Jim Balzotti's
Best Guest Ranches
And Horseback Riding Vacations

Second Edition

BALZOTTI PUBLICATIONS, INC.
5 BARKER STREET
PEMBROKE, MASSACHUSETTS 02359

PUBLISHER:
JIM BALZOTTI

MANAGING EDITOR:
ED MCCOLLUM

ASSISTANT EDITOR:
KRISTAN LANOUE

GRAPHIC DESIGNER:
CHERRY DAHLGREEN
GRAPHICS PLUS
DUXBURY, MASSACHUSETTS

COVER PHOTOGRAPHY:
JEFFREY GOODMAN
PHOENIX, ARIZONA

CARTOONIST:
FRED LEARY
LEARY CARTOON ASSOCIATES
PEMBROKE, MASSACHUSETTS

This book can be purchased through the American Automobile Association, numerous catalogues and bookstores across the country, or directly from the publisher. The retail cost through the publisher is $22.95 U.S. plus Priority Mail charge. Personal check, money order, or major credit card can be used. Please call our toll-free number for ordering and/or more information:

Inside US: 1-800-829-0715
Outside US: 1-781-829-0710

All rights reserved. No part of this work may be reproduced or transmitted in any form or by any means, electronic or mechanical, including photocopying and recording, or by any information storage or retrieval system, except as may be expressly permitted by the 1976 Copyright Act or in writing from the publisher.

Jim Balzotti's Best Horseback Riding Adventures logo image from: *All Roads Lead to Buffalo Bill's Wild West* with permission from Buffalo Bill Historical Center, Cody, WY.

Back cover/inside front cover photos were shot on location at the beautiful White Stallion Ranch, Tucson, Arizona.

Copyright © 1999 by Balzotti Publications, Inc.

Printed in Canada

ISBN No.: 0-9655278-1-6
ISSN No.: 1091-1642

Jim Balzotti's
Best Guest Ranches
And Horseback Riding Vacations

Dedication

I WOULD LIKE TO DEDICATE THIS BOOK TO
BOB O'HARA,
AKA THE BOSTON WRANGLER.
BOB AND I RODE MANY TRAILS TOGETHER.

IN MEMORY
ROBERT G. O'HARA

My very good friend and horseback riding buddy, Bob O'Hara, died in June, 1997 of cancer in Tucson, Arizona. Although Bob grew up and lived in the Boston, Massachusetts area most of his life, his heart was in the West and he moved to Tucson in 1992 with his wife, Pat, and his children, Mike, Terry, and Dawn. Bob and I shared years of horseback riding adventures all over the United States, some of which he had written about in "Confessions of the Boston Wrangler" which was published in the previous edition. Bob had appeared in small roles in a number of Hollywood western movies and has even written an, as yet, unproduced screenplay about the Wild West!

Bob is missed dearly everyday, especially by his wife and children.
He was a special person and I know that my life is
missing a part that will never be replaced.

Contents

From the Author	vii
Acknowledgments	ix
Location maps	2
Alaska	4
Crystal Creek Lodge	4
Northland Ranch	6
The Dude Ranchers' Association	8
Arizona Dude Ranch Association	9
Arizona	10
Circle Z Guest Ranch	10, 194
Crown C Ranch	12
Elkhorn Ranch	14
Flying E Ranch	16
Grand Canyon Bar 10 Ranch	17
Grapevine Canyon Ranch	18, 195
Iron Horse Ranch	22, 197
Lazy K Bar Ranch	24, 199
Muleshoe Ranch Preserve	26
Spirit Horse Ranch	28
Sprucedale Guest Ranch	30
White Stallion Ranch	32, 200
Arkansas	34
Horseshoe Canyon Ranch	34, 202
Lost Spur Ranch	36
Panther Valley Ranch	38
Scott Valley Resort and Guest Ranch	40, 203
California	42
Alisal Guest Ranch and Resort	42, 204
Coffee Creek Ranch	44, 206
The Homestead	46
Howard Creek Ranch	47
Hunewill Guest Ranch	48, 208
Rankin Ranch	49
Ricochet Ridge Ranch	50
Shannon Ranch	52, 209
The Wild Horse Sanctuary	54, 211
Colorado Dude & Guest Ranch Association	56, 212
Colorado	58
Aspen Lodge	58, 213
The Bar Lazy J Lodge	60, 214
C Lazy U Ranch	64, 216
Capitol Peak Ranch	66, 217
Casa Milagro Bed & Breakfast	68
Cherokee Park Ranch	69
Coulter Lake Guest Ranch	70, 220
Deer Valley Ranch	72, 221
Echo Canyon Guest Ranch	74
Elk Mountain Ranch	76, 222
Elkhead Ranch	78
Focus Ranch	80, 223
King Mountain Ranch	82, 224
La Garita Creek Ranch	85
Laramie River Ranch	86
McNamara Ranch	88
Mountain Meadows Ranch	90, 226
North Fork Guest Ranch	92
Peaceful Valley Lodge & Guest Ranch	94
Powderhorn Guest Ranch	96
Rainbow Trout Ranch	98
Rawah Ranch	100, 228
Red Feather Guides & Outfitters	102
Roubideau Western Adventures	104
San Juan Guest Ranch	106, 230
San Juan Outfitting	108
7 W Guest Ranch	110
Sky Corral Guest Ranch	112
Skyline Guest Ranch	113
Sylvan Dale Guest Ranch	114, 233
T Lazy 7 Ranch	116, 234
Weminuche Wilderness Adventures	118, 235
Wilderness Trails Ranch	120, 236
Wind River Ranch	124, 239
Florida	126
Ace Of Hearts Ranch	126, 240
Georgia	128
Victoria's Carriages & Trail Rides	128
Hawaii	129
Princeville Ranch Stables	129
Idaho	130
Bar H Bar Ranch	130, 241
Diamond D Ranch	132, 242

Hidden Creek Ranch	134, 244
Idaho Rocky Mountain Ranch	136, 243
Kingston 5 Ranch Bed & Breakfast	137
Moose Creek Ranch	138
Shepp Ranch	140
Western Pleasure Guest Ranch	142, 253
Iowa	144
Garst Farm Resorts	144
Kentucky	146
First Farm Inn	146, 254
Maine	147
Foothills Farm	147
Speckled Mountain Ranch	148
Michigan	150
Double JJ Resort Ranch	150
Missouri	152
Meramec Farm B&B	152, 257
Turkey Creek Ranch	154
Montana	156
63 Ranch	156
B Bar Guest Ranch	158, 260
Bar N Ranch	160
Bear Creek Lodge	161
Bear Tooth Ranch/JLX Outfitters	162, 261
Beaver Creek Trail Ranch	164
Bonanza Creek Country	166
C-B Cattle & Guest Ranch	167, 262
Circle Bar Guest Ranch	168
Covered Wagon Ranch	170, 263
DJ Bar Ranch	172
JJJ Wilderness Ranch	173, 265
Hargrave Cattle & Guest Ranch	174, 264
Horseshoe Hills Guest Ranch	176
Leffingwell's G Bar M Ranch	178
O Spear Guest Ranch	180
Monture Face Outfitters	182, 266
Parade Rest Ranch	184, 267
Powder River Wagon Trains	186, 268
RJR Ranch	188
Roundup Cattle Drive, Inc.	190, 270
Schively Ranch	192, 272
Sweet Grass Ranch	346
Triple Creek Ranch	348, 275
Two Medicine Teepee Adventures	350, 279
X Bar A Ranch	352
Adventure Camping Network	193
Color Photos Of Ranches	194-331
Feature Story— Posse Week At The N Bar Ranch, NM	332
The Photgraphy Of Jeffrey A. Goodman	342
New Hampshire	354
Horse Haven Bed & Breakfast	354
New Mexico	355
Cedar Crest Country Cottage & Stable	355
Dripping Springs Guest Ranch & Spa	356
The Galisteo Inn	358, 281
Los Pinos	360
N Bar Ranch (Special Feature Story—Page 332)	362, 282
New York	364
The Bark Eater Inn	364
Diamond Horseshoe Ranch	366, 283
Pinegrove Dude Ranch	370, 285
Ridin-Hy Ranch	373, 283
Rocking Horse Ranch	374
Roseland Ranch Resort	376, 288
North Carolina	380
Clear Creek Ranch	380, 290
North Dakota	382
Dahkotah Lodge	382, 292
Ohio	384
McNutt Farm II/Outdoorsman Lodge	384
Smoke Rise Ranch Resort	386
Oklahoma	388
Lotspeich Cattle Co.	388
Oregon	389
Ponderosa Ranch	389
Rock Springs Guest Ranch	390, 293
Pennsylvania	292
Buck Valley Ranch	392
Mountain Trail Horse Center	394, 294

- South Dakota ... 396
  - Triangle Ranch B&B ... 396
  - Triple R Ranch ... 398, 295
  - Western Dakota Ranch Vacations ... 400, 296
- Tennessee ... 402
  - Gilbertson's Lazy Horse Retreat ... 402
  - Namaste Acres Barn, Bed & Breakfast ... 404
  - Seraphim Stables ... 405
- Texas ... 406
  - The Bald Eagle Ranch ... 406, 297
  - Leon Harrel's Old West Adventures ... 408, 298
  - Prude Ranch ... 410
  - Y.O. Ranch ... 412, 299
- Utah ... 414
  - All 'Round Ranch ... 414
  - Best Western Ruby's Inn ... 416, 300
  - Dalton Gang Adventures LLC ... 418
  - Rocky Meadow Adventures ... 420
- Vermont ... 422
  - Firefly Ranch ... 422
  - The Jackson House Inn ... 424, 301
  - Kedron Valley Stables ... 426, 303
  - Mountain Top Inn ... 428, 304
  - Vermont Icelandic Horse Farm ... 430, 305
- Virginia ... 432
  - The Conyers House Inn & Stable ... 432
  - Fort Valley Stable Horse Campground ... 434
  - The Inn At Meander Plantation ... 435
- Washington ... 436
  - Hidden Valley Guest Ranch ... 436
  - The K Diamond K Guest Ranch ... 438, 301
  - North Cascade Outfitters ... 440, 307
- Wyoming Dude Ranchers Association ... 442
- Wyoming ... 443
  - A. Drummond's Ranch B&B ... 443
  - Absoroka Ranch ... 444
  - Bill Cody Ranch ... 445
  - Blackwater Creek Ranch ... 446, 308
  - Brush Creek Ranch ... 448, 310
  - Cheyenne River Ranch ... 451
  - Darwin Ranch ... 452
  - David Ranch ... 454, 314
  - Deer Forks Ranch ... 455
  - Diamond L Guest Ranch ... 456, 315
  - High Island Guest Ranch & Cattle Company ... 457
  - Flying A Ranch ... 458
  - Gros Ventre River Ranch ... 460, 316
  - The Hideout At Flitner Ranch ... 461, 317
  - Kedesh Ranch ... 462
  - Lazy L & B Ranch ... 464, 319
  - Lozier's Box "R" Ranch ... 466, 320
  - Paradise Guest Ranch ... 468, 303
  - Ranger Creek Guest Ranch ... 469
  - Rimrock Ranch ... 470, 321
  - Rocky Mountain Horseback Vacations ... 472
  - Savery Creek Thoroughbred Ranch ... 474
  - T Cross Ranch ... 476
  - Triangle C Ranch ... 478. 322
  - Triangle X Ranch ... 481, 324
  - Two Bars Seven Ranch ... 482
  - Two Creek Ranch ... 483
- Alberta, Canada ... 484
  - Warner Guiding & Outfitting Ltd., ... 484, 326
  - Saddle Peak Trail Rides ... 486
- British Columbia, Canada ... 488
  - Springhouse Trails Ranch ... 488, 327
  - The Hills Health & Guest Ranch ... 489
  - Three Bars Guest & Cattle Ranch ... 490
- Ireland ... 491
  - Black Valley Equestrian ... 491, 328
- Mexico ... 492
  - Saddling South ... 492
- Humor With Horses by Artist Fred Leary ... 494
- Interview With A Farrier ... 508
- Interview With A Veterinarian ... 511
- A Cowboy's Guide To Life ... 513
- On The Road Again, Trailering Tips For Safe Hauling ... 514
- Campfire Recipes ... 516
- U.S. Stabling Guide ... 520

From the Author

For all of you people out there who are expecting me to say that as a young boy watching Westerns I wanted to grow up and be a cowboy — sorry, it just "ain't" so. Actually, growing up in the inner city of East Boston, I always wanted to be a policeman. You see, in the city we really had never seen a horse. Pigeons, yes. Horses, nope.

I still remember the day I laid eyes on my first horse, as if it were a year or two ago. My big brother Bob had just gotten his driver's license and decided to take a road trip out of the city, and he was good enough to take his little brother along. He was great like that. We ended up driving to Maine. We were driving around aimlessly on some back country roads when we came up along an old farmhouse. On both sides of the road stretched the farmer's fields, lush green with the spring rains. Out my window, I could see in one of the fields a black horse, a magnificent creature with a black flowing mane. As we began to pass, he ran alongside us, kicking up his heels, tail straight out. I was absolutely awestruck, a witness to something magical that I had never seen before. I did not wish to ride that beast, oh no. Nevertheless, I couldn't take my eyes off him. As we drove past his boundary fence, he watched us until I could see him no more. That was my first encounter with horses, and it would be many years before I would actually ride one.

Without boring you with a lot of in-between stuff, a few years ago, I bought a ranch in Arizona, strictly as a vacation spot, a place to get away and do some serious horseback riding. As I was preparing to leave from my home in Massachusetts, traveling cross-country with four horses in tow, it dawned on me — where the heck was I going to board these guys overnight? I knew that I could always grab a room anywhere, even sneak my dog up to the room, but four horses? There was just no good, accurate information available to the traveling horseperson. So for the next few years, I criss-crossed the country, filling up a black book of stables that would board horses overnight, as well as accumulating a wealth of knowledge about traveling with horses and a lot of good stories to tell around the campfire, or coffee shop.

These experiences led to the founding of Balzotti Publications and to the U.S. Stabling Guide, which is the bible for those people traveling with their horses. Over the years, we have received numerous questions about horses and horse-related activities. The most common questions are "Where can I go on a horseback riding vacation?" and "Where can I take my horse with me and go riding?" Answers to those questions led to the first edition of Jim Balzotti's Best Horseback Riding Vacations. Over the last three years, since the first edition, I have "discovered" many other wonderful vacation destinations across the country. I wanted my book to be as complete as possible and offer additional choices for my readers.

I believe the places listed in this book represent the cream of the crop of horseback riding vacation destinations. There is something for everyone in this book. You'll find not only guest and resort ranches but also cattle drives, outfitters, a wild horse sanctuary, an inn that features fox hunts, and bed and breakfast facilities from Vermont to California.

The facilities are both small and large, and accommodations can be rustic (as on the cattle drives) or luxurious (as in the finer resorts), with every level in between. Some of the ranches are family oriented, while some cater to adults only. Some of the facilities are close or easily accessible to large cities, while others may be a 4-6 hour drive from the nearest large city or town. You can go riding in the surf along the Pacific or Atlantic coast, explore the back roads of New England, ride through the vast open spaces of the West, or climb above the tree line in the mountains. There are facilities where you can bring your own horse or facilities where you can ride the wonderfully trained horses that are provided. And there are places that can accommodate those who aren't as enthusiastic about horses as others in the party.

In the age of the Internet, this edition also includes website addresses of ranches that will enable you to view additional photos and information, and E-mail addresses to put you in contact with ranches when it's convenient for you — probably around 10:00 pm, if you are like me!

This was a really fun project, and the various ranches, inns, and B&Bs were just great to work with. Whichever one (or ones) you choose will provide you with memories that will last a lifetime. Enjoy, and safe riding.

Jim Balzotti

Acknowledgments

I want to take this opportunity to thank everyone who helped me put this book together. It was a major project, and I could not have done it without the counsel and hard work of dozens of people. A few of those deserve special thanks:

Tony Daniels, Alan Borne, and the other people at the American Automobile Association, who have given their help and support to the millions of horse owners and horse lovers out there.

My contributing professionals — Bruce Chase and Joe DeLowery — both experts in their field and horse lovers.

Fred Leary, a most talented cartoonist.

Ed, my senior editor and right hand man, and Cherry, my award-winning graphic artist. Without either of them, I could not have done this book. Their expertise and endless energy propelled them to work through countless nights, on usually nothing but too many pizzas and gallons of coffee! They were the two people behind the scenes that made this book happen.

To Russell and Gail and the True Family for allowing me to shoot the cover photographs for the Second Edition, on their beautiful White Stallion Ranch in Arizona.

To Jeffrey Goodman, a professional photographer who specializes in ranch and outdoor photography, for his expertise and spectacular photographs on our covers and in the Second Edition.

And finally to all the people represented in this book who open up their homes, ranches, and inns to those of us who want to take a horseback riding vacation.
These good folks epitomize the best in American hospitality, and it was wonderful working with them.

Publishers Notice

Every effort is made to compile and print this book as accurately as possible. However, errors or omissions may occur. In the event of an error or omission, Balzotti Publications assumes no obligation to correct the same in this book or by any special notice of any kind to the individuals and businesses listed herein.

Key

On the ranch description pages, dollar-sign symbols are used instead of actual amounts to indicate the rate charged by the facility. Many places, however, charge a weekly rate. All rates were broken down to "per person per day" to enable you to make some comparisons among the places included in this book. Rates are, of course, subject to change, and you should check with the facility itself to determine the actual, current rate.

$	means	$0 - $100 per person per day
$$	means	$100 - $200 per person per day
$$$	means	$200 - $300 per person per day
$$$$	means	Over $300 per person per day

The Guest Ranches, Inns and Adventures

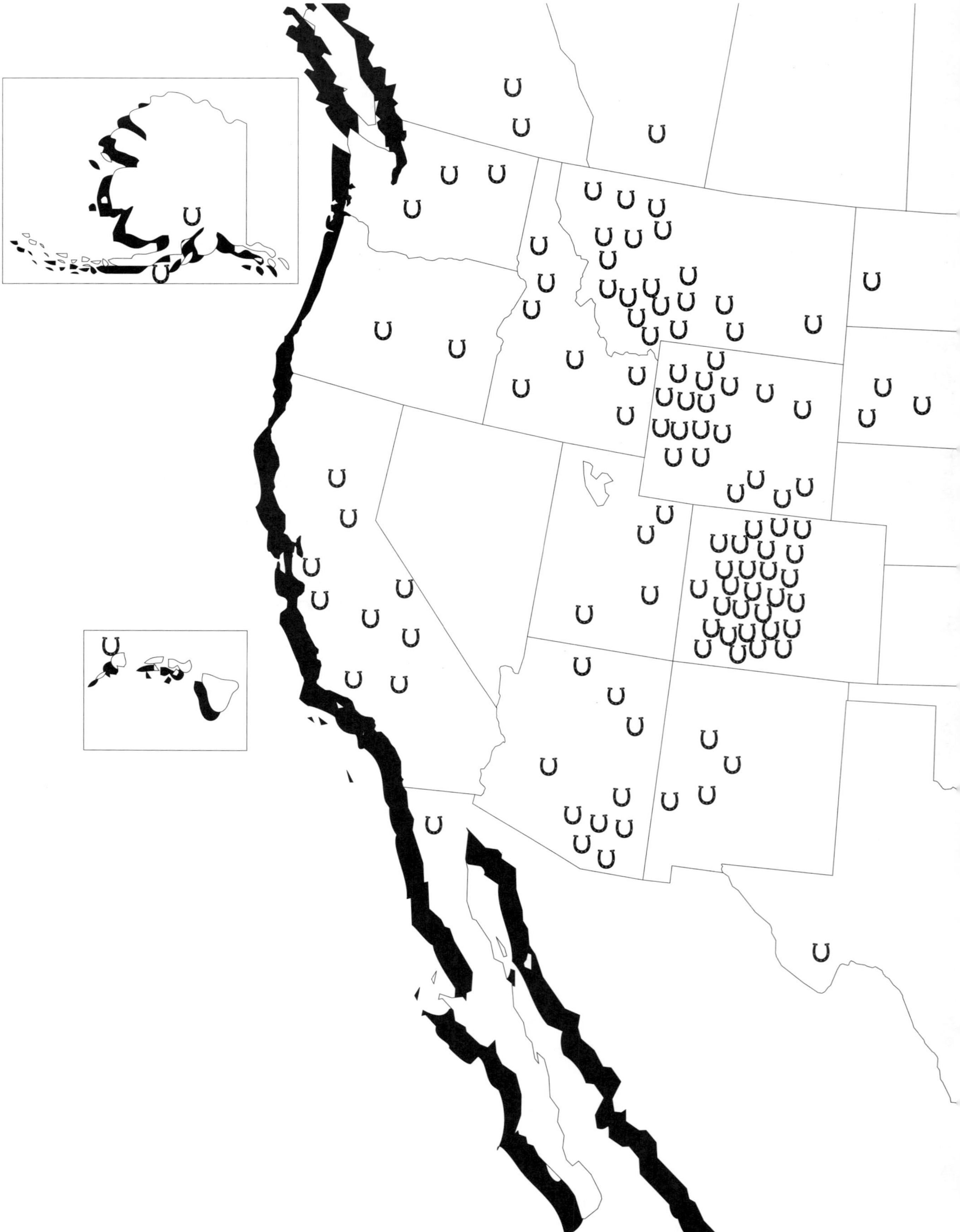

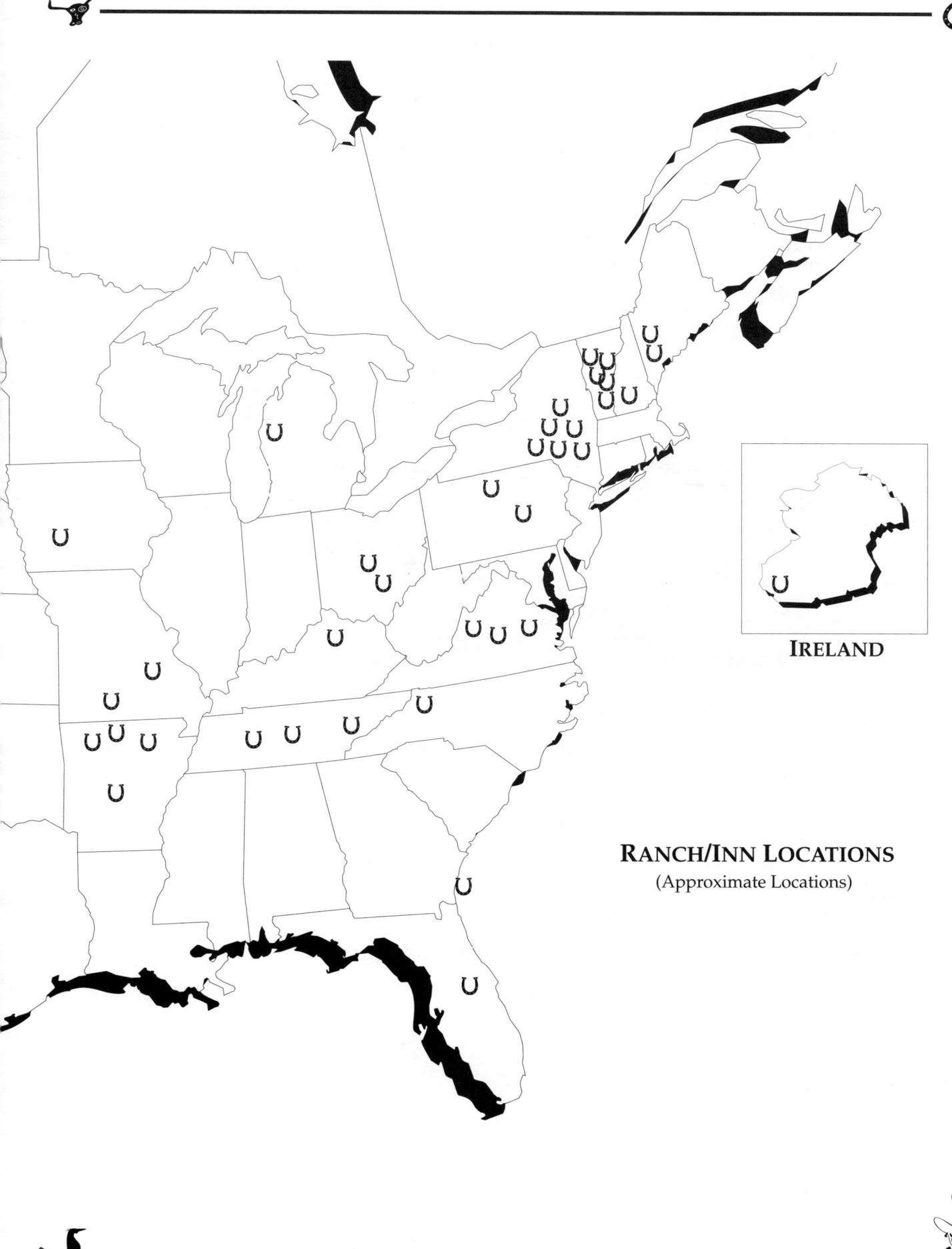

RANCH/INN LOCATIONS
(Approximate Locations)

IRELAND

ALASKA

CRYSTAL CREEK LODGE
ANCHORAGE, AK

Crystal Creek Lodge is located 350 miles southwest of Anchorage near Dillingham, Alaska, in the center of a world renowned region known as the Bristol Bay area.

Each morning at Crystal Creek Lodge, you'll board one of our aircraft. A fleet of meticulously maintained float planes and a helicopter are ready and waiting to whisk you to some of the most acclaimed fishing rivers in North America, including the Nushagak, Wood, Togiak and their tributary streams. The 4.3 million acre Togiak National Wildlife Refuge and the 1.5 million acre Wood-Tikchik State Park are both within sight of the Lodge. Each of these giant wilderness preserves are laced with hundreds of miles of productive rivers, streams, and lakes. Throughout your stay you'll visit different rivers and streams, each possessing individuality and unique fishing opportunities. Five Pacific Salmon species of Chinook (King), Chum (Dog), Sockeye (Red), Pink (Humpback) and Coho (Silver), run from the ocean into numerous area rivers. Six resident freshwater species are common throughout the area. These species are Rainbow Trout, Arctic Char, Arctic Grayling, Dolly Varden, Lake Trout, and Northern Pike.

Your daily fishing excursions are conducted by one of our dedicated and knowledgeable guides. All Crystal Creek Lodge guides are registered with the State of Alaska, certified U.S. Coast Guard Captains, expert outdoorsmen and bring years of individual experience to help you catch any of the Alaska fish species with fly or conventional tackle. An extensive selection of Orvis rods, reels, flies and tackle are available for complimentary guest use.

A fleet of over 25 boats are located either at the Lodge or at several of our best fishing rivers. All are jet-powered river boats, able to navigate shallow water with ease. Once you've arrived on the river by aircraft, the boats become invaluable for classic western-style drift fishing, reaching fish in deeper holes or mobility between wade-fishing areas.

The earliest waterfowl season in the United States opens in Alaska September 1, and from September 1 until the end of the season we offer guided waterfowl hunts to complement your Alaskan fishing adventure. All licenses, guns, ammunition and equipment are provided. For those who wish to exclusively fish, the participation in guided hunts is optional.

When the evening concludes a day of fishing or hunting, return to one of the word's great sporting lodges. We've anticipated your every need. Hang up your waders to dry in our" hot room" on your way to warming yourself in our spa and sauna room. Relax in front of the fireplace with a drink before dinner, then enjoy a delicious meal. Tie a couple of flies at our Tying bench before retiring to your room and a very comfortable bed.

The dining experience at Crystal Creek Lodge rates with the finest restaurants anywhere in Alaska. Breakfast and dinners, are offered from a daily menu then prepared to order by our chef, are served in our lakeside dining room. Delicious field lunches are prepared and packed to order.

Whether you're planning the fishing adventure of a lifetime or the best fishing vacation of the season, the excitement of Alaska fishing and warm hospitality await you at Crystal Creek Lodge.

 Mailing Address: P.O. Box 92170, Anchorage AK 99509-2170

Telephone: 907-245-1945; **Fax:** 909-245-1946
E-Mail: crystalc@alaska.net
Web: www.crystalcreeklodge.com

Location: 350 miles southwest of Anchorage

Season: June — September

Policy on Children, Pets: Children over 8 welcome. No pets.

Rates: $$$$

Credit Cards: Visa, Mastercard

Alaska

ALASKA

NORTHLAND RANCH
KODIAK ISLAND, AK

Located on beautiful Kodiak Island, Northland Ranch is not only one of the largest ranches in Alaska, but also the only Alaskan guest ranch. Here you will be able to enjoy the magnificence of Alaska and one of its most scenic and historic spots, Kodiak Island, also known as Alaska's Emerald Isle. Northland Ranch is not for everyone. It is no Club Med. We have no swimming pool, no paved tennis courts and no hot tubs or jacuzzis. What we do have is world-class fishing for five species of salmon and three species of trout; wildlife viewing for eagles, whales, fox and deer; miles of beaches with spectacular mountain views; and of course, horseback riding over some of America's wildest and most scenic terrain.

Your visit to Northland Ranch will begin with a 30 mile drive from the town of Kodiak, complete with local stories and beautiful scenery. When you arrive at the ranch, don't be surprised when you see your horses already saddled and waiting for your arrival. After you've enjoyed a complimentary drink or hot beverage, you'll be greeted by one of our friendly wranglers ready to take you on your first riding tour of the ranch. Back at the ranch, you'll be invited to join Omar Stratman, owner and operator of the ranch for over 20 years, for a little bit of history surrounding his cattle ranching on the island. Or you may want to go to your room and relax before dinner.

Our accommodations are comfortable but not plush. We are, after all, a working ranch. Dinnertime at Northland Ranch is always "family" style. After all, while you're a guest on the ranch you are part of the family. This is always the best time to talk to other guests and the cowboys to find out where the best fishing holes are and to plan the following day's adventures. You may want to plan an all-day or half-day ride. If so, we can pack lunches and bring you to some of the most beautiful trail-riding for all riding levels.

If you have a little cowboy or girl, they can ride alone on one of our gentle quarterhorses and be led by the wrangler or, if required, can ride double with you. If it's fishing that brings you to Kodiak, we can provide you with a ride to the "hottest" fishing spots. What we can't provide you with is the gear so don't forget your poles and tackle.

Your days here are all your own. We follow no schedules. You ride when you want to ride and fish when you want to fish, but if you miss a meal you might be heating up your own dinner!

Again, we are the first to admit that Northland Ranch is not for everyone. You won't find butlers and doormen, just a friendly bunch of cowboys and there are a few things you may have to do for yourself. You may have to bait your own hook, saddle your own horse, and well...if you step in the wrong place, you will have to clean your own boots!

We are a ranch for that special family who wants to experience a bit of Alaska. If you are that family, give us a call. We'll be happy to share our piece of Alaska.

Mailing Address: P.O. Box 2376, Kodiak, Alaska 99615

Telephone: 907-486-5578
Fax: Call
Web: www.jimbalzotti.com

Location: 45 minutes southeast of Anchorage

Season: Call

Policy on Children, Pets: Call

Rates: Call

Credit Cards: Call

Alaska
NORTHLAND RANCH

SEE THE BEAUTY OF ALASKA FROM THE SADDLE OF ONE OF OUR QUARTERHORSES.

THE SCENERY AT OUR RANCH.

RIDE the WEST with the BEST

THE DUDE RANCHERS' ASSOCIATION
Established 1926

Since 1926,
The Dude Rancher's Association
has been setting the standards for western ranch vacations. Assure yourself a quality ranch vacation by booking with the
Best, a member of
The Dude Ranchers' Association!
All members have gone through a two year selection process
(not all are accepted) and are publicly committed to the highest standards.

For an Official Directory of over 100 **inspected and approved** ranches or for assistance in finding
your perfect ranch, contact:
The Dude Ranchers' Association!
at
Post Office Box 471-H
LaPorte, CO 80535
or call (970) 223-8440
or E-mail
duderanches@compuserve.com
and Bobbi or Jim Futterer will
be happy to answer your questions
and help you plan
your best vacation ever!

Visit our Web Site at
http://www.duderanch.org
for great vacation tips and a description
of member ranches.

ARIZONA
DUDE RANCH ASSOCIATION

Our ranches, as diverse as our state, range from the more traditional smaller, rural ranches where riding and ranching are the primary focus to our larger resort-style ranches with all the amenities. No two dude ranches are alike; each is a reflection of the family who operates it.
Our guests – families, couples, and singles of all ages – come from every part of the country and the world. They share the common desire to experience the Western life-style, to be a cowboy or cowgirl (or at least ride with some), to slow their pace and relax and enjoy a vacation that is anything but ordinary. This wholesome environment also works wonders for corporate retreats, family reunions, and other group functions where a more intimate atmosphere is desired.
A horseshoe ∪ beside the ranch's name indicates the ranch has a detailed description and photos elsewhere in this book.

- **Ironhorse Ranch** ∪
- **Flying E Ranch** ∪
- Merv Griffin's Wickenberg Inn
- **Elkhorn Ranch** ∪
- **Circle Z Ranch** ∪
- Horseshoe Ranch
- Price Canyon Ranch
- **Grapevine Canyon Ranch** ∪
- Rancho de los Caballeros
- Kay El Bar Ranch
- Rancho de la Osa
- **Sprucedale Guest Ranch** ∪
- **White Stallion Ranch** ∪
- Tanque Verde Ranch
- **Lazy K Bar Ranch** ∪

Select your vacation destination with confidence when you use the Arizona Dude Ranch Association brochure as your guide. All member properties meet association requirements for their riding programs, lodging facilities, and amenities. All offer American Plan rates.

For your free color brochure describing these ranches, write to the Arizona Dude Ranch Association, P.O. Box 603, Cortaro, AZ 85652.

Arizona

Circle Z Guest Ranch
Patagonia, AZ

Founded in 1925, the Circle Z is Arizona's oldest working guest ranch. The ranch lies on over 6,000 acres, at an elevation of 4,000 feet in the foothills of the historic Santa Rita and Patagonia mountains. The unique, ever-flowing Sonoita Creek meanders throughout the ranch and is bordered by magnificent cottonwood and sycamore trees. *Red River*, starring John Wayne, and Spencer Tracy's *Broken Lance* were filmed here, as were more recent television films such as *The Young Riders*.

Riding is the principal activity at Circle A. The ranch is proud of its well-trained, ranch-bred Quarter Horses. Riding instruction is available, and guests may keep the same horse for their stay. Rides are scheduled twice daily (except Sunday), and variety rides include all-day, picnic, breakfast, and trailer rides to beautiful, remote locations. Trail rides are available for all levels, and the pace is either slow (walking) or medium (trotting and loping). With foreman approval, children over 5 may join the trail rides.

An overnight pack trip is offered once a week to the experienced rider, all equipment is provided. The destination is San Raphael Valley (where *Oklahoma* was filmed).

Hiking is also popular, and maps, water bottles, and back packs are provided. The ranch can even pack a sack lunch for hikers and sightseers. This is one of the foremost birding areas in the country. Hundreds of species may be seen, including great blue herons, phainopepla, hawks, and hummingbirds of many varieties. Adjacent to the property is the Patagonia-Sonoita Creek Nature Preserve, where guided tours are given on Saturdays.

The heart of the ranch is the large main lodge, where guests can relax by the fire and enjoy varied reading material. The extensive library features books of Western lore, novels, and reference material describing the plant life and unusual animals (such as chollo and javelinas) that they are likely to see in this high desert. Also in the lodge is the sun-filled dining room, where sumptuous meals are served buffet style. The ranch features home-baked breads and desserts, varied menus, and plenty of fresh fruits and vegetables. Returning guests look forward to Fridays, when mild authentic Mexican food is the fare. Adults look forward to each evening's hosted social hour. The Cantina is B.Y.O.B., but the ranch provides snacks, mixers, soda, and juices.

There are accommodations for singles, couples, and families in seven adobe cottages, each attractively furnished in Southwestern decor. A porch or patio is off each bedroom, and one 3-bedroom cottage has a fireplace in the living room. There are no TVs or telephones in the rooms.

Children are welcome at Circle Z. Although there is no organized children's program, they often ride with their own wrangler and "compete" for badges at the pole-and-barrel obstacle course. They dine with the wranglers and staff, and parents may join them. Children also have their own "cantina" with pool table, video games, and juke box. Children — and adults — can enjoy tennis, Ping-Pong, shuffleboard, and swimming in the heated pool.

Within an hour of the ranch are many fascinating places to visit including Tombstone, home of the famous OK Corral; the historic town of Bisbee, where there are guided tours of the Queen copper mine; and Tubac, a 1752 Spanish town with many shops and galleries. Nearby there are ghost towns and caves. Nogales, Mexico, with its many shops and restaurants, is about half an hour away. Three championship golf courses are nearby, and Lake Patagonia, where guests can rent rowboats and canoes, is only 4 miles from the ranch.

Circle Z offers excellent horses, varied trails, comfortable surroundings, and down-home hospitality.

Address: P.O. Box 194, Dept. BP, Patagonia, AZ 85624

Telephone: 520-394-2525
Web: www.jimbalzotti.com

Location: 65 miles southeast of Tucson, 4 miles southwest of Patagonia, 15 miles northeast of Nogales

Guest Capacity: 40

Season: November 1 to May 15

Policy on Children, Pets: Children are welcome but are the responsibility of parents. No pets.

Rates: $$ American plan. Three-night minimum stay. Children's and youth rates. Pack trips and trailer rides extra.

Credit Cards: None. Personal checks or cash accepted.

(See Pages 194 for Color Photos!)

ARIZONA
CIRCLE Z GUEST RANCH

OUR MONDAY MORNING COMMUTE.

ARIZONA

CROWN C RANCH
SONOITA, AZ

The Crown C Ranch is an "a la carte" guest ranch. We facilitate a variety of options in accordance with the desires of our guests. It's our belief that most people come here to find privacy and serenity in a unique setting of oak-studded grasslands and mountain vistas. We want our guests to feel as if they were in their own home with the freedom to choose their own activities, meal times and to sleep as late as they wish. At the same time, we want to make our guests comfortable. To this end, we provide fresh linens, equipped kitchens, and firewood in clean surroundings.

The ranch lies in country that has been the site of raising cattle for centuries. In the 1930's, the Crown C was split off from the famous Empire Ranch that stretched from outside Tucson to the Mexican border. Today, the original ranch house, built of immense, hand-crafted adobes, welcomes you.

The sprawling main ranch house consists of two wings that lead into a spacious living room, dining room, kitchen, porch and patio. The adjoining guest house is a cozy diminutive of the main house. These can be reserved as a whole, separately or by the individual room.

You'll enjoy relaxing in front of the large stone fireplaces or sunning on the patios. There is swimming in-season and year-round hiking to the historic ruins of Camp Crittenden or the small wildlife refuge located at Cottonwood Spring.

The Santa Rita Mountains and the surrounding country provide some of the most spectacular horseback riding in the world. Ride through lush, rolling grasslands, drop into canyons studded with oaks, sycamore and cottonwoods, or head for the peaks of the Santa Rita Mountains. The Arizona Trail, mapped from the Mexican border to the northern border of the state, has been completed in this area with miles and miles of maintained trails for riding, hiking and biking pleasure.

The Crown C Ranch is a working cattle ranch and guests are invited to participate in cattle drives, shipping and branding.

Local diversions include paramutual horse racing in April and May, AQHA show in June, a professional rodeo in September and the county fair in September.

The Sonoita-Patagonia area is best known for its ranches and historic ruins and, more recently, for its vineyards, world renowned bird sanctuary, spectacular scenery and hiking trails.

The surrounding area boasts many attractions including trips to historic towns such as Tombstone, shopping in Old Mexico, visiting arts and crafts centers, exploring museums such as the internationally known Arizona Sonoran Desert Museum, the movie location of Old Tucson, and the Biosphere 10. There are old Spanish missions to visit, a scenic railway to ride and bird and game hunting in season.

Our goal is to help our guests create their own adventure and enjoy all that the Crown C and the surrounding country has to offer. We want you to leave feeling that some of your dreams have come true.

Address: P.O. Box 984 Sonoita, Arizona 85637

Telephone: 520-455-5739
Web: www.flash.net/~crownc

Location: 50 miles southeast of Tucson

Guest Capacity: 16

Season: Year-round

Policy on Children, Pets: Children welcome with parental supervision, no pets

Rates: $$$

Credit Cards: None

ARIZONA
CROWN C RANCH

MOVING THE CATTLE.

THE WIDE OPEN SPACE OF THE WEST.

ARIZONA

ELKHORN RANCH
PATAGONIA, AZ

Since 1945, when the Miller family ventured south from Montana, guests have enjoyed the perfect winter climate of southern Arizona where the sun seeps into the very marrow of your bones. Elkhorn Ranch continues to offer unexcelled horseback riding and comfortable living for about thirty-two guests. The third generation of the Miller family and their crew provide the care you need to enjoy rest, quiet and stimulating activity around the ranch and on the miles of mountain and desert trails.

You will find a friendly welcome at Elkhorn Ranch and abundant wildlife and Sonoran desert vegetation will greet you each day. The spirit of Elkhorn is that of a considerate family group where guests enjoy fun times together or quiet hours alone.

Living accommodations provide comfortable space for singles, couples or families. Cabins are equipped with private baths, electric heat and good reading lights and some have separate sitting rooms or fireplaces—no television or phones. Accommodations are scattered throughout the mesquite trees to provide privacy for rest, sunning on chaise lounges, writing, reading or socializing. You'll find a bird feeder outside your door and all the birdseed you care to use. The variety of birds and wildlife will delight you!

Home cooked meals, buffet style, are served three times a day and you can order a hot breakfast of your choice. Meals are well-balanced, nutritiously prepared and the variety appeals to most appetites. You can sit indoors in the dining room or outdoors on the patio where frequently you will enjoy barbecue smells on the breeze, coming from our outdoor grill. A picnic lunch cooked over an open fire on the desert is a weekly treat. The ranch has no bar. Rather, those who wish may entertain in their own cabins. The Long House sitting area provides a place to gather by the fireplace, play games or browse the ranch's well stocked lending library.

Horseback riding at Elkhorn provides plenty of opportunity for adventure and a great way to see the incredible scenery of the Baboquivari Mountains. Gentle or loping rides through the desert and more vigorous mountain trips are offered every day except Sunday when the horses enjoy a day off. The ranch's ten thousand acres plus neighboring lands provide an unlimited area of remote canyon and mountain country to experience true wilderness. For those who wish to see the country, there are all-day adventures with sandwiches or a mule-packed hot lunch.

With six guides, groups are small so that you can have the length and pace of ride that you desire. Beginning riders have the chance to build skills and confidence while more experienced riders find plenty of challenge. People of all ages have learned to ride at Elkhorn (children must be six or older to ride). The right horse and saddle remain yours throughout the visit. Horses are raised and trained here for years and we have a variety that makes for fun and safety.

Elkhorn provides plenty of country for relaxed walks or vigorous hikes. The tennis court and heated swimming pool supply fun for kids of all ages, along with shuffleboard, Ping-Pong, horseshoes and a pistol and rifle range. Bring your camera and binoculars—the wintering birds are numerous and colorful, views are magnificent and spring flowers can be glorious. For the less energetic, there are always lazy hours in the sun, the pleasure of burying your nose in a good book and picnics on the desert where you can go by truck out to meet the riders for lunch or for supper beneath the full moon. Nearby sites within an hour or so drive provide the chance to learn about southern Arizona's natural and cultural history—the Arizona-Sonoran Desert Museum, Kitt Peak National Observatory, the Buenos Aires National Wildlife Refuge, Tubac and Nogales or the old Spanish missions at San Xavier del Bac and Tumacacori.

Address: HC1 Box 97 Tucson, Arizona 85637

Telephone: 520-822-1040

Location: 23 miles from Tucson Airport

Guest Capacity: 32-40

Season: Mid-November through the end of April

Policy on Children, Pets: Children welcome but must be age 6 or older to ride. No pets please.

Rates: $$

Credit Cards: Not accepted

Web: www.jimbalzotti.com

ARIZONA
ELKHORN RANCH

ON THE TRAIL.

Arizona

Flying E Ranch
Wickenburg, AZ

The Flying E Ranch is nestled in rolling hills at the foot of the Vulture Mountain range near Wickenburg — in dude ranch country. Wickenburg is a charming "real" Western town and the Flying E is a world all its own.

The Flying E was born in 1946, the dream child of Lee Eyerly, from Salem, Oregon. Lee purchased the original 3,000 acres and built the lodge, eight guest rooms, a picturesque Oregon-type barn, corrals, and a 3,200-foot-long airstrip. He also fashioned the brand ~E~ adding wings to the first initial of his surname. In 1947, Lee opened a "guest ranch."

Vi and the late George Wellik of Bellflower, California, discovered the ranch from the air in their private plane. They visited often and in 1952 found themselves the owners of the Flying E. In 1953, they added 17,000 acres to the ranch, making a total of 20,000. In 1960, the Wellik family moved to Wickenburg and assumed responsibility for the ranch's daily operations.

After George's death in 1983, Vi continued to house and entertain guests from all over the country and the world. One senses that spirit of continuity. There have been many additions and improvements, yet the Flying E remains essentially the same. It is not a "resort" but rather a homey spot where guests can enjoy privacy, relax in their blue jeans, and be themselves. Guests often refer to the ranch as "Camelot," a respite from the real world to which many return year after year. Some guests have returned more than twenty-five times!

The Flying E is "the riding ranch," although you don't necessarily need to ride to enjoy the ranch. The wranglers, good horsemen, share this enthusiasm. Guests are divided into beginner, intermediate, and experienced groups; horses are available for all types of riders. Rides are offered twice daily, once on Sunday. The scenery is picturesque . . . and the high desert trails are many. A number of activities are built around riding. There are breakfast rides where food is prepared on the desert floor, lunch rides to famed Robber's Roost, and chuckwagon campfire feeds. Games on horseback known as "dudeos" (gymkhanas) are held in competition with guests from neighboring ranches. The Flying E is also the home of Murray-Grey cattle, an Australian breed. Guests ride among the roaming cattle and witness the cattle work.

The Flying E accommodates approximately 34 — families, couples, and singles. Everyone knows everyone else, and a single person never feels alone. Rooms have a comfortable Western decor and are impeccably clean. All have a private bath, wet bar, and TV. There are also 3 separate suites.

The lodge is a cozy gathering place for dining, relaxation, and entertainment. Dedicated women ranch cooks take much pride in serving delectable, gourmet, home-cooked foods "family style." The coffee pot is always on and the cookie jar filled. Guests can help themselves to iced tea or lemonade. The "adult" saloon is a place for sips 'n' dips at cocktail hour, B.Y.O.L.

After a day in the saddle — or just basking in the ambiance and atmosphere of ranch life — the heated pool and hot spa are welcome retreats. Guests can also take advantage of a sauna/fitness room, tennis court, garden chess, volleyball, basketball, table tennis, and shuffleboard. The "hay loft" in the barn is ideal for occasional western dancing and organized entertainment. An 18-hole championship golf course is nearby; guests can take advantage of ranch membership.

The friendly, dedicated staff exudes the Flying E spirit . . . it is truly a heart-warming experience.

 Address: 2801 West Wickenburg Way, Dept. BP, Wickenburg, AZ 85390

Telephone: 520-684-2690; **Fax:** 520-684-5304; **E-mail:** flyinge@primenet.com

Location: 60 miles northwest of Phoenix, 4 miles west of Wickenburg

Guest Capacity: 34

Season: November 1 to May 1

Policy on Children, Pets: Children welcome. No pets.

Rates: $$ American plan. Children's rate if staying in parents' room. Horseback riding extra.

Credit Cards: None.

ARIZONA

GRAND CANYON BAR 10 RANCH
NEAR ST. GEORGE, UT

The grandeur, the mystery, the Old West flavor of the Grand Canyon and the Colorado River are yours at the Bar 10 Ranch — a working cattle ranch located about 9 miles from the north rim of the Grand Canyon on the fabled Arizona Strip! Whether it's a stay at the Bar 10 Lodge, trip-of-the-lifetime river run down the roaring Colorado River, a breathtaking helicopter or flight to the Grand Canyon, a four-day horseback pack trip, or a genuine cattle drive, you'll love the Western atmosphere and country hospitality at the Bar 10 Ranch!

For years, the Bar 10 has been a starting and ending point for guests on Colorado River rafting trips. There is no experience in the world that equals the grandeur and history of the Grand Canyon. From towering vistas and majestic canyons, you will descend back in time as you traverse the heart of the Colorado River, leaving each day wanting for more.

Any of the following Western ranch tours may be selected as an add-on to a river trip or booked as a separate adventure. You may begin your Bar 10 Overnight Experience with a breathtaking scenic flight from Las Vegas, Nevada, over the Grand Canyon. (Most guests of the Bar 10 fly from Las Vegas.) You'll be greeted by the Bar 10 cowboys and welcomed to the Bar 10 Lodge. After a buffet lunch and orientation, there will be a ranch demonstration that includes history and stories about the area and the cattle business. Ranch hands will demonstrate ranch chores such as branding a calf, shoeing a horse, or setting a coyote trap. Guests are welcome to watch or roll up their sleeves and help out! Throughout the day, ranch activities include skeet shooting, horseback riding, hiking, pitching horseshoes, playing table tennis or volleyball, browsing through the Bar 10 Trading Post for souvenirs, or just relaxing and enjoying the wide-open country and clean air.

The delicious Dutch-oven evening meal consists of potatoes, lean roast beef, tossed green salad, cooked vegetables, the famous Bar 10 homemade rolls, and dessert. After satisfying those city-slicker appetites, enjoy live Western entertainment, which could include singing, dancing, trick roping, and some good ole' cowboy humor. The two-story Bar 10 Lodge has comfortable dormitory-style rooms with bunk beds. Or if you're a bit more adventurous, try the Conestoga wagons positioned around the hillside. The next day...wake up to the wonderful aroma of cowboy coffee and enjoy a hearty cowboy breakfast before your flight back to Las Vegas.

If your idea of fun is getting out of a sleeping bag at dawn, eating camp food, riding a horse, singing cowboy songs around a campfire, and sleeping under a big sky full of stars, the Bar 10 Genuine Cattle Drive is for you! The final afternoon and evening are spent at the Bar 10.

Experience an authentic, Grand Canyon, Western ranch adventure on a Four-Day Horseback Pack Trip. Ride the trails of the early explorers, settlers, and outlaws of the Arizona Strip. An additional bonus is a trip to the fabulous overlook of the magnificent Grand Canyon. Enjoy two days of outdoor cowboy fun with meals, entertainment, and a day and night at the Bar 10 Lodge.

For those on a limited time schedule, the Bar 10 also offers various part and full day tours, which combine a short stay at the ranch with a scenic flight over the Grand Canyon.

The Bar 10 Ranch provides a unique setting for family reunions, vacations, business retreats, youth groups, church groups, or other excursions. Hearty country meals, varied ranch activities, and Bar 10 hospitality provide guests with an unforgettable Grand Canyon Western ranch experience! Let them customize a tour for you today!

 Mailing Address: P.O. Box 1465, Dept. BP, St. George, UT 84771-1465

Telephone: 801-628-4010, 800-582-4139; **Fax:** 801-628-5124; **E-mail:** bar-10@infowest.com

Location: 80 miles south of St. George, Utah; 200 miles east of Las Vegas, Nevada

Guest Capacity: 60

 Season: mid-May through mid-September. Off-season dates available for groups.

Policy on Children, Pets: Varies.

Rates: $ - $$ American plan. Helicopter rides extra.

Credit Cards: MasterCard, Visa

Arizona

Grapevine Canyon Ranch
Pearce, AZ

For a real Western experience, this working cattle/guest ranch is hard to beat. Nestled in an enchanting canyon in the remote, rugged Dragoon Mountains of southeastern Arizona, Grapevine Canyon Ranch enjoys a worldwide reputation for its friendly, personalized attention, good home-cooked food, first-class accommodations, and the best of horseback riding programs. Grapevine is not a dude ranch; it's a place where guests come to stay and return again and again as friends.

Grapevine prides itself on its quiet, tranquil atmosphere but also offers action-packed days. Whether you take the all-day ride or spend your day sightseeing, hiking, bird watching, or quietly lazing by the pool, your stresses and cares melt away as if by magic.

Eve Searle, the hostess of Grapevine, ensures that guests enjoy all the ease and comfort her well-trained staff can provide. The free-standing accommodations are carefully spaced round the 90-acre headquarters complex so that each is hidden in wooded groves that afford total privacy. The interiors are each individually decorated, cozy and comfortable with a true homelike touch. Large fluffy towels, patterned bed sheets, bath robes, refrigerators, coffee makers, and hair dryers are just some of the little comforts that make Grapevine your home away from home.

The food is all you'd expect it to be. Plentiful, tasty, well cooked and varied, the Grapevine menu does not repeat for a full six weeks. That's because some of the guests come and stay that long . . . "and we want them to enjoy as much as possible of our great home cooking," says Eve.

Gerry Searle, Grapevine boss, has been with horses all his life — training, riding, cowboying, and stunt riding in the movies, and what he doesn't know about a horse isn't worth knowing. " See this saddle," he says. "I've done over 100,000 miles on this saddle — and probably another 100,000 on my other one."

The Searles and their wranglers are very proud of their horses, who are accorded the same rights and respect as are the other staff members. Grapevine horses are almost unionized — their work hours are strictly monitored. They work five days and get a day off, work three weeks and get a week off to run free on the range. You can see they're happy horses. Walk up to the enormous corral, lean on the rail, and horses come meandering from all directions for a horsy chat or a friendly pat on the nose. It's obvious they like people — and why not? They experience only kindness and fair treatment at the hands of their humans. Grapevine has horses for all types of riders, and guests come from a variety of riding experiences, from novice to professional.

The trails vary from flat country and rolling foothills to steep mountains and rugged rocky paths winding high up where only eagles fly. Rides take you to places such as the Cochise Stronghold, hidden camp of the Chiricahua Apaches whose land this was only a hundred some years ago. Their most famous chief, Cochise, is buried somewhere in these mountains, his grave a secret to this day. Ferocious war chief Geronimo walked these same trails, and the echoes of the past seem to float on the still air as you ride the canyons and peaks once inhabited only by this proud and fierce people.

The rides are divided according to rider experience and skill; you are certain to get the type of riding you want. As Grapevine is also a working cattle ranch, there are many opportunities to join the crew at roundup and test your Western skills. One thing is certain — the Grapevine wranglers are eager to please you, and their aim is to give you the experience of your lifetime.

One only has to read the many enthusiastic comments in the ranch Guest Book to see that this is the place where dreams come true. As one Belgian couple put it, "As children we dreamt it, as adults we lived it at Grapevine Canyon Ranch."

Address: P.O. Box 302, Dept. BP, Pearce, AZ 85625

Telephone: 520-826-3185; **Reservations:** 1-888-BE-A-DUDE; **Fax:** 520-826-3636; **E-mail:** egrapevine@earthlink.net; **Internet:** http://www.gcranch.com or www.beadude.com

Location: 80 miles southeast of Tucson, 35 miles southwest of Willcox

Guest Capacity: 30

Season: Year round (Closed December 1-15)

Policy on Children, Pets: No children under 12. No pets.

Rates: $$ American plan. Minimum stay required.

Credit Cards: American Express, Discover, MasterCard, Visa

(See Pages 195 and 196 for Color Photos!)

ARIZONA
GRAPEVINE CANYON RANCH

LUXURY ACCOMMODATIONS.

SCENIC TRAIL RIDE.

Arizona
Grapevine Canyon Ranch

SOME OF OUR FINE HORSES.

GUESTS ENJOYING SEASONAL CATTLE WORK.

ARIZONA

Arizona

Iron Horse Ranch
Tombstone, AZ

Ironhorse Ranch is a working cattle ranch with a reputation for making people feel at home in an "Old West" atmosphere. The ranch offers visitors tours, entertainment, food, lodging and a swimming pool that is open May through October. Its unique 1880's town, "Tumbleweed'" provides a perfect backdrop for the ranch's many activities and special events.

The ranch is located along the San Pedro River, ringed by the Huachuca, Dragoon and Whetstone mountains. The area enjoys a year-round mild climate with plenty of sunshine, punctuated by stunning sunrises and sunsets.

The San Pedro Riparian Area provides a refuge for more than 100 species of breeding birds and another 250 species of migrant and wintering birds. There are also frequent sightings of deer, javelina (wild pigs), mountain lions and bobcats.

Daily horseback rides visit a wealth of nearby historical sites from the Indian and Spanish eras. One of the most interesting sites is the Presido Santa Cruz De Terrenate, located on the steep cliffs overlooking the San Pedro. The Presidio was established in 1775 for the Royal Spanish Army but later abandoned due to Apache raids.

The nearby ghost town, Contention, was established because of the abundant water supply needed for processing raw ore into silver bullion. Its name resulted from an argument between Dick Gird and Ed Schieffelin of Tombstone.

The Boquillas ruins near the ranch headquarters are the remains of an old railroad section station. These stations were built during the 1880's every 12 miles along the railroad for maintenance. It was here that the ranch got its beginnings. The property was acquired through the Homestead Program and the deed was signed by President Calvin Coolidge in 1927. It remained in the Kellar family until 1992 when it was remodeled to accommodate guests.

Today, the Ironhorse Ranch offers its visitors the newly constructed western town of "Tumbleweed" consisting of 18 buildings each offering very comfortable sleeping accommodations.

The Boquillas House, the newly remodeled 1903 railroad section house built on the ranch, offers large rooms, private baths and a spacious sitting porch. The quaint room above the Winchester Hotel in "Tumbleweed Town" has a veranda with wonderful views.

Ironhorse Ranch guests enjoy a wide variety of activities including guided historical horseback rides, wagon rides and other horseback activities. The Ironhorse Ranch has its own rodeo arena and offers a trail ride to Tombstone, aka "The Town Too Tough To Die," every week. The ranch also has a swimming pool to wash off the trail dust.

Ironhorse Ranch offers 16 rooms including 8 suites, 4 standards and 4 singles. The rooms at the ranch all appear to be buildings in a 1880's town. Inside, they are very comfortable sleeping rooms, each with its own western decor and personality. All of the rooms have heat, air conditioning, and private baths. Some rooms even include Jacuzzi tubs.

The Longhorn Saloon and Dining Room with its 4,000 square feet offers spacious dining with meals prepared by our accredited chefs. The Saloon with its full service bar and sitting area is a fun place to finish your long day in the saddle.

Address: P.O. Box 536, Dept. BP Tombstone, Arizona 85638

Telephone: 520-457-9361
Fax: 520-457-2471
Web: www.jimbalzotti.com

Location: 22 miles East on I-10 from Tucson.

Guest Capacity: 32

Season: Year-Round

Policy on Children, Pets: Cater to children over the age of 12. No pets.

Rates: $$

Credit Cards: Yes

(See Page 197 and 198 for Color Photos!)

Arizona

Iron Horse Ranch

ENJOY OUR OLD WEST TOWN!

YOU NEVER KNOW WHO'S COMING TO TOWN.

Arizona

Lazy K Bar Ranch
Tucson, AZ

No one ever quite recovers from their first exposure to the American Southwest: the sky, the dry clarity of the air, the flat terrain of the deserts colliding with the last outcroppings of the Rockies. The vistas are both harsh and magnificent, tolerant but not indulgent of man. It's a land you want to lose yourself in, match if you can. One way to do both — a dude ranch vacation. Since 1936, the Lazy K Bar Ranch (altitude 2,300 feet) has hosted tenderfeet of all ages and riding abilities. Like any ranch worth its salt, the Lazy K offers a unique vacation experience because it seems to exist in a world of its own.

Owned by Bill Scott and managed by Carol Moore, the ranch has something for everyone — horseback riding, swimming, tennis, mountain biking, hiking, or just plain relaxing. There are no TVs or telephones in the rooms to jangle your nerves; however, there is a big screen TV and VCR in the main lodge if you want some indoor entertainment. The bar is B.Y.O.B., and the nightly "happy hour" provides an opportunity to get acquainted with fellow guests. Evening programs consist of nature shows, country-western dance lessons, astronomy programs, local cowboy singers, cookouts, and of course, hayrides.

The main lodge is the hub of all social activities. All of the guest rooms are within an easy stroll of the lodge and pool, and each of the rooms has a full, private bath as well as individual heat and air conditioning. There are four bedroom-living room suites with fireplaces. All buildings are ground level, and some offer private patios. Gorgeous mountain views come with all accommodations!

As with any guest ranch operation, the star performers are the horses. Beginners or just-occasional riders shouldn't be anxious. All the horses are well trained, and the wranglers are adept at evaluating your experience (or lack of it) right off to assure you the most comfortable ride. The tall tales they spin don't hurt either! There are morning and afternoon rides with an option of walk-only or a ride that includes some loping. Scenic walk rides head up rocky paths in the foothills of the Tucson Mountains through stands of saguaro, barrel, and cholla cacti. The pace is slow due to the terrain, but there's a built-in excitement as the panoramic view unfolds beneath you. The desert flatlands offer the opportunity for some wonderful loping trails for the more experienced riders. A specialty of the Lazy K is team penning. This event takes place in the arena and is a timed team activity involving horses and cattle. This is a great opportunity to improve your horseback riding skills while you're having fun.

For those who want to enhance their riding skills, lessons are available at a nominal charge. Other riding activities that are frequently available but not included in your nightly rate are: moonlight rides, sunrise rides, and all-day excursions to surrounding mountain ranges.

If you want to take a break from riding — no problem. There is plenty to do around the ranch, or you can enjoy some of the local sites such as the Sonoran Desert Museum, Old Tucson Movie Studios, Tombstone, or a little jaunt into old Mexico. Excellent golfing is within a 20-minute drive.

Most folks find it hard to leave the serene, relaxed atmosphere of the ranch. You'll meet the friendliest staff, ride some of the best horses, and fill yourself with delicious food. Put it together, the sky, the desert and hills, wildlife, riding, an easy almost bygone pace. What it all spells… is the Lazy K Bar Ranch. There you'll experience a vacation that truly rejuvenates your soul.

Address: 8401 N. Scenic Drive, Dept. BP, Tucson, AZ 85743

Telephone: 520-744-3050, 800-321-7018; **Fax:** 520-744-7628; **E-mail:** lazyk@theriver.com; **Internet:** http://mmm.arizonaguide.com/Lazy.k

Location: 16 miles northwest of Tucson

Guest Capacity: 50

Season: September 15 to June 15

Policy on Children, Pets: Children must be 6 to trail ride. No pets.

Rates: $$ American plan.

Credit Cards: American Express, Discover, MasterCard, Visa

(See Page 199 for Color Photos!)

Arizona

Lazy K Bar Ranch

RIDING THROUGH THE ARIZONA HILLS.

Arizona

MULESHOE RANCH PRESERVE
WILLCOX, AZ

The Muleshoe Ranch Preserve is located on 49,000 acres in the foothills of the Galiuro Mountains, one of the most remote ranges in Arizona. In fact, access to the Preserve is a 27-mile dirt road passable by two-wheel-drive vehicles. Straddling the transition from Sonoran to Chihuahuan desert, the Muleshoe Ranch Preserve is owned and jointly managed by The Nature Conservancy, the U.S. Forest Service, and the U.S. Bureau of Land Management.

Elevation on the Preserve ranges from 4,100 feet to over 7,600 feet in the Galiuros. The landscape ranges from the high desert to the rugged uplands of the Galiuros to the lush areas lining the Preserve's seven perennial streams to the canyons and their cottonwood and willow forests. This varied ecosystem results in a diverse plant and animal life. Rare gray hawks, zone-tailed hawks, and black hawks nest along the streams. The streams themselves are a perfect habitat for native fish species like the Gila chub and speckled dace. Bobcat, mountain lion, deer, and bighorn sheep roam the hills and canyons.

Imagine spending a day — or more — riding through this natural setting! Riders must provide their own horses, and they are generally on their own once they arrive at the Preserve. There are several marked trails, and maps are provided at the Visitors Center. Riders can also use the BLM and Forest Service Galiuro Wilderness Areas north of Muleshoe Headquarters. Trails are rocky and hilly, so horses (and their riders!) should be in good condition.

In October, the Preserve arranges a special 3-day trail ride. Naturalists accompany guests on two trail rides and provide insight into the surroundings and the conservation efforts in the area. The 12-mile "High Lonesome Trail" ride features both rugged mountain views and lush canyons. The 6-mile "Beth Woodin Scenic Vista Trail" meanders through the uplands and provides an unrivaled panorama of the San Pedro River flyway. A special "sit-down" barbecue dinner in the Commons dining room gives guests a chance to relax, unwind, and share their experiences.

Hiking is also a popular activity at Muleshoe. There are both developed and undeveloped hiking trails near the Headquarters area. Other recreational opportunities include backpacking, photography, and birding.

Guests can camp or stay in a mix of original and renovated historic buildings that date from the early 1900s. The five "casitas" can accommodate up to 21 guests. They are charming and comfortable and completely furnished. Guests prepare their own meals. Casita guests are also welcome to use the natural hot springs soaking tubs. Three large corrals constructed of mesquite wood provide overnight accommodations for horses. If corrals are full, however, horses may have to be tied to trailers, hitching rails, or picket lines.

If you like getting away from it all and riding in a rugged, untouched area, you may find that the Muleshoe Ranch is just what you're looking for.

 Address: RR #1, Box 1542, Dept. BP, Willcox, AZ 85643

Telephone: 520-586-7072

Location: 30 miles northwest of Willcox, 110 miles east of Tucson

Guest Capacity: 21

 Season: Year round, day visits only June-August

Policy on Children, Pets: Children are welcome. No pets.

Rates: $ No meals included. Corral charges extra.

Credit Cards: American Express, MasterCard, Visa

Arizona

Arizona

Spirit Horse Ranch
Pearce, AZ

Ride quietly into ancient Apache lands, once home to Cochise, Geronimo and Buffalo Soldiers. Listen carefully and you may still hear the echo of charging horses or a blaring bugle. Then again, given the vastness of this land, you may for the first time hear only the night's falling stars as you soak in the outdoor hot tub. Come to the Real West, home of Tombstone, lost gold mines and endless cattle range. Well-trained horses provide both safe and adventurous rides suiting any level of experience. You will enjoy our "discovery" rides to places like Cochise Stronghold, Chiricahua National Monument, Fort Bowie, etc. A limited number of guests allows for time to see, touch, and learn about the land and its history. You won't find yourself lost at the back of a long line here.

Spirit Horse Guest Ranch rests at the foothills of the spectacular and historic Chiricahua Mountains of southeastern Arizona. As "Sky Islands", these mountains and adjacent areas are home to more than half of North America's bird species. Deer, hawks, coyotes, owls, and other wildlife dwell on our ranch. At an elevation of 5000 feet and set along Rock Creek, our ranch is nestled in a grove of ancient oak trees. Surrounded by tens of thousands of acres of National Forest land and cattle range, our location allows endless opportunities to discover this magnificent land. Spend the day riding into "secret" canyons or ride out onto vast cattle range. "Wow" becomes an often heard word on the trail. Our setting offers ideal year-round weather with warm sun-filled days and clear, cool nights. We are located only 90 miles southeast of Tucson.

The main house is where we meet for dining and relaxation. Sit on our verandah and sip a cool drink while you take in our scenery. Or, shake the dust from your boots and kick up your feet in our fireplaced lounge. After dinner, join us for good stories, a game of checkers, Native American flute music or some piano music.

Our newly refurbished guest rooms are located in separate bunkhouses that contain a queen sized bed, futon, a bath with shower, hair dryers, linens and bathrobes. Guest rooms can accommodate 1 to 4 people. All meals are served family style with picnic lunches provided for the trail rides. Transportation to and from Tucson's International Airport can be arranged.

If you love to ride horses or just want to lie quietly in a shaded hammock and are looking for a vacation that will truly restore you, then Spirit Horse Guest Ranch is the place for you. Cross through our gate and leave it all behind. You can do that here. So pack those jeans, boots, and hat and head on down!

 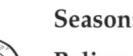

Address: HC 1 Box 360, Pearce, Arizona 85625

Telephone/Fax: 520-824-3667
E-mail: trailboss@vtc.net
Web: www.spirit-horse-ranch.com

Location: 90 miles SE of Tucson, 40 miles south of Wilcox, Arizona

Guest Capacity: 10

Season: Year-Round

Policy on Children, Pets: No children under 12. No pets.

Rates: American Plan. Minimum stay required.

Credit Cards: Visa, MasterCard

ARIZONA
SPIRIT HORSE RANCH

AN ARIZONA TRAFFIC JAM!

COME VISIT THE SOUTHWEST.

Arizona

Sprucedale Guest Ranch
Alpine, AZ

Sprucedale is a family-owned and -operated guest ranch located high in the cool White Mountains of eastern Arizona. Owned by the Wiltbank family since 1941, Sprucedale has made many wonderful memories for guests from all over the world. In fact, the Wiltbanks are into welcoming the second and third generation of returning guests.

The beautiful, peaceful surroundings are just one of the things that bring guests back year after year. The ranch is surrounded by the Apache National Forest, with its stands of ponderosa pine, aspen, and, of course, spruce trees. The grassy meadows are decorated by a wide variety of wildflowers and crystal-clear streams, with deer and elk grazing nearby.

Of course, the wonderful horseback riding is a major reason people come to the ranch. The Wiltbanks raise and train their own string of horses, so they know each horse and its capabilities. They can match up each rider with a horse that is well suited to the rider's needs. Whether beginner or experienced, everyone enjoys the horse rides and other horse-related activities. There are two trail rides a day, with the morning ride separated into three ability levels. There's also a picnic and swimming ride and a gymkhana (guest) rodeo once a week. From rounding up cattle to branding to milking the cows to working with the young colts, there are lots of hands-on activities for guests of all ages. The annual lovable, furry kitten litter is always a big hit with both children and adults.

Sprucedale is off the beaten path, with a road that ends at its gate. The ranch setting makes for a safe place for children to roam and experience life on a working cattle/horse ranch. When not on a horse, children love to splash in the creek or play basketball or volleyball. Parents love to sit on the porch and relax or read or just drink in the awesome view. Everyone enjoys the evening activities of bonfires, western dancing (they'll even teach you how!), and hayrides. You can also bring along your mountain bike and fishing rod and try your hand in the nearby streams and lakes. There is plenty of time to spend with your family, and many opportunities to make some great new friends.

The accommodations are rustic, cozy, and clean cabins. Each cabin has a full bath and a wood-burning stove for those cool evenings. And the beds are covered by handmade quilts. The front porches on each cabin are great places for reading, relaxing, or perhaps entertaining newly made friends.

The Wiltbank family and their capable and friendly kitchen crew serve up great home-cooked meals family style. Guests look forward to the homemade breads, delicious soups, tasty jams, hearty meals, fresh salads, and of course the baked goodies of all kinds. Yep, you definitely won't go away hungry!

Many a family reunion has been held at Sprucedale. Families can come and enjoy each other's company, all the while having a great ranch vacation. Corporate groups and other organizations come (usually in June or September) to really get away and do some quality planning and peaceful regrouping.

The Wiltbank family continue the tradition of offering a unique family-oriented experience. Sprucedale is a haven for those who want to get away from it all and to have a great time while doing it.

 Address: Summer: HC61 Box 10, Dept. BP, Alpine, AZ 85920; Winter: P.O. Box 880, Dept. BP, Eagar, AZ 85925

Telephone: 520-333-4984

Location: 24 miles southwest of Alpine off Route 191

Guest Capacity: 45-50

 Season: Memorial Day to mid-September. Specialty groups and seminars after that.

Rates: $ American plan. Children's rates. Riding extra. Six-day minimum stay.

Credit Cards: None. Personal checks and cash accepted.

Arizona

ARIZONA

WHITE STALLION RANCH
TUCSON, AZ

This charming, informal guest ranch truly gives you a feeling of the Old West. Only at White Stallion Ranch can you find 3,000 acres of such wide-open land at the foot of the ruggedly beautiful Tucson Mountains. The ranch has an excellent location, sharing 2 miles of common fence with the Saguaro National Park and access to its thousands of acres. Yet the city of Tucson is just 15 minutes away on the other side of the mountains. The ranch seems isolated but is close to "civilization."

The ranch has been the home of the True family since 1965, and the third generation of Trues is now growing up there. Ranch informality is blended perfectly with the comforts of a top resort. You'll enjoy the western-style horseback riding with the wranglers, fine horses, and scores of scenic trails. Whether horseback riding is new to you or you are an experienced rider, the varied riding program will fit your needs. From slow scenic rides through the saguaro cactus to brisk, fast rides to exciting mountain rides, there is something for everyone. For the rugged few, the weekly all-day rides are a challenge, and for the cowboy in all of us, team cattle penning is a taste of work on a cattle ranch.

This is a working cattle ranch, and it's great fun to watch the longhorn cattle herd coming in to water. The weekly rodeo that is held on the ranch has ranch owners, staff, and local cowboys and cowgirls roping, steer wrestling, barrel racing, and working cattle. The announcer talks about the realities of rodeo today and gives some of the history of rodeo.

Many of the ranch guests prefer petting horses to riding them. They enjoy the sparkling heated pool, indoor redwood hot tub, guided hiking, tennis courts, shuffleboard, horseshoes, pocket billiards, table tennis, basketball, volleyball, and the extensive collection of books in the library. The petting zoo will delight young and old alike with its miniature horses, llamas, pot-bellied pigs, and pygmy goats. Guests can spend some time relaxing on the patios, in the cozy and comfortable living room, or in the Happy Hour Saloon, which offers a full bar. The ranch is noted for delicious meals, cookouts, steak barbecues, hayrides, and breakfast rides.

The White Stallion Ranch is a great place for children. The Trues say they have never heard a kid complain of being bored. White Stallion Ranch takes the family approach to children. There are no specific children's programs. Activities, meals, and riding are available for the whole family to do and enjoy together.

Southern Arizona has a lot to see, and the ranch offers sightseeing trips for an additional cost. Rental cars are also available at the ranch. The ranch gift shop has a unique collection of Southwest gifts and many of the necessities you may have forgotten.

The White Stallion Ranch truly offers the best of both worlds — an authentic ranch experience with the convenience of being close to sightseeing and shopping.

Address: 9251 West Twin Peaks Road, Dept. BP, Tucson, AZ 85743

Telephone: 520-297-0252;
Reservations: 888-977-2624;
Fax: 520-744-2786
Web: www.wsranch.com or www.jimbalzotti.com
E-mail: wsranch@flash.net

Location: 17 miles northwest of Tucson

Guest Capacity: 75

Season: September 1 to May 31

Policy on Children, Pets: Children welcome, but no organized children's program. No pets.

Rates: $$ American plan. Riding instruction extra.

Credit Cards: None

(SEE PAGE 200 AND 201 FOR COLOR PHOTOS!)

ARIZONA
WHITE STALLION RANCH

LOPING ALONG A TRAIL.

WHITE STALLION RANCH

Arkansas

Horseshoe Canyon Ranch
Jasper, AR

Horseshoe Canyon Ranch is 350 beautiful acres located in the high country of the Natural State...the pristine Buffalo National River wilderness of the Ozark Mountains near Jasper, Arkansas.

This ranch, owned and managed by two generations of the Johnson family, offers guests a ÒWestern experience with Southern hospitalityÓ. Guests stay in comfortable, rustic log cabins and have spectacular views of the rolling meadows and high cliffs. All cabins have excellent cooling and heating systems, but for those nights with a nip in the air, there are also wood-burning stoves on rock hearths that provide a romantic glow as well as additional warmth. Imagine "settling in" at the end of an incredible day...a day filled with new adventures in the clear Ozark mountain air. Our guests rarely have trouble sleeping!

Ranch fare at its finest is served family style in the main lodge from early spring to late fall. Guests respond quickly when they hear the bell ring signaling another delicious meal. For "snackers" and those with special needs, each cabin is equipped with a microwave and small refrigerator.

A six-day stay at the ranch includes all meals, horseback riding (including a great breakfast, moonlight, and cookout rides!), canoeing, a day trip to Branson, Missouri, ranch rodeo, hayride, hiking, climbing and entertainment. Pond fishing is available with river fishing nearby. Table tennis, horseshoes, volleyball, and other activities are provided for use during unscheduled time. The pool is seasonal, but the hot tub is not; either one sure feels relaxing after a day in the saddle! A short ride to one of the Buffalo River swimming holes is a unique and fun experience, too.

Families are encouraged to participate in ranch activities together, but special events for youngsters include riding in the arena, camping, roping and nature walks. Each evening there is a special activity, and guests are welcome to do as much or as little as they like.

Riding is the main attraction at the Horseshoe Canyon Ranch, but for those who prefer simply a quiet interlude from the hustle and bustle of "real" life, Horseshoe Canyon Ranch is the place to be. Every season of the year has a different appeal. The sky is magnificent when the moon is full, and without the city lights to obscure them, the stars seem close enough to touch from your cabin porch. Several streams cross the meadow surrounding the cabins and provide pleasant sights and sounds during early morning or evening walks. A quiet porch, a comfortable rocker, no fax or phone...what a wonderful vacation!

The 7,000 square foot barn is the hub of ranch activity and guests are welcome to just hang around or help with the stock. Animals in the petting zoo need care too, and children are invited to feed, water, and groom them under the supervision of the staff.

Because the Johnsons operate the ranch themselves, they are able to arrange activities to suit the particular desires of their guests. Wranglers and other staff members will guarantee that though you may come as strangers, you'll leave as friends.

Address: HCR 70 Box 261, Jasper, Arkansas 72641-9723

Telephone: 1-800-480-9635 or 1-870-446-2555
Web www.jimbalzotti.com
E-mail: hcr@aristotle.net

Location: 50 miles from Branson, Missouri

Guest Capacity: 52

Season: Open Mid-March to Late Fall

Policy on Children, Pets: Children welcome. No pets.

Rates: $-$$

Credit Cards: Visa/MasterCard

(See Page 202 for Color Photos!)

ARKANSAS

HORSESHOE CANYON RANCH

THE LODGE AND CABINS AT HORSESHOE CANYON RANCH.

OUR BACKYARD VIEW.

ARKANSAS

LOST SPUR RANCH
HARRISON, AR

Nestled in the beauty of the Ozark Mountains you'll find the Lost Spur Ranch. Owned and operated by Lynn and T.J. Hunter, the Lost Spur is a wonderful place for folks who want to experience the good life on a honest-to-goodness ranch.

Saddle up and follow the wrangler out on an early morning trail ride. It's not unusual to spot deer, wild turkey, fox, beaver-even an eagle circling high above the bluffs. Enjoy a hearty cowboy-style breakfast cooked over an open fire.

At the Lost Spur, you'll enjoy horseback riding, sharpening your riding skills in the arena and roping lessons. A gigantic tree house, farm critters and playing in the creek will delight kids of all ages. Our covered wagon rides, pulled by our matched pair of Belgian Mares, is a treat for everyone. The clear waters of Crooked Creek invite you to fish for small mouth bass, take a refreshing afternoon swim, or try a peaceful canoe outing. You might just relax and while away a lazy afternoon in an old-fashioned porch swing. And who could resist a moonlight hayride or a romantic time in the cedar hot tub nestled away in the woods before turning in?

The Lost Spur is able to host up to 34 guests in four large cedar cabins and the homestead cottage. A beautiful wooded setting offers panoramic water and mountain views. Uniquely decorated cabins offer heat and air conditioning; queen, twin and bunk beds; as well as private baths, small refrigerators, and coffee makers. Two cabins equipped with kitchens are available November 1 through March 1. Although no activities are offered these dates, you are free to enjoy the serenity of our countryside.

We all know that great food is one of the best parts of a vacation. Join the rest of the Lost Spur family at the big dining barn three times a day for mouthwatering meals prepared by Lynn Hunter herself. Family-style buffets feature homemade breads, pastries, fresh vegetables and the finest cuts of meat. Enjoy homemade ice cream and fruit cobblers every week.

At Lost Spur, there's something for everyone no matter what your age, or feel free to sit back and soak up the tranquility. Go for a long hike. Take a horseback ride: we offer rides twice a day on trails that wind through woods, meadows, and creekside. Relax in a hot-tub, swim, canoe, or fish in Crooked Creek (bring your own tackle). Try volleyball or a good old-fashioned game of horseshoes. Kids can spend as much time as they like on the playground or in the game room.

We offer a weekly authentic Chuckwagon dinner cooked western-style, complete with a live country music show. Branson, Missouri, Eureka Springs, and numerous caves are within an hour's drive of the ranch. Special tour packages are available. Harrison and surrounding areas offer great antique shops and flea markets.

Come join us. You'll feel like family before the sun sets.

Address: Route 3 Box 93 Harrison, Arkansas 72601
Telephone: 1-870-743-7787
Fax: 870-743-6686
Web: www.jimbalzotti.com
E-mail: lostspur@yournet.com
Location: 45 miles south of Branson, Missouri
Guest Capacity: 32

Season: March 1 - October 30th, Cabins available for rent in off-season
Policy on Children, Pets: Children welcome. No pets.
Rates: $-$$
Credit Cards: Yes

Arkansas
Lost Spur Ranch

A COUPLE OF YOUNG COWBOYS HEADIN' OUT FOR A RIDE.

Panther Valley Ranch
Hot Springs National Park, AR

An authentic western horse ranch that is over one hundred years old, Panther Valley Ranch is just five miles from the city of Hot Springs and nestled in the beautiful Quachita Mountains. Roger Stanage and his wife Jerrie, who own and operate Panther Valley, especially cater to horseback riding vacationers who want to escape the hustle and bustle of their daily routine and do some of the finest riding in Arkansas. At Panther Valley, you might catch a glimpse of a deer, wade in crystal-clear streams, experience the lush fern forest, pick wild blueberries, or generally just enjoy the natural beauty of this the "Natural State."

At Panther Valley, there's a spacious ranch house that sleeps ten. It has king, queen, and bunk beds; a full kitchen; a den complete with air conditioning and a wood heater; two bathrooms; and a sprawling, long lazy front porch that offers you a cozy, rustic retreat. The large porch swing is the perfect place to sit and watch all of the ranch activities. The Loft sleeps eight and has a large entry/den/kitchen area. The deck overlooks the corral. The unique Barn Room is inside the 25-stall stable and has a fireplace and a 25-foot beamed ceiling.

There is a large barbecue and plenty of picnic tables under the three giant cedar trees near the house. You can fish in the ranch's small private lake or try your luck on one of the area's excellent bass lakes. There are many area marinas that rent fishing boats, ski boats, pontoon barges, or anything else you might need.

The horseback riding on the ranch is wonderful. You can go on daily rides or try the overnight horseback campouts, where our seasoned guides do all the work — leaving you to have all the fun, roasting marshmallows over the open flames, and listening to scary ghost stories! Panther Valley also offers hayrides, rock hunting, horseshoes, and a terrific place to just get away from it all!

The overnight adventure is a favorite of ranch guests. You'll saddle up your favorite horse and ride to a nearby scenic mountaintop campsite. There you'll find the campfire and a wrangler cooking a big juicy rib eye steak, just the way you like it. Roger's special twice-baked potato, secret salad (we would tell you what's in it, but you'll have to come down and find out for yourself!), and specialty desserts round out the meal. Topped off with a fresh, hot pot of coffee, you'll settle right in and enjoy the sunset. After a refreshing night's sleep (in a tent you didn't have to pitch), you'll awaken to the aromas of a big hearty ranch-style breakfast before mounting up and heading back down off the mountain.

Come and be a part of Panther Valley Ranch, and you'll be one of the many friends who come back again and again!

Address: 1942 Millcreek Road, Dept. BP, Hot Springs National Park, AR 71901

Telephone: 501-623-5556;
Fax: 501-624-2998
Web: www.jimbalzotti.com
E-mail: pantheval@prodigy.net

Location: 5 miles from Hot Springs

Guest Capacity: 60

Season: Year round

Policy on Children, Pets: Children welcome. No pets except horses.

Rates: $ - $$ American plan. Guided trail rides and overnight adventure extra.

Credit Cards: Discover, MasterCard, Visa

Arkansas
Panther Valley Ranch

YOUR HOST, ROGER STANAGE, LEADING A TRAIL RIDE.

ARKANSAS

Scott Valley Resort & Guest Ranch
Mountain Home, AR

Owners Tom and Kathleen Cooper have fashioned a riding resort that will meet the criteria of every level of rider. If you've never tried a ranch vacation before, this one is perfect because it offers a huge variety of activities at an extremely reasonable rate. Named as one of *Family Circle*'s "family resorts of the year" for three years in a row, Scott Valley is nestled on over 1,200 acres in a peaceful valley in the Ozark Mountains. The Ozarks are not huge snow-capped mountains, but they are some of the oldest on earth. You might even find some Cambrian and Precambrian rocks and you won't have to worry about altitude sickness.

Scott Valley has a fine line of gentle horses, most of which were bred, born, and gentle-trained right there on the ranch. They specialize in Missouri Fox Trotters, a horse that is people-oriented and willing to please. Wranglers will match you with a horse that suits your riding level and temperament. Trail rides are separated into three categories — novice, intermediate, and advanced. Parents can even ride double with their little tikes!

There's something for everyone, even that "not so enthusiastic" horseperson. You'll find world-class trout fishing on the White and North Fork rivers or, if you prefer, canoes for a quiet scenic trip down the river. There are plenty of hiking trails, plus tennis, volleyball, shuffleboard, badminton, or horseshoes for those who are looking for a little more action. After a long ride, you can relax and cool off in the swimming pool or whirlpool spa. The recreation room has table tennis, a pool table, and electronic games. Or, you might just want to curl up with a good book. For those looking for relaxation and terrific riding, the spring and the fall are great times to come for unlimited riding plus peace and quiet.

As a family-oriented resort, there are lots of things for children to do. There's a full playground complete with slide, swings, rocking horse, and sandbox. Kids can help out in the petting zoo. And one night, there's a hayride and a cookout that is always a hit with the young ones.

During the summer months, there is nightly entertainment. One night, you can test your athletic skills in a ballgame. On another night, Scott Valley charters a ferry for a tour of Lake Norfolk, and guests have dinner on the ferry. There's wrangler night, and one night reserved for country line dancing. A magician will entertain the young at heart, and the ice cream social is fun for guests of all ages.

If you'd like to go farther afield, Blanchard Springs Caverns are nearby and Branson, Missouri, is just 1-1/2 hours away. Tom and Kathleen will provide you with a map showing the best places for crafts shopping, antiques shopping, or just plain shopping.

Accommodations at Scott Valley are clean and comfortable, and with maid service you don't even have to clean up after yourself — much. There are no telephones or televisions in the rooms. But there are TV lounges for those unable to go cold turkey! The resort is known in the area for its great food, and it's all-you-can-eat for breakfast, lunch, and dinner. Meals are "down-home, good cookin'" things like homemade chicken and noodles, biscuits and gravy, and desserts the like of which you've not seen! Salads and fresh fruits and vegetables are a-plenty for those people who prefer a low-fat diet.

Tom and Kathleen Cooper are here to see that you thoroughly enjoy your ranch experience — whether you're a first-time visitor or a veteran. As they say, "Our guests become our friends, and we place a very high value on friendship."

Address: P.O. Box 1447, Dept. BP, 223 Scott Valley Trail, Mountain Home, AR 72653

Telephone: 870-425-5136;
Fax: 870-424-5800;
Web: www.scottvalley.com or www.jimbalzotti.com

Location: 7 miles southeast of Mountain Home, 156 miles north of Little Rock

Guest Capacity: 75+

Season: March to November

Policy on Children, Pets: Family-oriented resort; children most welcome. Small, house-trained pets welcome for additional charge.

Rates: $ - $$ American plan. Children's and group rates.

Credit Cards: MasterCard, Visa

Arkansas

Scott Valley Resort & Guest Ranch

SOUTHERN RIDING AT ITS BEST.

SADDLED UP AND READY TO RIDE

CALIFORNIA

ALISAL GUEST RANCH
SOLVANG, CA

The Alisal Guest Ranch and Resort is a horse lover's paradise, an unspoiled 10,000 acre expanse of Santa Barbara County countryside where the romance and majesty of the Old West continues today in resort activities as varied as trail rides, rodeos, cowboy poetry recitals and more.

The Alisal has hardly changed since the days of the vaquero. Its history began 39 years after the Spanish padres established Mission Santa Ines in 1804, when conquistador Raimundo Carrillo was given a land grant in the Santa Ynez Valley for his service to the Mexican government. Carrillo used the grant to raise cattle, and it has been a cowboy's land ever since.

In 1946, a century after the original land grant for "Rancho Nojoqui" (after the Chumash name for the Santa Ynez Valley), The Alisal entered a new era and opened as working cattle ranch/guest ranch combination - and remains so today. It was an immediate success, soon achieving a national reputation. People come from all over and return year after year.

With countless trails to explore, The Alisal offers horseback adventures into the undisturbed Santa Ynez foothills, where cattle, horses and deer roam freely. On certain days of the week, the morning rides begin with a campfire breakfast and cowboy poetry at the historic Adobe Camp.

The Alisal's own spring-fed lake offers a tranquil setting for sunning, canoeing, bird watching, or fishing by rod and reel or fly - the lake is teeming with bass and bluegill. Private guided fishing excursions are available, too, and nature walks highlight wildlife from bald eagles to bobcats to deer.

The Alisal also offers two scenic championship golf courses to challenge even the best players. Each course has a distinctive personality shaped by the area's unique terrain. Test your skills on the secluded guests-and-members-only Ranch Course, or tee off on the more recently-built River Course. Both courses are ideal for private tournaments, have temptingly-filled pro shops and inviting club houses.

A complete seven-court tennis complex in one of the most scenic areas of the property, replete with a well-stocked pro shop. The tennis professionals, one of which is nationally ranked, offer clinics and private lessons by appointment. They also coordinate group tournaments or individual games.

Children are always welcome at The Alisal. A full program of supervised activities fill their days and create lots of memories. Among the activities are a daily Arts and Crafts program, horseback riding program, Petting Zoo, lake activities, golf and tennis, numerous outdoor games and the pool.

To sum up The Alisal experience, some come for the horses, others for golf, tennis or fishing. But everyone comes for the great food and easy going atmosphere. Just a two-and-a-half hour drive north of Los Angeles, The Alisal Guest Ranch and Resort offers an early California life-style nearly as valued for what it doesn't have as for what it does.

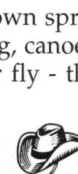 

Address: 1054 Alisal Road, Solvang, California 93463

Telephone: 805-688-6411 800-4-ALISAL
Fax: 805-688-2510; **E-Mail:** sales@alisal.com
Web: www.alisal.com

Location: 35 miles NW of Santa Barbara

Guest Capacity: 200

Season: Year Round

Policy on Children, Pets: Very "child friendly". No pets.

Rates: $$$

Credit Cards: Amex, MC, Diners Club, Visa

(SEE PAGE 204 AND 205 FOR COLOR PHOTOS!)

CALIFORNIA
ALISAL GUEST RANCH

RANCH COURSE — BRIDGE IN FOREGROUND.

CALIFORNIA

COFFEE CREEK RANCH
COFFEE CREEK, CA

Picture-postcard views await you in the Trinity Alps Wilderness Area of snow-capped mountains, cascading waterfalls, lush meadows, and sparkling lakes. In the heart of it all is northern California's finest guest ranch, Coffee Creek — or, as the guests affectionately call it, "home away from home."

The ranch lies along Coffee Creek, an excellent fly-fishing stream for wild trout. If that isn't enough, there's a pond that awaits young and old looking for those 3-pound "whoppers." Or hike or ride to an alpine lake for some tasty brook trout!

The chefs prepare nutritious meals laden with fresh fruits and vegetables. Only the choicest, leanest beef is served alongside home-baked breads from old family recipes. The menu is planned and prepared keeping in mind both children's tastes and the hearty appetites of the adults.

The secluded cabins at Coffee Creek Ranch may look rustic from the outside, but the atmosphere inside is cozy. And each cabin has a theme based on its brand; Circle B, for example, is decorated with bears, Bigfoot, and butterflies! All fourteen cabins have tall pines and ceiling fans to help keep them cool. One ranch house room and one cabin are equipped for the handicapped. Most family cabins have wood-burning stoves where a view of the fire sets the mood for those "romantic weekend getaways" for couples. Adult/singles weeks are offered during the spring and fall.

Coffee Creek has every imaginable dude ranch amenity, with the most popular still being horseback riding. The ranch breeds its own Thoroughbreds and Appaloosas. Typically, two rides a day are scheduled, and rides are divided by ability level. Private lessons can also be scheduled. Breakfast rides are the most popular, with the all-day "wilderness" ride to a lake the most spectacular! For a unique riding experience, the all-day ride can be extended into an overnight pack trip returning the next day. Wilderness pack trips led by our licensed and bonded guide come complete with cook for roughing it in style! The biggest riding event of the week is the gymkhana, where adults and children compete for the prized blue ribbon in various events.

When guests aren't out riding in the Trinity Alps, there's plenty to keep them busy back at the ranch. They can swim in the heated pool, soak in the waterfall spa, or work out in the exercise room. They can try their hand at badminton, volleyball, horseshoes, table tennis, shuffleboard, canoeing, or basketball. Guests can test their skills at the rifle and archery range. Hiking trips can be planned, and mountain bike trails parallel the Wilderness Area.

Summer evening events are organized by the ranch staff. Bonnie and Clyde, the ranch Belgians, head up the Sunday night hayride and bonfire. The rest of the week is sprinkled with bingo, square and line dancing, a talent show along with the ranch band, the "Rattlesnakes."

Supervised summer youth programs are specially designed by age groups: Bronc Busters for 13 to 17 year olds, Junior Wranglers for ages 8 to 12; and Cowboys and Cowgirls for ages 3 to 7. Kids enjoy a fun-filled schedule of pony rides, nature walks, crafts, and panning for gold. Ranch weddings are special whether it be just the two of you or 250 guests. The staff helps you arrange everything from the ceremony to the reception! Small business conferences and family reunions are also welcome. During the winter months, the ranch offers cross-country skiing. Other winter activities include snowshoeing, hiking to the pond for some ice fishing, tubing down Shotgun Hill, sleigh rides, and dog sledding. You can be sure that you'll find just what you're looking for at Coffee Creek Ranch.

 Address: Coffee Creek Road, Coffee Creek, CA 96091; Mailing Address: HC2 Box 4440, Dept. BP, Trinity Center, CA 96091

Telephone: 530-266-3343, 800-624-4480; **Fax:** 530-266-3597

Location: 72 miles northwest of Redding

Guest Capacity: 50; 30 in winter

 Season: Year round; dude ranch, April through October; cross-country ski resort, mid-November through February

Policy on Children, Pets: Children welcome; organized children's program by age group. No pets.

Rates: $ - $$ American plan. Riding is extra in summer; included in spring and fall.

Credit Cards: American Express, Discover, JCB, MasterCard, Visa

(SEE PAGE 206 AND 207 FOR COLOR PHOTOS!)

CALIFORNIA
COFFEE CREEK RANCH

OUR SUNDAY DINNER BUFFET BRINGS KIDS AND COWBOYS TOGETHER.

RANCHHOUSE OVERLOOKING OUR POOL.

CALIFORNIA

THE HOMESTEAD
AHWAHNEE, CA

The Homestead is not for everyone. What owners Larry Ends and Cindy Brooks have created from the ground up is a special getaway for couples who want to ride their own horses during the day, yet come back in the evening to luxury and pampering. What you'll find at The Homestead is absolute peace and quiet, a place of serenity where you can forget the "real world" and have a special treat not only with your horse, but also with that special person in your life.

Larry and Cindy looked long and hard for the perfect property that combined the best of both worlds from a horse lover's perspective — breathtaking, wide-open country filled with memorable riding adventures and private, comfortable accommodations for both you and your equine friend.

In 1990, Larry and Cindy purchased 160 oak-studded acres and spent a year planning the site to preserve nature yet provide ultimate views. Then the fun began — creating four wonderfully self-contained country cottages, sparing no details of comfort and design. Each cottage has a master bedroom and bath, a comfortable living room, and a full kitchen stocked daily with coffees, teas, muffins, and fruit. The horses are comfortably housed in 12 x 12 foot partially covered pipe corral stalls.

Larry and Cindy have researched and prepared a full range of riding possibilities — complete with maps and helpful information. They will help you plan your rides and will share with you their own very special spots, many just outside of Yosemite in the Sierras where you will most likely not see anyone else! Although there is no direct access to the trails from The Homestead's property, trailering to the trailheads is close and easy. You'll see such spectacular sights as the Giant Sequoias and panoramic views of Bass Lake. The possibilities are endless. Every few weeks, Larry and Cindy ride the trails to get the latest update on trail conditions. Many endurance riders take advantage of training on the logging roads that crisscross the Sierras and can keep you trotting on for miles and miles.

The Homestead is also quite well known as a destination for non-equestrians who want to enjoy the peace and solitude of the mountains. Many have spent a memorable honeymoon, anniversary, or special birthday in the serenity of the cottages and peaceful grounds. The Homestead is run in tune with nature's quiet rhythm so that guests may enjoy the ultimate in peace and quiet.

At The Homestead, you'll feel like you're a million miles away. With a fully equipped kitchen and fabulous outdoor barbecue, you can relax in the evening preparing your own meals and gazing at the incredible stars. Or, just 10 minutes to Oakhurst for a choice of over fifteen restaurants including world-renowned and 5-star Erna's Elderberry House.

Antique stores, art galleries, feed and saddlery shops make a fun out-of-the-saddle afternoon, or you could drive to Bass Lake for a refreshing swim. Relax while away from The Homestead knowing your horses are in a safe environment and caring folks are keeping an eye on them. And, of course, there is always the massage therapist who will come out to your cottage and soothe those aching muscles.

Larry and Cindy are always available to assist you, but they also respect your privacy to just rest and relax. Try The Homestead, a relaxing retreat for horses and their owners at the gateway to Yosemite.

Address: 41110 Road 600, Dept. BP, Ahwahnee, CA 93601

Telephone: 209-683-0495; **Fax:** 209-683-8165; **Web:** www.sierranet.net/~homestead

Location: 60 miles from Fresno

Guest Capacity: 10

Season: Year round

Policy on Children, Pets: Children limited to a maximum of two. No pets.

Rates: $$ Continental breakfast provided.

Credit Cards: American Express, MasterCard, Visa

CALIFORNIA

HOWARD CREEK RANCH
WESTPORT, CA

Imagine yourself riding horseback on a private beach while Pacific Ocean waves crash along the shore. That could be *you* if you visit the rustic Howard Creek Ranch on the scenic Mendocino coast of northern California.

You'll find a bit of New England in the large 1871 Victorian farmhouse filled with antiques and set in the middle of wide, green lawns. Award-winning flower gardens surround the inn. Horses and cows graze the pastures. And they're all snuggled in the valley alongside Howard Creek, which flows to the ocean two hundred yards away. It has been called enchanting, "a private theme park of rural pleasures."

Unmistakably California, however, are the inn's hot tub and sauna and the buildings, which are constructed of virgin redwood from the ranch's own forest. There is also swimming in the solar-heated pool after an adventure-filled day of exploring the redwood forest by foot or by riding the famous Skunk Train, a 40-mile journey on a historic logging line into the redwood forest. Tide pooling and whale watching are also favorite pastimes. A 75-foot swinging footbridge over Howard Creek connects the main house to this working farm's barns, cabins, and the beach. The ranch is home to two giant Percherons that pull a farm wagon for day and moonlight rides. You might want to combine that with a barbecue on the beach.

Horseback riding is available through arrangements with the neighboring ranch — Ricochet Ridge Ranch and its owner Lari Shea. Her Akhal-Teke/Russian Orlov crosses, Arabians, Appies, and Quarter Horses are well trained and suited to any riding level. Rides range for miles along the Pacific beaches, through a working cattle ranch, and up into the coastal mountain range. Private rides are available if arranged in advance.

Nearby Mendocino and Fort Bragg can provide a change of pace if you'd like to explore their art galleries, museums, and fine shops. Deep sea fishing is available from Noyo Harbor, and you might also want to go hiking in the nearby mountains.

Nominated to the National Historic Register, Howard Creek Ranch is set on 40 acres with sweeping ocean and mountain views — an oasis of peace and beauty. The ranch is also located near the "Lost Coast," a 60-mile-long wilderness along the California coast.

Accommodations include the main house and cabins, with suites and rooms furnished with antiques, large, comfortable beds, and handmade quilts — all with views of the ocean, mountains, creek, or gardens. While the richly restored decor exudes a feeling of authenticity, it also conveys a feeling of serenity.

A hearty country breakfast is prepared each morning on the ranch's original wood stoves by your hostess, Sally Grigg. The menu changes each day and depends on what is farm fresh that day. A typical breakfast might include fresh juice and fruit, omelets, country style potatoes, extra-thick bacon or locally made sausage, fruit and spice muffins, and strong coffee served early. Tea is available anytime. There are also microwaves and refrigerators on the premises for guests who get the "munchies."

Howard Creek Ranch caters mainly to adults seeking a leisurely, relaxing, and serene "get-away-from-it-all." It's even a favorite destination for honeymooners. Pets are welcome if arrangements are made in advance.

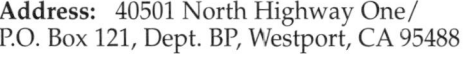

Address: 40501 North Highway One/ P.O. Box 121, Dept. BP, Westport, CA 95488

Telephone: 707-964-6725; **Fax:** 707-964-1603

Location: 170 miles north of San Francisco

Guest Capacity: 22

Season: Year round

Policy on Children, Pets: By prior arrangement only.

Rates: $ Gourmet breakfast provided.

Credit Cards: American Express, MasterCard, Visa

CALIFORNIA

HUNEWILL GUEST RANCH
BRIDGEPORT, CA

It starts when you drive out the gate.
Where the one lane blacktop splits a green meadow
in a valley where the Sawtooth rise to the sky.
That's when the memories start.

Hunewill Guest Ranch offers some of the greatest riding in the world. Wide open meadows beneath Yosemite's lofty Sierra peaks are the backdrop for the kind of riding you've only dreamed about. At this family-owned and -operated ranch, you can unwind from your day's ride with a leisurely stroll down the ranch road, take a hayride on a cool mountain evening, or you can watch a western sunset from your cabin porch. If you want evening activities, you can try your hand at roping, take part in square and round dancing, share your talents at skit night, watch the young foals as they are worked with for the first time, and enjoy barbecues on the creek. N. B. Hunewill homesteaded this ranch 135 years ago when the West was young. You'll be happy to find that time and custom, at least at the ranch, still run at that slower Old West pace.

Horseback riding is the premier attraction at Hunewill Ranch. Horses and riding groups are available for every riding ability. You can saunter slowly across the meadows with the "buckaroo" (beginning) ride or lope across thousands of acres of wildflower-covered meadows with the advanced ride. Splash your way a-horseback through one of the ranch's many puddles on a hot summer day, ride to breakfast on Robinson Creek, take an all-day ride into high mountain meadows that approach both Yosemite and the Hoover Wilderness, or have fun at our horseback play day. Ride one of the ranch's 120 horses, or bring your own for the best riding you have ever had.

Hunewill Ranch is a working cattle ranch with 1,200 head of cattle. You will have an opportunity to learn "cowboying" skills such as how to gather and sort cattle. Special cattle events include a cattle working week (in June and in September), desert and mountain gathers with other ranches, a ranch roping clinic, and our traditional 60-mile fall cattle drive.

For a vacation suited to families, singles or couples, you won't find a friendlier, more down-home atmosphere with all the elements for the greatest experience of your life. If you want home-style meals, wide open space, great riding, and lasting friendships, then Hunewill Guest Ranch is the destination for you.

First come the dreams;
a herd of horses thunders
through a shallow bog
and the cadence of the spray
marks time across your child's smile.

As you wend you way home the memories grow
and for the next 51 weeks they will not end.
A time when existence was relegated
to a time between bells
and the most complex decision
was to ride, or not to ride.
David Vassar, Ranch Guest

Address: P.O. Box 368, Dept. BP, Bridgeport, CA 93517

Telephone: Summer: 760-932-7710, Winter: 702-465-2201
Fax: Summer: 760-932-7933, Winter: 775-465-2056

Location: 120 miles south of Reno

Guest Capacity: 45

Season: May to October 1

Policy on Children, Pets: Children included in all activities; separate rides when needed. No pets.

Rates: $$ American plan. Children's rates.

Credit Cards: None

(SEE PAGE 208 FOR COLOR PHOTOS!)

CALIFORNIA

RANKIN RANCH
CALIENTE, CA

The Quarter Circle U Rankin Ranch has long been known for its history, horses, and Western hospitality. Rankin Ranch headquarters are in the beautiful mountain valley of Walker's Basin, deep in the Tehachapi Mountains at an elevation of 3,500 feet. Just 2-1/2 hours north of Los Angeles in central California, you'll find no smog, long lines, or crowds at Rankin Ranch. Guests from all over the world love staying at this working cattle ranch where they can leave the hectic city pace behind. With no telephones or televisions, the stress level at Rankin Ranch is zero!

The Quarter Circle U was founded in 1863 by Walker Rankin, Sr. It is one of the oldest and largest family-owned ranches in the West. Six generations of Rankins have lived and worked this 31,000-acre ranch. Rankin Ranch has been featured in *Sunset* and *Better Homes and Gardens* magazine, *U.S. News & World Report*'s "10 Best All-American Vacations," and *Adventure West*'s "Kidstuff."

Guests can be as busy or as relaxed as they wish. There are lots of activities — hiking, swimming, tennis, trout fishing, bicycling, shuffleboard, horseshoes, and more. Of course, the twice-daily horseback rides are the highlight of the day! Guests enjoy riding through mountain and meadow cattle country and sharing a way of life that continues the warm hospitality and exciting traditions of early California.

Each season brings something special. Springtime wildflowers lure guests back year after year. The gorgeous peaceful scenery and century-old barns and buildings lend themselves to our annual "Art Week" for sketching and painting. "Western Week" is booked a year in advance. Guests help with the roundups, weigh and ship cattle and live the long hours of a cowboy's life . . . it's a favorite time for many. Summertime and holidays are filled with returning families. Parents and grandparents agree that Rankin Ranch is the perfect family vacation; they can rest and relax yet there are plenty of activities to keep the children busy and happy. During the "regular" summer season, the ranch has a special supervised children's program for ages 4 to 11. Counselors are on hand to make sure children enjoy their time at the ranch. There are arts and crafts, nature hikes, treasure hunts, boat and kite making, swim meets, and much more. Children are never bored at Rankin Ranch!

Lunch rides, barbecues, and picnics along the trail are enjoyed by everyone. Late evening walks are routine for some, while others enjoy the glory of starlit evenings from the comfortable chairs outside their cabins. Autumn brings a slower pace with longer horseback rides and cooler nights. Couples and groups especially enjoy these quieter months.

The comfortable duplex cabins are designed for privacy and quiet. Each has a connecting private bath and large picture windows overlooking the mountain valley. The small number of guests adds to the personal, charming atmosphere . . . somewhat like that of a private house party. Each evening before dinner, adults enjoy gathering in Lightner Square for the patio party, where dips, chips, and mixes are provided. When the dinner bell rings at 6:30, everybody is ready for another hearty ranch meal . . . complete with homemade breads and desserts! The family antiques used in the dining room add to the personal, friendly atmosphere.

Evening activities for everybody include hay wagon rides and meadow barbecues, square and western line dancing, bingo and pool tournaments with prizes, talent shows, and " horse races" complete with stick horses and Rankin Ranch fun money for your choice of winners!

Families . . . couples . . . singles . . . and *groups* are all welcome at the Quarter Circle U Rankin Ranch.

 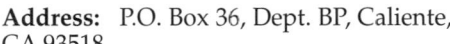

Address: P.O. Box 36, Dept. BP, Caliente, CA 93518

Telephone: 661-867-2511; **Fax:** 661-867-0105

Location: 13 miles east of Caliente, 42 miles northeast of Bakersfield

Guest Capacity: 25 — 45

Season: March to October

Policy on Children, Pets: Children are welcome; supervised children's program. No pets.

Rates: $$ American plan. Children's and group rates.

Credit Cards: American Express, MasterCard, Visa, Discover

CALIFORNIA

RICOCHET RIDGE RANCH
FORT BRAGG, CA

Canter along deserted ocean beaches, ride the bluffs while the Pacific crashes against rocks far below, and meander on mossy trails through the magnificent redwood forest during a destination vacation on northern California's rugged Mendocino Coast. This is an area of great beauty, rich in wildlife, where time flows at its own pace; the ghosts of Russian, Spanish, and English adventurers tenaciously cling to this isolated part of the world.

Lari Shea, the owner of Ricochet Ridge Ranch, has won dozens of endurance events, including the prestigious Tevis Cup 100 Mile Endurance race. On her ranch, Lari breeds Akhal-Teke, Russian Orlov crosses, and Arabians, many of whom have excelled in dressage and jumping, as well as on competitive trails. Her Appies and Quarter Horses round out the stable, 80 in all, of wonderfully well-trained horses, for both English and Western style enthusiasts.

Both novice and expert riders will find horses matched to their skill level. The ranch can accommodate six-year-old children with comfortable ponies or mount adults on docile beasts that walk placidly when and where they are told. For the serious, experienced riders, there is the chance to ride endurance-trained horses — spirited Arabs and graceful, majestic Orlovs. You can live the fantasy of cantering past crashing waves on the beach toward a misty horizon.

At Ricochet Ridge Ranch, you can create your own riding adventures. You can ride for a day, a weekend, or a week. Your vacation is there for your planning. There are regularly scheduled trail rides on the beach lasting about 1.5 hours. Custom rides are offered year round. California's unique climate offers unmatched riding days even in the dead of winter. Rides of 1.5, 3, or 4 hours or all day (which includes lunch in some scenic spot) are available on the beach and in the forest.

On the Virgin Redwood Wilderness Weekend, you'll experience the grandeur of the virgin redwood forest, where individual trees may be over 2,000 years old! You'll ride surefooted horses through the world's oldest forests under the towering canopy of giant Sequoia Sempervervins, the "ever-living redwoods." You'll camp at Cuneo Creek, which features solar showers and meals on redwood slab tables.

The Redwood Coast Riding Vacation is a week-long adventure that ranges for miles along ocean beaches, through a working cattle ranch, and up and into the coastal mountain range, offering views that stretch to eternity. You'll ride in the ancient redwood forests and along the Noyo River to its headwaters. trails. Along the way, you'll learn about the conditioning of sport horses. The last day's riding fulfills a dream for many people, concluding in a canter for many long miles along the surf. Free time during the week allows for exploring the many art galleries and quaint shops of picturesque Mendocino, the town used as the location for many movies and TV shows, including "Murder, She Wrote." The coast has many secluded beaches where you can watch the seals, sea lions, and whales.

Rest and repast are to be found at the end of the day's trek at the luxurious and historic Mendocino Hotel and at two of Mendocino's unique bed and breakfast inns. You will sample the fares at various acclaimed local restaurants. Nightly entertainment ranges from the melodic rhapsodies of classic piano and guitar to folk music to the usual fun of a hot tub or impromptu song-fest by the fireplace.

Perhaps the most treasured memories from your trek will be of a personal nature; the friendships that develop between fellow riders sharing this experience and the sense of self-accomplishment each feels for meeting the challenge of the trail.

Address: 24201 North Highway One, Dept. BP, Fort Bragg, CA 95437

Toll-free: 888-873-5777
Telephone: 707-964-7669; **Fax:** 707-964-9669
E-mail: larishea@shorse-vacation.com;
Web: www.horse-vacation.com or www.jimbalzotti.com

Location: 2 miles north of Fort Bragg

Guest Capacity: 24

Season: Year round

Policy on Children, Pets: Children, if accomplished riders, must be at least 6 years old. No pets.

Rates: $$ - $$$ American plan

Credit Cards: None

CALIFORNIA

California

Shannon Ranch & Golden Trout Wilderness Pack Trains
Porterville, CA

Shannon Ranch on Deer Creek near Porterville, California, offers a little of everything to anyone interested in horses, cattle, or terrific scenery. The Shannon Ranch has been herding cattle and driving them to lush summer rangeland in the Sequoia National Forest in the Sierra Mountains, then escorting them back to their fenced pastures for the winter since 1917. The home ranch spans 12,000 foothill acres and is adequate to feed the 320 head of cattle during the wet winter months.

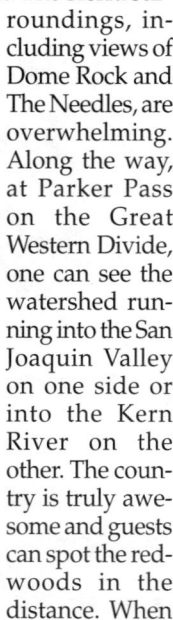

To move such a herd entails horses and cowboys. Since 1996, the Shannon Ranch and Golden Trout Wilderness Pack Trains have been taking guests who want a *real* outdoor adventure and the chance to work with animals on the spring (June) and fall (September) drives. They experience the history and excitement of the search and roundup of the cattle, a better understanding of a vanishing way of life, and the solitude of true wilderness — cooking over a campfire, sleeping in a bedroll, bathing in mountain streams. The wranglers supervise the city folk along an 80-mile drive that starts at an elevation of about 700 feet at the ranch, climbs to about 8,000 feet, then drops to about 6,200 feet in the Sequoia National Forest. Along the way, they ride through pristine meadows across rivers, through old-growth forests, past mountain cabins, cow camps, and ranger stations and finally to the top. The guests work along with the ranch's regular cowboys and learn by doing. There are long hours spent sitting on a horse and lots of time to think.

The Shannon Ranch cattle drive is one of the oldest continuing drives in California and one of the longest. The drive is a week long and begins on Monday before daybreak. With breakfast comes a condensed education on horse and people safety when handling cattle. This is a working vacation where each person is expected to carry his or her own weight — an opportunity for hands-on experience. Everyone lends a helping hand, and that includes the guests.

The scenery along the way is magnificent. The Sierra surroundings, including views of Dome Rock and The Needles, are overwhelming. Along the way, at Parker Pass on the Great Western Divide, one can see the watershed running into the San Joaquin Valley on one side or into the Kern River on the other. The country is truly awesome and guests can spot the redwoods in the distance. When the cattle drive reaches the Grey Meadow cow camp in mid-June, guests can explore the wilderness area — provided the weather cooperates.

The nightly campfire is the center for entertainment. There's cowboy poetry, and the guitar playing and old cowboy songs are lively. Story telling and swapping tall tales are on the menu — it's a time to relax and ease sore muscles. Being outdoors all day in the fresh mountain air guarantees a good and untroubled night's sleep.

When the top of the Sierras has been reached and the herd is feeding on the green pastures in the grazing lands, you'll have a sense of having done something a lot of people have never done, a lot of people will never do. You have accomplished something. This cattle drive fills the bill for a grand adventure!

Address: Rte. 4, Box 107, Dept. BP, Porterville, CA 93254

Telephone: 209-782-8836; **Fax:** 209-535-4673

Location: 80 miles from Fresno, 60 miles from Bakersfield

Guest Capacity: 12

Season: Cattle drives in June and September

Policy on Children, Pets: No children. No pets.

Rates: $$ American plan.

Credit Cards: None. Personal checks accepted.

(See Page 209-210 for Color Photos!)

CALIFORNIA
SHANNON RANCH

DAWN BREAKING ON A CATTLE DRIVE.

MOVING THE HERD.

CALIFORNIA

WILD HORSE SANCTUARY
SHINGLETOWN, CA

Track wild horses through the ruggedly beautiful foothills of Mt. Lassen in Northern California on the first natural wild horse sanctuary refuge. Ride out of a base camp to observe and photograph wild horses and burros in their natural, protected habitat shared with black bear, bobcat, mountain lion, wild turkey, deer, badger and bald eagles as well as over 150 species of song birds.

Each day, ride the trails created by the horses and burros through pine- and oak-studded hills - a bit of paradise dotted with meadows, woods, creeks and ponds. Bordering Vernal Lake, the base camp consists of frontier-style sleeping cabins; a "cook" house featuring a main kitchen, a wood-burning stove, hearty meals, and bathroom facilities complete with hot shower. Evenings are spent relaxing around a warm campfire and enjoying a peaceful night's rest after a day of adventure.

The horses on the Sanctuary are unique in that they represent the diverse herd types of wild horses found throughout the Western United States. During the evenings, upon request, informal presentations by wildlife experts will be given explaining and illustrating the various herd types as well as complex social behavior and mountain ecology of the wild horses.

Come share in the spirit of the Old West as you experience a way of life from the past. Seeing the horses running free in the beauty of their protected environment is both exhilarating and inspiring. Your ride will not only be educational, but more importantly, your participation will help the Wild Horse Sanctuary preserve "America's Living National Treasure."

If you haven't experienced the beauty of the wild horses here first-hand on one of our weekend rides, we'd like to encourage you to think about joining us this season. If you have ridden with us, we'd love to ride with you again! We have had so many wonderful experiences with our various guests we'd like to share some of these comments with you. One "city" dweller said, "I felt like we were your city cousins who came to visit. You treated us like family and made sure we had a good time."

As we ride home from the camp on Sundays we often stop at one of the springs to eat lunch under the shade trees. This is a beautiful spot where we sometimes surprise wild horses who've gathered there to drink or to languish in the shade. On one of these lunch stops I asked how our rides compared with everyone's expectations? All the comments were favorable but the one I remember most came from a visitor from Los Angeles who said, "It's everything I hoped it would be but was afraid to think it could be."

We're still excited to be able to invite you to ride the trails created by the wild horses themselves through this rugged piece of natural range that is home to almost 300 wild horses and burros. It's the best way we've found to share their beauty and the inspiration they evoke and at the same time raise the needed funds to ensure that we can continue to provide this home for rescued wild equines. We do hope, if you're able, you can ride with us this year.

Address: Box 30, Wilson Hill Road, Shingletown, CA 96088

Telephone: 530-474-5770; Fax: 530-474-5728

Location: Guest Capacity: 3

Season: May, June, July, September, and October

Policy on Children, Pets: 14 years or older, under 14 with management approval. No pets.

Rates: $$

Credit Cards: None

(SEE PAGE 211 FOR COLOR PHOTOS!)

California
Wild Horse Sanctuary

A LONE WILD HORSE

RESCUED HORSES AT THE SANCTUARY
RUNNING FREE.

Colorado

Colorado Dude Ranch Association

Colorado Dude & Guest Ranch Association

Since 1934

See page 212 for color photo and information.

(See Page 212 for Color Photos!)

COLORADO

COLORADO DUDE RANCH ASSOCIATION

COLORADO DUDE & GUEST RANCH ASSOCIATION
Since 1934

OUR PLEDGE

For over 60 years the CDGRA (Colorado Dude and Guest Ranch Association) has maintained high standards and strict requirements for membership. Each ranch is regularly inspected for cleanliness, facilities, hospitality, quality of riding program, and honest representation in promotional messages. This assures that guest expectations are consistently fulfilled.

OUR SECRET

For pure value, CDGRA ranches offer America's best kept travel and vacation secret. Base prices typically include nearly every cost for an entire ranch visit – all lodging and meals, riding and fishing, and a smorgasbord of ranch activities – like Western entertainment, cookouts, hay rides, square dancing and more. Compare this to the typical pay-as-you-go ordinary vacation. At a cost you can confidently budget for and live within, your ranch hosts and new life-long friends provide you with virtually everything you will need for your entire stay. Plus, wonderful Western hospitality and lifetime memories.

CDGRA member ranches have proudly identified themselves in this book.

COLORADO DUDE & GUEST RANCH ASSOCIATION
P.O. Box 300
Tabernash, CO 80478
970-887-9248 • www.coloradoranch.com

Colorado

Aspen Lodge
Estes Park, CO

Estes Park in Colorado has long been considered one of the most scenic places in North America and a favorite vacation destination. It also happens to be the site of the Aspen Lodge Ranch Resort, one of Colorado's premiere horseback riding vacation destinations. Located in the spectacular high country adjacent to the Rocky Mountain National Park and surrounded by the Roosevelt National Forest, you can come and celebrate life in this exhilarating paradise. Select superb accommodations within Colorado's largest log lodge or in the cozy comfort of cabins framed by aspens. Awaken to a stunning view of Longs Peak, deep forests of evergreen and aspen trees, and shimmering mountain lakes. At over 9,000 feet, the Lodge is set against a sky too blue to be believed. Whatever your Western dream, you will find it at the Aspen Lodge, in its 50th year as a top guest ranch, where our professional staff is dedicated to serving you.

The Lodge's finely developed, year-round activities program offers a multitude of choices including fishing in the lake, beautiful hiking trails, mountain biking, tennis, racquetball in the Sports Center with a weight room, sauna, hot tub, and outdoor heated seasonal swimming pool. Extensive winter activities include cross-country skiing, snowshoeing, romantic sleigh rides, and the only snowmobiling on the Front Range of the Rocky Mountains. Wildlife viewing is, of course, abundant any time of year.

The summer children's program lets you choose to be with your children or relax on your own. Supervised activities for ages 5 to 12 include pure fun as well as educational experiences of Indian lore, local pioneer lifestyles, and nature rambles.

Aspen Lodge offers outstanding outdoor barbecues, hayrides, van tours, pack trips, overnight tent camping, singalongs, square dancing and two-stepping, and live entertainment. Complete your day in the lodge's friendly Western bar or the wonderful deck overlooking the Rocky Mountain's highest peaks.

At Aspen Lodge, you don't have to compromise on food because the meals range from traditional Western to gourmet continental. They are served in the award-winning casual dining room without the restrictions of rigid meal times. Varied menus, prepared with imagination are dedicated to your satisfaction.

Aspen Lodge offers its guests two different types of horseback rides. The Lodge rides are one- and two-hour scenic walking trail rides on both the Aspen Lodge property and the adjacent Rocky Mountain National Park. There is an advanced ride for more experienced riders. The unique, open-country Ranch rides give guests the opportunity to get in some trotting and loping on Aspen Lodge's 3,000-acre working ranch on the other side of Twin Sisters. On the Ranch rides, you'll explore the gorgeous open country and, on the all-day ride, the 33,000 acres of Homestead Meadows in Roosevelt National Forest.

There's also a "trip of a lifetime on top of the world" deluxe pack trip. You'll cross the Great Continental Divide going from Allenspark to Winter Park. Following the Divide trail is possibly the most breathtaking, thrilling ride on the earth. At night, you'll settle in to a roaring campfire and a terrific cowboy supper. You'll sleep soundly serenaded by the evening sounds and wake to a piping-hot Buckaroo breakfast. The trail is considered "moderate" in difficulty and is not for the beginning rider.

Come celebrate life in this exhilarating mountain paradise. Leave Aspen Lodge refreshed, energized, and revitalized by the warmth of good fellowship and inspired by country that will stir your soul. Let Tom and Jill Hall and their friendly staff share the beauty and the power of this unique place with you.

Address: 6120 Highway, 7 Longs Peak Route, Dept. BP, Estes Park, CO 80517

Telephone/Fax: 970-586-8133, 800-332-6867

Location: 7 miles from Estes Park, 80 miles from Denver

Guest Capacity: 175

Season: Year round

Policy on Children, Pets: Children always welcome; special children's program in summer. No pets.

Rates: $$ American plan. Children's rate; childen under 2 free. Riding extra.

Credit Cards: American Express, Diner's Club, Discover, MasterCard, Visa

(See Page 213 for Color Photos!)

Colorado

Colorado

Bar Lazy J Lodge
Parshall, CO

Opened in 1912, Bar Lazy J is considered the oldest continuously operating guest ranch in Colorado. We believe our ranch to be ideally located just southwest of Rocky Mountain National Park and nestled in a peaceful valley on the Colorado River. With a guest capacity of 38 we offer our guests a friendly, informal atmosphere and a true "western experience".

We have 12 comfortable guest log cabins clustered along the Colorado River - all with screened porches overlooking the river. At 7500 ft. elevation the crisp, clean mountain air will stimulate your appetite for the homestyle food and fresh baked goods served in our historic log lodge.

Our private 3/4 mile stretch of Colorado River (designated Gold Medal Fishing) offers some of the best Rainbow and German Brown trout fishing in the state. Not only is it a challenge for the "old time" fisherman, it is great fun for the beginner or in-between. We have a free fishing clinic once a week for anyone interested in learning more about fishing our stretch of the river. We also have a stocked trout pond for everyone to enjoy. Our horseback rides take you on sage covered hills and majestic mountains with sapphire skies and crystal clear views. We offer horseback riding for all levels of riders. Our horse rides takes you through the Arapaho National Forest's sage-covered hills andmajestic mountains, with it's sapphire skies and crystal-clear views. Twice a day we have slow, medium and fast rides on numerous riding trails covering a variety of scenic terrain. Once a week we offer a breakfast ride and an all-day ride. Our wranglers carefully consider experience and ability before pairing you with your horse for the week. Our wranglers will be glad to give you personalized riding instuction for those who wish to feel more comfortable in the saddle. Other activities (all packaged in one price) include jeep rides, hiking, mountain biking and sightseeing van trips. Golf, tennis and white water rafting are available options.

The children's' Ranch Fun Program is flexible and designed to meet the needs of children in various age groups so they may be kept safe, busy and entertained. Our counselors lead activities for children (ages 3-12) which include hiking, fishing, swimming, arts & crafts, in addition to many games and a well equipped play area. Children 8-12 are assigned a horse for the week and trail ride every day. Children 3-7 are led on horseback in our arena.

Every evening we have planned activities including hay rides, campfire sing-a-longs, rodeo, staff show, and western dancing. Of course you can play your own activities by starting a card game, soak in the hot tub, visit with the people in the next cabin, walk under the stars or sit in front of your cabin and enjoy the peaceful, cool nights with the river rushing by.

Whether you are a family, a couple or traveling alone your every vacation need will be taken care of at Bar Lazy J. Relaxation is what we offer. The pace is slower...or not at all. We offer many things to do...or not at all. The choice is always yours.

Address: 447 County Road 3, P.O. Box N, Parshall, CO 80468

Telephone: 970-725-3437;
Toll Free: 800-396-6279;
Fax: 970-725-0121

Internet: www.barlazyj.com
E-Mail: barlazyj@skymtnhi

Guest Capacity: 38

Season: May through September

Policy on Children, Pets: Children welcome. No pets.

Rates: $$

Credit Cards: Visa, MC

 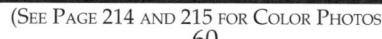

Colorado

Bar Lazy J Lodge

ENJOY OUR FINE FISHING.

Colorado
Bar Lazy J Lodge

A RELAXING AFTERNOON.

COLORADO BEAUTY.

Colorado

C Lazy U Ranch
Granby, CO

Located 100 miles northwest of Denver, high on the western slopes of the Colorado Rockies, this Mobil 5-Star, AAA 5-Diamond ranch provides a wealth of opportunities for family fun. The 2,000-acre C Lazy U Ranch has been serving guests since 1925. Because of its personal service, outstanding food, comfortable accommodations, and beautiful surroundings, guests from all over the world visit — and returning time and again. The C Lazy U is a guest ranch that melds rustic elegance and old fashioned informality.

The many miles of scenic mountain trails are a rider's dream. The ranch has a variety of horses, ranging from gentle to spirited. Each guest is matched to a horse that fits her or his level of skill and keeps that particular horse for the entire stay. If you don't know how to ride or would like to ride better, you'll find expert instruction by the friendly wranglers. Trail rides are led by experienced wranglers and vary from instructional, to slow scenic, intermediate, or fast. There's also an indoor riding arena for those (few) rainy days.

Children too learn to appreciate the "ways of the West" under the watchful eyes of experienced counselors. There are separate programs for 3-to-5 year olds, 6-to-12 year olds and teenagers, each designed around their interests. Those might include horseback riding, outdoor and indoor sports, hiking, fishing, swimming, tennis, or hayrides. The Saturday afternoon "Shodeo" enables children and adults to show off the horsemanship skills they've learned or to compete in a variety of equestrian events. Everyone wins a ribbon, but only a few win the coveted blue!

The main lodge of the ranch is a splendid log building that houses the main living room, dining room, library, bar, and card room. The Patio House, which is next to the heated, spring-fed swimming pool, is the ranch's recreation center. There you'll find a whirlpool, sauna, exercise equipment, racquetball court, game room, bar, TV room, and a large all-purpose activity room. It also has facilities for Western cookouts twice a week. In addition to riding, you might want to try hiking, tennis, trout fishing in Willow Creek (which runs through the ranch), or skeet and trapshooting. Evening entertainment might consist of a staff talent show, country and western music, square or line dancing, and singalongs. Or, you might want to just rest and relax in conversation and fellowship around the lodge's huge fireplace. There is even a library for the readers.

Guest accommodations surround the main lodge and range from one to three rooms with private baths, most with fireplaces and some with Jacuzzis. All meals are served family style. After breakfast, children go off with their counselors and are served separately for lunch and dinner. Dinner is relaxed and informal, and the food is outstanding.

In October, the ranch offers the Buck Brannaman Clinics. If you're serious about improving your riding skills and using them on the trail, this clinic is for you. You can bring your own horse or ride the ranch's horses and take advantage of Buck's expertise. Only graduates of the clinic are permitted to help bring the herd to pasture.

The fall season, when children are getting ready and going back to school, is for adults only. You'll still find all of the summer's activities, but you'll also have a greater chance to find peace and quiet. The Ranch is also open for winter activities, including skiing and sledding.

The ranch family — staff and owners alike — are eager to make your stay at C Lazy U one of the most memorable experiences of your life. Come and enjoy!

 Address: P.O. Box 379, Dept. BP, Granby, CO 80446

Telephone: 970-887-3344; **Fax:** 970-887-3917

Location: 8 miles west of Granby, 100 miles northwest of Denver

Guest Capacity: 110

Season: Summer through winter

Policy on Children, Pets: Excellent children's program; children under 3 require a nanny. No pets.

Rates: $$ - $$$ American plan. Children's rates.

Credit Cards: None

(See Page 216 for Color Photos!)

C Lazy U Ranch

POWDER DAY AT THE RANCH.

Colorado

Capitol Peak Ranch
Granby, CO

Capitol Peak Outfitters offers a full range of horseback opportunities, from hourly rides to all-expense wilderness pack trips, to fully guided hunting and fishing expeditions. We are located near the resort town of Aspen, Colorado. Our trips center around the Maroon Bells/Snowmass Wilderness, a setting unsurpassed with 14,000 foot peaks, mountain lakes, wildflowers, and breathtaking panoramas.

A successful vacation is measured in many ways. We want the following to be a part of the rewards of your vacation: fondly remember our area as the most scenic, pristine and enchanting countryside in the Rockies; recall home-ur stretch of river. We also have a stocked trout pond for everyone to enjoy.

We offer horseback riding for all levels of riders. Our horse rides take you through the Arapaho National Forest's sage-covered hills and majestic mountains, with it's sapphire skies and crystal-clear views. Twice a day we have slow, medium and fast rides on numerous riding trails covering a variety of scenic terrain. Once a week we offer a breakfast ride and an all-day ride. Our wranglers carefully consider experience. Take pictures of wildlife in their natural habitat and horseback ride off the beaten trail. We'll show you country you've only seen in pictures. All you need is your personal gear and an adventuresome spirit - we will do all the rest.

Our pack-in base camp is nestled in a secluded basin with trout filled beaver ponds and mountain meadows. The camp is comfortable with wall tents, wood stove, cots, tables and chairs. You will wake up in the morning to the smells of fresh brewed coffee and breakfast food that will make your stomach growl with the anticipation of digging in. Our cooks prepare special, fresh meals that you won't forget.

Bring the whole family and just relax, fish, hike or ride. Create your own vacation by using one of our fully-equipped camps for as long as you wish. We will pack you in and out and you may choose to ride or walk. The beauty of Capitol Peak Outfitters is that we will tailor your vacation to accomodate you. If these services don't suit you, inquire about our Aspen to Crested Butte trail rides, full-day fishing excursions, fly fishing lessons, or guided and drop-camp hunting trips.

Most people go their whole lives without ever seeing firsthand the majestic splendor of the Rocky Mountains. Come let us be your guide and allow us to show you this country the best way to see it, on the back of one of our fine horses. We promise you a unique vacation you'll never forget.

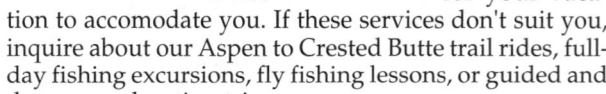

Address: 0554 Valley Road, Carbondale, CO 81623

Telephone: 970-963-0211; **Fax:** 970-963-0497
E-Mail: capitolpk@aol.com

Location: Approx. 10 miles from Aspen

Guest Capacity: 85

Season: May through September, until November (hunting)

Policy on Children, Pets: Children over 7 years old welcome. No pets.

Rates: $

Credit Cards: MC, Amex, Visa

(See Page 217 for Color Photos!)

Colorado

Capitol Peak Ranch

THE MANY VIEWS OF CAPITOL PEAK OUTFITTERS.

COLORADO

CASA MILAGRO BED & BREAKFAST
HOT SULPHUR SPRINGS, CO

Casa Milagro B&B is the closest log home bed and breakfast to Denver that offers a setting on the river in the pines and aspens, and home-cooked, family-style dinners with home made bread and dessert as part of a relaxing getaway. We serve a hearty gourmet breakfast such as baked rellenos, homemade tortillas and honey, and Colorado's best bacon at the 'crack of nine.' In the afternoons, there will be coffee and tea and a variety of snacks including popcorn with Lynn's outrageous spices (upon request) or Paul's homemade fudge.

You'll enjoy a room with a queen-sized bed with our comfortable "cuddle ewe" mattress, gas log stove, private bath, comfortable reading chairs, books galore and a TV/video player for enjoying movies from our extensive collection. After a late evening movie, relax and soak under the stars in our hot tub. Our log home is decorated with native American southwestern arts and crafts, reflecting Lynn's family history of trading with the Navajos, Hopis and Pueblo Indians of New Mexico since 1908.

Bring your horses and ride our many nearby forest trails. Our large outdoor corral was built with care by Paul, who took care that there are no sharp edges anywhere. We recommend you bring their feed so they don't experience any problems from sudden changes in their food. Horses from warm climates might appreciate your bringing a blanket for our cool, mountain evenings. Our cats, Maxine and Bill, ask that you leave your other pets at home.

Come to Casa Milagro for a relaxing getaway! Our river will soothe you, our forest will energize you, our birds will delight you, the mountain activities will invigorate you, our cats will charm you and WE WILL PAMPER YOU! See four dozen ways we pamper you on our site, www.casamilagro.com

Call us at 888-632-8955 to find out more about our 2 + 2 getaway package which includes two nights, breakfast both mornings, dinner for 2 your second night, and 2 all-day passes to the newly renovated, nearby Hot Sulphur Springs Mineral Baths & Spa. Our fall "Pampered Shopping Getaway" includes one or two nights at Casa Milagro with breakfast and dinner, and coupon savings for the Silverthorne Factory Outlet Stores at I-70, and a massage and facial at the hot springs. Prices on the "Pampered Shopping Getaways" range from $217-$386 for two. Make your reservations now and enjoy the fall color of nearby aspen groves.

Be sure to reserve your holiday getaway at Casa Milagro! Enjoy our family traditions, decorations, and special holiday cooking along with your favorite winter activities! We're less than 50 minutes from Winter Park, Silver Creek, Keystone, Breckenridge and Copper Mountain! After the hustle and bustle of the slopes, come back to the Casa and enjoy a relaxing soak in the hot tub! Dinner is only footsteps away and is served family-style in a great room!

Address: 13628 County Road 3, Hot Sulphur Springs, CO 80468

Telephone: Office, 970-725-3640
Reservations: 888-632-8955; **Fax:** 970-725-3617
Web: www.casamilagro.com or www.jimbalzotti.com
E-Mail: casamilagro@rkymtnhi.com

Location: 45 minutes from Grand Lake, Winter Park, Frisco and Keystone

Guest Capacity: 6

Season: Year round

Policy on Children, Pets: No Pets

Rates: $

Credit Cards: MC, Visa, Amex, Discover

Colorado

Cherokee Park Ranch
Livermore, CO

Cherokee Park Ranch invites you to experience a true Western adventure. Nestled in the Rocky Mountains, the ranch is just a two-hour drive from Denver. The history of Cherokee Park is unique among local dude ranches. The log lodge was a stagecoach stop between Fort Collins and Laramie, Wyoming in the 1880's. Incoming guests had to chance Indian confrontations as they traveled 15 miles down the ranch's dirt road! Since then, the ranch has grown to become a haven from the hustle-and-bustle of everyday life.

Owners Dick and Christine Prince describe themselves as "the guests that never went away." After vacationing at Cherokee Park Ranch with their children for four consecutive years, the Princes purchased the ranch... and a dream was fulfilled. What makes Cherokee Park Ranch such a special place is 22 young adults who dedicate themselves to this 300-acre spread. The genuine warmth and friendliness of the staff is evident in all facets of the ranch from the corral to the kitchen to the kid's program.

The ranch has a fine herd of 82 horses to assure your riding enjoyment. Horseback riding at Cherokee Park Ranch is truly a nonstop favorite - you pick and choose what type of riding you would like to do each day. Whether you were born in the saddle or have never mounted a horse, the ranch will meet your riding needs and exceed your expectations. The kid's rides are undeniably popular as the young cowhands venture out to ride as a group, building forts and playing games with their counselors after lunch. Personal instruction is available each day in the ranch's arena for you to gain skill and confidence on horseback.

An overnight pack trip, covering 28-miles, is offered mid-June through August for those seeking a camp-out experience. Late September and October are reserved for those "true grit" riders interested in the ranch's cattle drives.

Cherokee Park Ranch is 100% kid-friendly. We offer an extensive children's program for 3-5 and 6-12 year-old age groups. Imagine: hiking, fishing, panning for gold, arts and crafts, kid's cookouts, trips to an old gold mine, and a kid's overnight in a bonafide Indian teepee! For young ones under 3, they offer a unique opportunity: bring your own full-time baby-sitter free of charge.

Accommodations are complete with front porches, hummingbird feeders, and blooming flowers. There are four spacious suites in the main lodge, varying in size from two to three bedrooms each. Five cabins overlook the river and can accommodate between 2-6 guests each. Home cooked meals are served family-style in our dining room and are designed to satisfy all appetites, both hearty and light. Several cookouts during the week add to the delicious menu offered at the ranch.

Although horseback riding is the favored activity, non-riding activities are plentiful at Cherokee Park Ranch. No activity is required, so you may do as much or as little as you like. Highlights include fly fishing with the ranch's Orvis-certified instructor, skeet shooting, hiking, whitewater rafting, bait fishing in their stocked trout pond, and sightseeing trips to Rocky Mountain National Park and the Territorial Park in Laramie, WY. The ranch offers nightly activities every evening including square dances, sing-a-longs, and hayrides.

Address: 436 Cherokee Hills Drive, P.O. Box 97, Livermore, CO 80536

Telephone: 970-493-6522, **Toll Free:** 800-628-0949; **Fax:** 970-493-5802; **Web:** www.jimbalzotti.com **E-Mail:** thomas@pageplus.com

Location: Approx. 115 miles from nearest airport.

Guest Capacity: 35

Season: May through September

Policy on Children, Pets: Children welcome. No pets.

Rates: $$ All meals included.

Credit Cards: MasterCard and Visa

Colorado

Coulter Lake Guest Ranch
Rifle, CO

Coulter Lake is a delightful small family ranch, nestled in a mountain valley deep in the White River National Forest at 8,100 feet. In operation since 1938, Coulter Lake is one of Colorado's oldest and most respected guest ranches. The ranch is surrounded by some of Colorado's most spectacular scenery, stretching for miles in all directions. Guest are likely to spot deer, elk and other wildlife, while riding through forest of quaking aspen and spruce with high mountain meadows abundant with wildflowers. The mood is casual and non-regimented, with plenty of great food and healthy outdoor activity for the entire family. The owners have purposely kept the ranch small and intimate with no televisions and no telephones to interrupt the serenity. We have eight cabins that stand in beautiful aspen trees nestled in the mountainside. They accomodate between two to nine people, with each having its own private bath. Some are equipped with fireplaces and porches. You can ride to your heart's content, swim, row across cool spring-fed waters in search of rainbow trout, play an invigorating game of volleyball or horseshoes, or just recline on the deck of the lodge letting the sun melt away your cares and worries. We offer short horseback rides to all-day rides to places like Long Park, Little Box Canyon and Pot Holes. We have limited children's programs with supervised kiddie rides. If you feel the need to get off the Ranch for a day, nearby you will find an 18 hole golf course, tennis, rafting, hot springs, shopping in Aspen or Vail, ballooning, mountain biking, or rock climbing. Summertime brings the famous Coulter Lake boat races which take place on Saturday's and is always lots of fun. Many guests like the intimate nature of the ranch and come to get away from it all. We offer a superior place to come hike in spectacular mountain country streching for miles in every direction. You may wish to go hiking after eating our delicious, home-cooked food. All meals are hearty, stick-to-your-ribs, served family style. Our cookouts are very popular and we feature a supper ride to 10,000 foot Coulter Mesa. On Wednesday we have our steak dinner, Friday you could be dining on fresh cooked trout, and on Saturday we offer a grand buffet. If you have special dietary needs, just let us know and we will be able to accommodate you. We have guests bring their own alcoholic beverages. Fall offers excellent hunting for elk, and mule deer, guided hunts, drop camps and cabin rentals available. Winter brings incredible snowmobiling, cross-country skiing or showshoeing. With access to over 80 miles of groomed trails and all the powder you can handle, we invite you to bring your own sled or rent ours. Overnight accommodations and meals provided. Call or e-mail us for more information. Promotional video is available.

Address: 80 C.R. 273, P.O. Box 906, Rifle, CO 81650

Telephone: 970-625-1473
Toll Free: 800-858-3046
Fax: 970-625-1473
E-Mail: coulterlake@sopris.net

Location: 2 miles from Denver International Airport.

Guest Capacity: 30

Season: Late May to October/Mid-December to Early April

Policy on Children, Pets: Children welcome.

Rates: $$

Credit Cards: Amex, Discover, MC and Visa

(See Page 220 for Color Photos!)

Colorado

Coulter Lake Guest Ranch

A BEAUTIFUL LAKESIDE RANCH!

RIDING THROUGH THE ASPENS.

Colorado

Deer Valley Ranch
Nathrop, CO

"There are many ways to enjoy a horse. However, riding in the beautiful mountains of Colorado surpasses all others." Harold and Sue DeWalt and John and Carol Woolmington welcome you to experience pronghorn antelope on a plains ride to our Wednesday cookout, elk wandering through high mountain valleys after your lunch on the High Country Ride or deer bedded down as you ride to Monday morning's Wrangler's Breakfast.

Deer Valley offers horseback riding for all age groups and riding abilities. A flexible program designed to accommodate every level of rider is our specialty. Riding instruction is for beginners to advanced riders. Trail rides are available from easy walk rides to those covering more difficult terrain. Our program surpasses the guidelines and standards established by the Colorado Dude and Guest Ranch Association. Safety is our forte! It is our goal to help each individual develop an understanding of the horse's inherent nature and to build a positive relationship between the horse and rider. Our horses have Quarter Horse breeding with easy going dispositions. Also featured is a small string of excellent riding mules.

Variety is the spice of life and at Deer Valley Ranch, our meals are "spiced" to your liking. Each morning you will begin your day with a yogurt/bagel bar, hot and cold cereal, juice, cocoa and coffee. Add to that a daily selection of eggs, cinnamon rolls, pancakes with strawberries and whipped cream or French toast and you are ready for a day in the saddle or a climb to the top of a 14,000 foot peak.

You will find the accommodations at Deer Valley neither "quaint" nor "rustic". Our fully carpeted, elegant, western cabins all have fireplaces and full kitchens. It would be hard to pass up one of Sue's fine meals for the pleasure of cooking your own, but you might want to keep the fridge stocked with snacks. When staying in our ten-bedroom lodge, you are automatically on the American Plan where meals are included.

Whether you come to Deer Valley Ranch to experience the outside of a horse or just enjoy all that we have to offer, you will want to be involved in our extensive activities program. Special counselors direct ages 4-11 in children's activities such as a tour of the local fish hatchery, evening hayride and Thursday pool party. Teens sign up for 3-on-3 basketball, an overnight campout and breakfast or mountain biking on nearby trails. Your whole family can take part in fishing on our private lake or in nearby streams and rivers. Also enjoy mountain climbing, white-water rafting, rockhounding, gold panning, horseshoe tournaments, tennis, hayrides, square dances and our hilarious staff Western Party. After a day of excitement and activity, relax in our hot springs swimming pools, hot tub, sauna and Jacuzzi.

Deer Valley Ranch has been in the same family for 45 years. Our purpose today is the same as it was when Parker and Clara Woolmington first started the ranch...to provide families with a western vacation in a warm Christian atmosphere. Our no alcohol policy and fine Christian staff add to the ambiance you feel as you start your vacation. Come and be a part of our family and let us offer you the vacation of a lifetime.

Address: 16825 C.R. 162, Nathrop, CO 81236

Telephone: 719-395-2353
Toll Free: 800-284-1708
Fax: 719-395-2394
Web: www.deervalleyranch.com
E-Mail: fun@deervalleyranch.com

Location: 135 miles SW of Denver, 100 miles W of Colorado Springs

Guest Capacity: 135-150

Season: Year round

Policy on Children, Pets: Children welcome. No pets.

Rates: $ American and Modified European Plans

Credit Cards: None

Colorado

Colorado

ECHO CANYON GUEST RANCH
LA VETA, CO

Welcome to Colorado's quality guest ranch - ECHO CANYON. We are proud to be small enough to afford you individual attention with quality activities. Echo Canyon Guest Ranch is situated among some of the most breathtaking scenery in the United States. Horse Program: We own and train every one of our horses and take great pride in their disposition and athletic ability. You'll enjoy a horse of your own during your stay, and a very diversified riding program. We provide a professional orientation for the novice through the most advanced rider. In addition to our excellent horseback riding program, you can participate in cattle gathering, roping lessons, and pack trips, if you so desire. Since our ranch includes 1000 acres of private property and 100,000 acres of federally permitted land to ride through, one can go mountain riding next to Echo Canyon Creek in the morning and ride the gentle slopes next to the Cucharas River in the afternoon, with views of mountain tops in all directions.

Accommodations: Our quality lodge is 12,000 square feet which includes two spacious lounges. You'll love to browse the art work that adorns each activity area. You will also enjoy the panoramic views and the ability to catch a glimpse of elk, deer, or wild turkey. On the main level of the lodge is a large dining area and adjacent to this is our Cantina. On the lower level, you will enjoy the game room which includes a pool table, juke box, ping pong table and arcade games. We also have a separate library reading room, a separate video movie area, an equipped fitness area and an extensive unique "wranglers gift shop." We have lovely lodge guest rooms with spacious private bathrooms and handmade aspen pole furnishings. We also feature four-bedroom cabins plus large bedroom accommodations with balconies located over the Longhorn Conference facility.

Other Activities: Some of our other onsite activities include: fishing for mountain trout in our four ponds; trap shooting; archery; gold panning; 4-wheel tours; cookouts; wildlife viewing; indoor game room; library; movie videos; and a hot tub. Plus, our most recent addition of nature walks with our on-staff naturalist. Fishing is great on the ranch! Four ponds on the property filled by Echo Canyon Creek, produce 16" to 24" trout for your eating pleasure.

Offsite Activities: River rafting, day tours to Pikes Peak or the Royal Gorge, plus mountain biking. Four miles from our gate you can play the #1 rated, 18-hole public golf course in Colorado designed by Tom Weiskoff. Dining: Family style service and western cookouts are part of the stay at Echo Canyon. Our food is delicious and the variety of cuisine will keep you pleasantly surprised.

Entertainment: Includes musical entertainment by a professional who has written over 500 country and cowboy songs and recorded his hit songs, plus a "Mountain Man" lecture by a true mountain trapper and hunter.

Longhorn Conference Center: Echo Canyon Guest Ranch has a newly completed state-of-the-art conference center with two meeting areas. The Longhorn Center seats 30 people and houses a kitchen and conference equipment. We customize our activities to fit your meeting schedule.

Address: P.O. Box 328, La Veta, CO 80155
Telephone: 719-742-5524; **Toll Free:** 800-341-6603
Fax: 719-742-5525
Web: www.guestecho.com   **E-Mail:** echo@rmi.net

Location: Fly into Colorado Springs, then drive 120 miles south on freeway. Pick-up is available.

Guest Capacity: 35

Season: May 23rd through October 2nd, 1999
Policy on Children, Pets: Children 11 years & older. No pets.
Rates: $$
Credit Cards: All major credit cards accepted.

Colorado
Echo Canyon Guest Ranch

YAHHOO!

MOSEYING ALONG.

Colorado

Elk Mountain Ranch
Buena Vista, CO

At Elk Mountain Ranch, you'll experience a real Western vacation high and remote in the Colorado Rockies. Your hosts, Tom and Sue Murphy, and their staff are dedicated to excellence in accommodations and hospitality and committed to upholding the deserved reputation for a traditional Western vacation experience that is unsurpassed.

Elk Mountain is unique for its spectacular location, horseback rides of unmatched beauty and variety, superb menu, and intimate capacity. The Ranch lies at a very high 9,535 feet — the highest elevation of any guest ranch in Colorado. It is also remote — located in the San Isabel National Forest near Buena Vista, Colorado. The ranch is nestled among aspen and evergreen trees on Little Bull Creek and is surrounded by a breathtaking panorama of snow-capped peaks.

The log cottages and lodge rooms are carpeted, tastefully furnished, with daily housekeeping. Flowers and fresh fruit await you in your room! The main lodge houses the dining room, game room, library, sun deck, and a fireplaced main room where guests can gather and get to know one another.

Horseback riding and instruction are unsurpassed. Elk Mountain offers highly personalized daily trail and off-trail riding. You'll ride through miles of aspen forests, not above the timberline, with spectacular vistas of the finest the Rocky Mountains have to offer, down through rock canyons, and across open high mountain plateaus. They purposely keep the trail ride groups small and try to tailor each ride to the particular interest and ability of the group. Experienced and competent wranglers accompany you as your instructors and guides. Included in the weekly full-rate package is a night under the Colorado stars "cowboy style." You'll ride to a secluded campsite, set up tents, roll out the sleeping bags, and tend to the horses. A roaring campfire provides the heat to fix up a "trail boss size" supper.

The wide variety of activities at Elk Mountain include trapshooting, archery, and marksmanship. They provide the guns, bows, competent supervision and instruction, and targets. The ranch is located near several excellent "fishin' holes," so bring your fly rod or spinning tackle. There are also two stocked ponds on the ranch property. Ride the rapids of the roaring Arkansas River on the ranch's thrilling whitewater raft trip. Travel through the beautiful canyon scenery and whitewater rapids that the Rocky Mountains have to offer. You'll thoroughly enjoy the vehicle trip over the Continental Divide to Aspen or Breckenridge. The spectacular drive to Aspen will take you across Independence Pass. Other ranch activities include hiking, Jacuzzi, horseshoes, volleyball, hayride, and library. On Tuesday, there's a square dance and a hammer dulcimer concert.

The Ranch offers a full children's program for ages 4-7. They ride in the arena improving their skills on the barrels, poles, and other gymkhana events. Children also enjoy fishing, hiking, softball, arts and crafts, and scavenger hunts.

At Elk Mountain meals are a main event. Superb, varied, full-course meals delight both hearty and discerning appetites — sure to be a highlight. Trail ride cookouts, steak fry, hors d'oeuvres, candlelight supper, and fabulous desserts. Enjoy!

One of the best features about Elk Mountain Ranch is the limited number of guests — 35. The Murphys believe that their guests do not want to be lost in the crowd or one among many. They want their guests to get the most out of their vacation experience in the Rockies and believe that can best be achieved in a more intimate atmosphere.

At Elk Mountain Ranch, you'll find that "the mountains, the sky, and the log walls seem to suggest that all is well with the world, and that there's enough time for everything."

Address: P.O. Box 910, Dept. BP, Buena Vista, CO 81211

Telephone: 719-539-4430

Location: 120 miles southwest of Denver, 90 miles west of Colorado Springs

Guest Capacity: 35

Season: May to September

Policy on Children, Pets: Family oriented with program for children 4-7. No pets.

Rates: $$ American plan. Children's rates.

Credit Cards: American Express, Master Card, Visa

(See Page 222 for Color Photos!)

Colorado
Elk Mountain Ranch

FAMILY TIME AT ELK MOUNTAIN.

ENTRANCE TO A WEEK OF FUN.

COLORADO

ELKHEAD RANCH
HAYDEN, CO

The Elkhead Ranch would like to extend an invitation to horse enthusiasts of all interests. Our historically designated 1883 cattle ranch is nestled in the Elkhead Valley. We're surrounded by stocked ponds, lush hay meadows, rolling sagebrush covered hills, aspen spotted ridges, and an abundance of wildlife, all awaiting your exploration. Our valley offers the chance for complete seclusion and privacy from the hurried pace of reality. What better place to indulge in the riding retreat of a lifetime? Bring your own horse or use one of ours and come on out to the Elkhead Ranch for the opportunity to experience Colorado's "Best Kept Secret."

We invite you and your horse for a vacation of unlimited Western adventure. Bring your gear and get ready to ride, rope and work on a real Colorado ranch in the Rockies. First homesteaded in 1883, Elkhead Ranch now encompasses 10,000 acres rising from the valley headquarters at 6,700 feet elevation through sagebrush covered hills to shimmering, aspen groves sprinkled among pines on the high mountain slopes reaching to 9,400 feet. Ride through lush mountain hayfields or through mature cottonwood trees growing along the creek.

Several original homestead sites are still visible around the ranch, waiting for exploration. The ranch recently received an historical designation from the Routt County Historical Society.

The ranch is prime habitat for wildlife as well as innumerable songbirds, waterfowl and a resident colony of nesting Sand Hill Cranes.

Elkhead Ranch is located 40 miles from Steamboat Springs, a world famous ski town rich in western history, and home of weekly PRCA rodeos from June through August. Winter activities include cross country skiing and snowshoeing based from our lodge.

Mount up and head off on your own for a day's riding or help us to move and work our 1,600 yearling cattle. Or meander down to the creek to wet a line and soak up a good book. Children and pets are generally not permitted. Special arrangements can be made for riding families. Childcare is not available at the ranch.

To ensure a quality riding holiday, we limit the number of guests to six. Accommodations range from our lodge, to the bunkhouse, a mobile sheepcamp and a tipi for those wanting seclusion.

Experienced English or Western style riders are encouraged to bring their own mounts. Or use one of our ranch-raised Quarter Horses. We have a full-service saddle shop and farrier on hand at all times and a veterinarian on call. We'll do our best to give you the vacation of a lifetime.

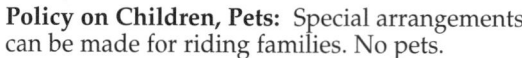

Address: 8102 RCR 56, Hayden, CO 81639

Telephone: 970-276-3920; Fax: 970-276-3377
E-Mail: elkhead@ranchweb.com

Location: 20 miles from Yampa Valley regional.

Guest Capacity: 8

Season: Year round

Policy on Children, Pets: Special arrangements can be made for riding families. No pets.

Rates: Varied packages range from $150 - $390 for a three day minimum stay.

Credit Cards: None

Colorado
Elkhead Ranch

YOUR HOSTESS, HEATHER STERLING.

COLORADO

FOCUS RANCH
SLATER, CO

Focus is a working cattle ranch that was homesteaded in the late 1800's on the Colorado/Wyoming border. The principal ranch function is pasturing 1,000 to 1,500 yearling cattle form May to October. In addition, the ranch looks after 50 cow/calf pairs (including two milk cows), 30 to 35 head of horses, a few bums lambs (orphans), and an assortment of barnyard animals year-round. Overseeing this operation are Terry and Maureen Reidy. The ranch setting is remote and lightly populated, so wildlife abounds. This is a traditional family ranch that remains so today for most of the summer, but increasingly, singles and couples come to participate in a challenging vacation. Everyone is allowed to participate according to their desires and abilities. At Focus, you are a guest in Maureen and Terry's home. The atmosphere is casual, and the program centers on meals, riding, and cattle work. Everything else is flexible according to the season, the guests, and the whims of nature.

Riding is the principal activity and it's unique - usually cross-country working with cattle. Yearly cattle are difficult to handle and must be moved through five to seven pastures in the Medicine Bow National Forest, and five on the ranch. The owners require using the guests to get the work done. Daily riding usually settles the cattle and constant attention increases their weight gain. Guests of all ages and riding abilities may participate - greater ability begets more challenging participation.

Children's Programs: Loosely structured. The ranch setting, chores, and barnyard animals provide entertainment for kids. Riding starts at age 5, usually with a special wrangler who stays with the children throughout the week. Children are allowed to participate in cattle work. With no TV, children rarely become bored. Historical "pop shop" with clubhouse atmosphere, games, and Ping-Pong table belong to the young. Dining: All meals are made from scratch and served family-style. Nutritional balance and quantity are the hallmark of ranch meals at Focus. Special dietary needs can be accommodated. BYOB.

Entertainment: For those who ride, the hot tub is very popular. A basketball court, volleyball net, horse-shoe pit, and a few other props are available. Evening entertainment depends on the mix of guest - Europeans, young, old, staff, and those with energy left to play. The Valley has roping twice a week and occasional dances with the rodeo in nearby Steamboat.

Summary: Working yearly cattle on a daily basis provides a purpose to riding that makes Focus unique. Each week is a little different, dictated by weather and season. Spring and fall cattle work is intense (best for seasoned and advanced riders only), while summer allows more diversity in riding. Whatever the ranch needs might be, guests may participate or not. Fishing, hiking, and relaxation are perfectly acceptable, as the intimacy of the experience leads to many perennial guests. Ideal for family reunions.

Address: P.O. Box 51K, Slater, CO 81653

Telephone: 970-583-2410; call for fax.

Location: Steamboat Springs, 1-1/2 hours away.

Guest Capacity: 25

Season: May through September

Policy on Children, Pets: Children 5 and older. No pets.

Rates: $$ - $$$ American Plan. Children's rates.

Credit Cards: None

(SEE PAGE 223 FOR COLOR PHOTOS!)

Colorado
Focus Ranch

DOING A LITTLE COW WORK IN COLORADO.

HORSEBACK RIDING IN COLORADO.

Colorado

King Mountain Ranch
Granby, CO

King Mountain Ranch offers guests an outstanding, memorable experience in the high Rocky Mountains of north-central Colorado. Built by a Denver oil tycoon and politician as a personal and corporate retreat, King Mountain Ranch has played host to presidents, astronauts, foreign dignitaries, and Hollywood celebrities.

Set on 320 acres, this famous Fraser Valley resort is totally surrounded by over a million acres of the Arapaho National Forest. The secluded setting is ideal for pure relaxation and enjoyment. The two 3-story lodges house 34 guest rooms, a comfortable lobby and library (both with fireplaces), a bar/lounge, and the beautiful Charlie Russell dining room with its tall windows and native stone fireplace.

Accommodations are warm and spacious. Some of the rooms have private fireplaces, and all rooms provide beautiful views of the valley. Creative western cuisine is prepared daily by the executive chef. Most meals are served in the dining room, with the exception of the weekly cookouts for guests during the summer months. The cookouts center around a large campfire on the edge of the lake and give guests the added plus of breathtaking views of the ranch and property.

Guests enjoy the numerous recreational facilities. Horseback riding is, of course, the focus of summer activities. The surefooted mountain horses are all trailwise, with varying degrees of spirit. The experienced wranglers will match you to a horse that fits your riding level. The trails into the National Forest offer a great variety of scenery and an opportunity to do some serious wildlife viewing. Rides usually last about two hours, with all-day rides offered weekly.

In between rides, guests can fish for trout in the ranch's 50-acre private lake, play tennis on the two lighted tennis courts, hike, try their hands at skeet and trapshooting on the outstanding hilltop range, or swim in the indoor heated pool. A hot tub, exercise room, billiards, table tennis, and two-lane bowling alley are also available to guests. And, of course, the Trails End Bar is always a relaxing meeting place and a great way to end an eventful day.

There are two professional award-winning golf courses nearby, and whitewater rafting and hot air ballooning can also be arranged. You'll find a number of boutiques and shops in Winter Park, just a short drive away.

The ranch is open in the winter for cross-country skiing, snowshoeing, and snowcat tours through the National Forest. Winter guests can also rent snowmobiles in nearby Grand Lake or take advantage of nearby ski areas. The ranch also has three meeting rooms for gatherings, family reunions, or retreats. Service here is always of the quality that its impressive setting suggests.

The King Mountain Ranch provides guests with a combination of beautiful seclusion and exciting activity.

Address: P.O. Box 497, Dept. BP, Granby, CO 80446

Telephone: 970-887-2511, 800-476-KING; **Fax:** 970-887-9511

Location: 2.5 hours from Denver

Guest Capacity: 75

Season: Year round. Horses available June 15 - September 30.

Policy on Children, Pets: Children's counselors on staff mid-June to mid-September. No pets.

Rates: $$ American plan. Children's rates.

Credit Cards: Discover, MasterCard, Visa

(See Page 224 and 225 for Color Photos!)

Colorado
King Mountain Ranch

OUR WRANGLERS KNOW ALL THE ROPES TO HAVING FUN.

Colorado

King Mountain Ranch

ONE OF THE MANY WINTER ACTIVITIES AT KING MOUNTAIN RANCH.

Colorado

La Garita Creek Ranch
Del Norte, CO

La Garita Creek Ranch, located at 8100 feet, is located on the Eastern slope of the San Juan Mountains and the Rio Grande National Forest in southwestern Colorado. In the late 1970's and early 1980's, the Ranch was known as the Balloon Ranch. Guests stayed here two weeks and could leave with a Hot Air Balloon Pilot's license. The Balloon Ranch was the only ranch of its kind in the country. La Garita Creek Ranch has combined the Balloon and Guest Ranch concept once again. If you have your own plane, landing on our dirt 2500' airstrip, tying up for the week, and having the time of your life at our ranch can be yours.

Activities: The weekly activities program is dependent on our guest's wishes. Unlimited horseback riding (not nose to tail) and quality horsemanship instruction are the main focus. However, the Ranch offers a diverse number of activities. Included in the rate is riding, fishing, skeet and target shooting, archery, hiking, mountain biking, swimming, hot tub, sauna, jeep rides, hay rides and more. Half day, all day, breakfast and dinner rides are experienced by all. The hot tub and sauna are most welcome after a full day of riding. A choice of whitewater rafting, technical rock climbing, golfing, riding the Toltec-Cumbres Narrow Gauge Railroad is available.

Children's Programs: Excellent supervised children's program with counselors for ages 3 - 5, 6 - 11 and teens. The 3 - 5 year olds ride ponies while the children 6 and up have their own horse for the week and take trail rides as well as participate in horsemanship classes. Nature hikes, picnics, fishing, exploring, swimming, treasure hunts, organized games and crafts are just a few of the activities provided. One night a week, all children 3 and up head out for a super campout while the adults have a quiet candlelight dinner with a champagne breakfast next to the creek the next morning.

Dining: Hearty meals of "Country Cookin with a Gourmet Flair" are served buffet or family-style, all you can eat. Everyone eats together, including most of the staff. Special diets, such as vegetarian or kosher meals can be accommodated with prior notice. Fruit, fresh baked goods, cookies and several beverages are available throughout the day. Entertainment: Campfires, Sing-a-longs, Western movies (with popcorn), Western Dance instruction Guest/Wrangler Talent Show, Hayrides, Local History talks, Showdeo, Awards Night and Boot Branding.

Accommodations: Six guest rooms are in the Stone Mountain Lodge with private baths and southwestern decor. Four duplex cabin rooms which can accommodate a total of 16 guests and the Cliff Dwelling Family Cabin suitable for a family up to 8 persons. The cabins are secluded in trees near the creek. Several have fireplaces and sitting decks out front.

 Address: 38145 County Road 39, Del Norte, CO 81132

Telephone: 719-754-2533; **Toll Free:** 888-838-3833; **Fax:** 719-754-2666
Web: www.lagarita.com
E-Mail: balzotti@lagarita.com

Location: 10 miles north of Del Norte

Guest Capacity: 16

Season: May — September

Policy on Children, Pets: Children 3 years and older.

Rates: $ - $$

Credit Cards: MC, Visa

Colorado

Laramie River Ranch
Glendevey, CO

The Laramie River Ranch is a small working ranch with a total capacity of 25 guests. Our small size allows us to give each guest our personal attention.

Horseback riding is the central activity at the ranch and our abundant terrain provides new vistas with each ride. Our wranglers take out small groups of riders so that you can enjoy riding at a pace suitable for your ability. We are happy to provide new riders with instruction in our arena. When you feel ready you can venture out and practice your new skills on the open range. Experienced riders can enjoy loping rides across meadows and over hills. Morning, afternoon, and all-day rides are available.

Fly fishing on the Laramie River is just super. Our professional instructors provide casting lessons once a week free of charge. You can learn the basics and try your luck in the river with equipment on loan from our shop. It's a great opportunity to fish for wild brown trout in a really beautiful environment.

In addition to riding and fishing, we have a full time naturalist who leads a variety of activities including hiking, birding, hunting for wild flowers, exploring beaver dams, and orienteering. Guests can entertain themselves with volleyball, horseshoes, croquet, and late night strolls to gaze at the stars. Do as much or as little as you like.

And when you are ready for some real adventure you can saddle up for an overnight ride. Escape into the mountains for steaks grilled over an open fire and great conversation. Then relax into your own private tent for a comfortable night's sleep. You will awake to beautiful views and a warm breakfast before riding back down to the ranch.

Outside activities and fresh mountain air create healthy appetites. Whether it's on the trail or at the ranch, meals are an event at the Laramie River Ranch. Our kitchen crew prepares homemade breads and desserts, fresh salads, fruits and vegetable, and delicious entrees. In keeping with informal western tradition, meals are served family style.

Our lodge at the Laramie River Ranch dates back to the 1890s and has a new dining room and kitchen, private bathrooms for each guest room, and both enclosed and open porches for relaxation after a long day of riding. Our five historic log cabins have been remodeled to provide guests with the comfort of modern accommodations.

Each summer guests come to the west to see high flying rodeo action. Only ninety minutes from the ranch, Cheyenne Frontier Days is the biggest outdoor professional rodeo and has been a tradition for over 100 years. Only fifty minutes from the ranch, Laramie Jubilee Days provides both men's and women's rodeo action as well as street dances, parades, art shows and more. Call us for dates.

The Laramie River Ranch prides itself on providing outstanding vacations for its guests. Come and experience our hospitality and the Rocky Mountain west.

Address: 25777 County Road 103, Jelm, WY 82063

Telephone: 970-435-5716; **Toll Free:** 800-551-5731; **Fax:** 970-435-5731
Web: www.jimbalzotti.com
E-Mail: vacation@LRRanch.com

Location: 45 minutes from Laramie airport, 3 hours from Denver airport

Guest Capacity: 25

Season: June through October

Policy on Children, Pets: Children welcome. No pets.

Rates: $$

Credit Cards: Mastercard and Visa

Colorado

Laramie River Ranch

COME AND GET IT!

Colorado

McNamara Ranch
Florissant, CO

If you visit McNamara Ranch near Florissant, Colorado, you'll enjoy unlimited trail riding personalized to suit your horseback riding abilities and desires. Almost due west of Colorado Springs — remote yet easily accessible — the small ranch offers hundreds of acres of pristine wilderness riding amid 14,000-foot snow-capped peaks.

Sheila McNamara, an avid horsewoman all her life, owns and operates the ranch. The bright and sunny ranch house is nestled in a valley of Saddle Mountain, with spectacular views of legendary Pikes Peak. Since McNamara Ranch can only accommodate up to four guests, your horseback vacation is completely individualized. Sheila will guide you through mountain trails and beautiful wooded areas to spacious meadows covered in wildflowers. You can ride to your heart's content or just relax at the ranch house.

A typical day at McNamara Ranch begins with a hearty breakfast to prepare you for a day of horseback riding. Then you'll select one of McNamara Ranch's excellent horses. There is a horse for every riding style and ability. You can ride either English or Western, and you can choose a different horse every day or become attached to just one horse. They are all pets and love their treats! Guests are welcome to help tend and tack the horses or just sit back and anticipate the wonderful day ahead. With the saddle bags packed, the horses are eager to go and respond readily to your commands.

Your ride will take you over wilderness trails, seldom seen by humans, that climb above the timberline for views from the "top of the world." What a terrific way to spend your day, away from the crowd, with horses, while searching for glimpses of the various and abundant wildlife in the area — deer, elk, antelope, hawks, eagles, and other wildlife.

After a few hours in the saddle, you'll be ready for the lunch packed in your saddle bag. Great care is taken to find a lunch location with unbelievable views. Once the horses are tied and given their treats, riders relax and enjoy their lunch. You can even lay back and bask in the warm sunshine and savor the quiet solitude.

Upon returning to the Ranch after an exhilarating day in the saddle, take a soak in the hot tub on the deck and enjoy the views of the surrounding snow-capped mountains.

Supper at McNamara Ranch will satisfy even the most ravenous appetite, with wonderful home-cooked meals. After your day outdoors, you'll probably be ready for a good book and your comfortable bed. For the more active, a stroll in the woods or a good-night visit to the horses might be in order. One last soak in the hot tub under the stars is also an option. This is your vacation, and here you do what you want to do! If you want more of a camping experience, McNamara Ranch offers a secluded teepee by a nearby pond. You can even try and fish for your dinner.

Another terrific day of horseback riding to a different area is in store for you tomorrow. You'll be amazed at how each ride is even more beautiful and spectacular than the day before. If you're willing, Sheila will trailer the horses to nearby wilderness areas for more riding in paradise.

McNamara Ranch is a working ranch that offers a unique vacation opportunity. If quiet solitude, nature, individual attention, and horseback rides tailored to your stamina sound appealing, make yourself at home at McNamara Ranch and stay for a day, week, or more. Experience the ultimate in horseback riding in the Colorado mountains and sample ranch life.

Address: 4620 County Road 100, Dept. BP, Florissant, CO 80816

Telephone: 719-748-3466

Location: 45 miles from Colorado Springs

Guest Capacity: 4

Season: Summer and 2 weeks at Christmas

Policy on Children, Pets: Children must be riders, otherwise little for children to do. No pets.

Rates: $$ American plan.

Credit Cards: None

Colorado

McNamara Ranch

A QUIET MOMENT.

GOING FOR A DRINK.

Colorado

Mountain Meadows Ranch
Gunnison, CO

What makes Mountain Meadows a special place for horseback riding enthusiasts is that it is truly a riding retreat, geared for both confirmed horsepersons who wish to challenge themselves and the weekend enthusiasts looking for a truly Western adventure and personalized service.

The Mountain Meadows Ranch was built in 1875 and is a reflection of Colorado's historic ranching past. The ranch was the homestead of Governor Dan Thorton and his wife Jessie in 1940, where together they developed their "triumphant type" Hereford cattle into one of the great herds of America. A collection of historic memorabilia is displayed in one of the rooms in the house.

Horseback riding is Mountain Meadows's speciality — not just another attraction. Owner Denise Fisher takes pride in the stable of well-trained horses, although guests are certainly allowed to bring their own horses. The highly personalized riding instruction will make even the novice rider comfortable in the saddle. There's even a great established kids program. Try your hand on Mountain Meadows's cross-country course. The lighted indoor arena allows year-round use — regardless of what the weather throws at you. They can even set up a custom pack trip for you. The best way to experience the West is on horseback, just like the old days!

Mountain Meadows sits in a beautiful valley tucked away between the San Juan and Elk mountains and affords year-round activity. Winter brings pristine and quiet serenity, abundant sunshine, and a slew of fun activities. With both Crested Butte and Monarch ski resorts only minutes away, guests can know the thrill of downhill and backcountry skiing. Or try cross-country skiing, ice skating, snowshoeing, ice fishing, snowmobiling, or even a dog sled ride! After a delectable dinner at one of the great restaurants nearby, end the day with a winter horse-drawn sleigh ride, a hot tub, a massage, and a good night's sleep.

In spring, the streams are plentiful, and the fish in them an irresistable lure for the fisherman (or woman). The surrounding mountains will appeal to the naturalist, hiker, or horse enthusiast. Visit Crested Butte, a National Historic District preserving our mining past with a Victorian architectural flair. It is renowned for its fine dining as well as unique galleries and abundant boutiques.

Summer brings wildflowers and festivals. Activities are limitless. Horseback riding is at its best in the summertime in Colorado. Exhilarating rock climbing, mountain biking, jeep tours, hiking and camping, boating, windsurfing, whitewater rafting, hang gliding, ballooning, golf, and tennis are just some of the other activities riders might enjoy when they're not riding. And, by all means, catch a piece of local history — don't miss Cattleman's Day in July, Colorado's oldest rodeo.

There is an explosion of color in the mountains in the fall — a paradise for photographers. Intense blue skies and warm days are great for hiking, fishing, visiting area ghost towns, and horseback riding to view the trees' splendor. It's a time also for relaxing and just reliving all the wonderful moments of the past seasons in those warm days before the snow flies. Every September, the ranch hosts the Mountain Meadows Classic, a fun horse show for all ages and levels. Everyone gets in the act.

Whatever time of year you visit, Mountain Meadows Ranch is a retreat, a safe place where you can leave behind your busy schedules. Relax, recharge, experience the atmosphere and down-home comfort.

Address: 237 County Road #48, Dept. BP, Gunnison, CO 81230

Telephone: 970-641-9501

Location: 3 miles north of Gunnison

Guest Capacity: 12

Season: Year round

Policy on Children, Pets: Children and pets both welcome.

Rates: $ Full breakfast included; kitchen available anytime. Riding extra.

Credit Cards: MasterCard, Visa

(See Page 226 for Color Photos!)

Colorado

Mountain Meadows Ranch

ALPINE MEADOW.

COLORADO

NORTH FORK GUEST RANCH
SHAWNEE, CO

If you are looking for a ranch that has it all — North Fork is it! A small ranch specializing in personal attention to each guest, it's perfect for families who love the outdoors and being together outdoors. Located on the North Fork of the South Platte River at an altitude of 8,100 feet, the ranch is surrounded by beauty. Gorgeous meadows blanketed with wildflowers, quaking aspens, awesome rock cliffs, and ponderosa pines that lead up to the timberline just above them fill the eye in every direction. Always visible a few miles to the west is the Continental Divide with its snow-capped peaks reaching over 14,000 feet.

Horseback riding is available for all levels of riders. Your week at North Fork Guest Ranch will begin with some basic instruction and walking rides. As the week progresses, the length of rides increases, and riders will have an opportunity to do some loping and cantering. Experienced wranglers will guide groups of 6 to 8 people on half-day and all-day rides into the high country wilderness surrounding the ranch. Overnight pack trips can be arranged, and a guest rodeo at the end of the week offers a chance to show off your skills in horseback games and events.

When you're not on horseback, you might want to try whitewater rafting down Brown's Canyon on the Arkansas River. That's an experience you won't want to miss! The trip combines the thrill of a roller coaster with incredible canyon scenery — for those calm periods when you actually have the time to look around! For the less adventuresome, you can fish right outside your back door on the North Fork of the South Platte River. The ranch also has fly fishing instruction and a stocked trout pond.

Other favorite activities include hiking, swimming in the heated pool, outdoor and indoor games and sports of all kinds, trap and target shooting, horseshoes, and just plain lounging around. In the evening, there are hayrides, square dancing, volleyball games, and campfire singalongs. If you want to soak away sore muscles or just relax, the hot tub's the place for you. On the last night of each week, there is a farewell candlelight dinner where guests can share — and create — fond memories of their time at North Fork.

If you want to see the country off-ranch, there's an alpine slide and great little shops in Breckenridge. Fairplay has the South Park City Museum.

The children's program offers pony rides, nature hikes, and crafts — all supervised by the terrific children's counselors. There's also an overnight campout in the teepee and the animals in the "petting farm" to be attended to. Babysitting can be arranged for infants. The children's program is designed for children under 6. Children over 6 are invited to enjoy all of the ranch activities with their parents.

Accommodations vary at the ranch and are suitable for any size family, as well as singles and couples. The Wildhorse Lodge is the main building and is centrally located. Here each room has its own private bath and is decorated with country quilts, Western artwork, and log furniture. Families will feel right at home in one of the other buildings or cabins. All have private baths; some have shared common areas, and some have private sitting areas with fireplaces.

Meals are served family style with all you can eat! Refreshing salads and homemade breads start every meal, and typical dinners feature steak, roast beef, pork loin, or trout with fresh vegetables and scrumptious desserts.

The terrific accommodations, great food, varied activities, and of course the wonderful horseback riding will be sure to make your vacation at North Fork Guest Ranch one you will remember for years to come!

Address: P.O. Box B, Dept. BP, Shawnee, CO 80475

Telephone: 303-838-9873, 800-843-7895; **Fax:** 303-838-1549 **Web:** www.jimbalzotti.com

Location: 50 miles southwest of Denver, 6 miles west of Bailey

Guest Capacity: 40

Season: May through September

Policy on Children, Pets: All children welcome; counselors on staff for those 6 and under. No pets.

Rates: $$ American plan. Children's rates.

Credit Cards: None. Personal checks, traveler's checks, and cash accepted.

Colorado

North Fork Guest Ranch

MAKE OUR HOME YOUR HOME THIS WEEK!

THE RUSHING SOUTH PLATTE RIVER RUNS THROUGH THE RANCH.

COLORADO

PEACEFUL VALLEY LODGE AND GUEST RANCH
LYONS, CO

Peaceful Valley has been putting guests in the saddle for over forty years. Experienced horsemen, dudes, and greenhorns — saddle up to view some of the most beautiful country ever created! There's a horse waiting here for you.

A quality resort with a tradition of old-fashioned hospitality, you'll surely appreciate the "Fun, Food, and Friendly Feeling" provided by Mable Boehm, Debbie and Randy Eubanks, and their terrific staff.

You'll choose what activities you want to participate in, or not. There are daily horseback rides along the scenic trails or through the surrounding Roosevelt National Forest. If you need a refresher course, riding instruction is always available. You can choose from one of several all-day trips, either by horse, vehicle, or hiking. When you're not on a horse, you can enjoy tennis, swim in the indoor solar-heated pool, refresh tired muscles in the sauna and hot tub, or fish for trout in the ranch's own ponds or in nearby streams and mountain lakes. Perhaps resting and loafing were more what you had in mind.

Sign up for the overnight pack trip and the thrill of a lifetime. The trip leaves the ranch and heads into the Roosevelt National Forest. You and your companions will fish, take in the wonderful scenery that you can only see by horseback, and share some real trail fellowship. What's more, you'll have breakfast on the mountain within sight of the snow-capped Continental Divide — a never-to-be-forgotten experience.

During the summer months, bring the children and let them enjoy the special programs designed for them. There are full-time trained counselors on duty throughout most of the day, one for each age group. Horseback riding is a big attraction, either on the trails or in the ring — or in a pony cart for those under 6. Children ride twice a day — one trail ride and one horsemanship lesson. The Pet Farm will draw the youngest to feel, pet, and love young kid goats, calves, and bunnies. Teenagers will enjoy a program that includes dancing, swimming, horseback rides, and llama hikes.

Riding a horse isn't the only way to see this magnificent country. Climb aboard the ranch's "horseless carriages" and ride off to check out old historic narrow-gauge railroad beds, gold and silver mines and towns, or the grandeur of Rocky Mountain National Park, all within a short distance.

In the evenings, guests of all ages enjoy the hayride; poolside barbecue; square, folk, and other dance parties; singalongs; weekly talent show; and late evening Kaffeeklatsch. In fact, like the "Old" West, the coffee pot is always on at Peaceful Valley Ranch.

Accommodations are varied, and range from cabins, to suites, to private rooms in the chalet or one of the lodges. All of the private cabins have stone fireplaces and hot tubs. Delicious home-cooked meals are served family style in the dining room. They'll even cook up any fish you catch!

Peaceful Valley also has a conference center that can accommodate up to 150 persons. Food, lodging, and recreational activities are all available for that special group.

Peaceful Valley Ranch has one more surprise for guests — a beautiful chapel where a church service is conducted each Sunday. For those saying goodbye after a week's vacation, it may provide a most inspirational experience. This serene, quaint chapel has also been the setting for many weddings.

Family . . . fellowship . . . food . . . fun. You'll find all of those at Peaceful Valley Lodge and Guest Ranch in Colorado's beautiful mountain setting.

Address: 475 Peaceful Valley Road, Dept. BP, Lyons, CO 80540-8951

Telephone: 303-747-2881; **Fax:** 303-747-2167

Location: 60 miles northwest of Denver, 22 miles south of Estes Park

Guest Capacity: 100

Season: mid-May through September

Policy on Children, Pets: Children very welcome, with on-staff counselors and supervised children's program. No pets.

Rates: $$ American plan. Children's rates. Overnight pack trip extra.

Credit Cards: American Express, Diners Club, Discover, MasterCard, Visa

Colorado

Peaceful Valley Lodge and Guest Ranch

THAT'S HORSEPOWER!

DON'T FORGET TO BRING YOUR CAMERA.

Colorado

Powderhorn Guest Ranch
Powderhorn, CO

Once you make the turn off the pavement onto the gravel road for the 10-mile drive to the ranch, you can feel yourself start to relax. The road follows the Cebolla Creek as it meanders through the beautiful Powderhorn Valley in southwestern Colorado. Cattle graze peacefully in the meadows — and sometimes on the road, but they'll lazily move out of the way of your car. Gradually, the road narrows into a little canyon where the creek tumbles over rocks an is framed by huge, pine trees.

Just as you're wondering if you'll ever get there, the big red barn appears around a bend, and it's like coming home. Pulling into the driveway and crossing the bridge, you can barely park your car in front of the office before Jim and Bonnie Cook are there to greet you and welcome you to a week in their very own paradise.

They clearly love it here; it's a lifelong dream come true for them, and the best part about it is sharing it with others. Jim and Bonnie say they can welcome the most uptight, burned-out executives on Sunday and by Tuesday evening have them relaxed, unwound, content, and born for the saddle. The biggest decision they'll have to make is whether to use a salmon egg or a fly for the evening's fishing.

You'll settle into a comfortable and cozy log cabin complete with a refrigerator for your snacks, pop, and perhaps a libation (B.Y.O.B.). The front deck is the perfect place to lounge and breathe in the pure air while you observe the kids on the playground or in the swimming pool. Here, you'll be able to get away from it all as there are no telephones or televisions in the cabins.

Before dinner, you'll have a chance to meet the other guests, who come from all over the country and Europe. You'll also meet the staff, mostly college students wanting to spend their summer on a ranch, learning all types of skills, and meeting wonderful people. The food is outstanding and plentiful, and seconds and thirds are just normal once you've spent any time in the clear mountain air.

Monday morning after your first night's sleep under blankets in the middle of July or August (maybe even with the furnace on) you're ready to meet your horse! He'll be chosen just for you, and you will keep him for the week. After a short orientation with the basics of riding, you're off on the trail. Groups are kept small so you have a chance to enjoy the scenery and views of snow-capped mountain peaks as well as being able to hear the wrangler as he points out wildlife or interesting things along the way. Rides vary from 2.5 hours to all day.

The days and nights speed by, filled with horseback riding, spectacular Jeep rides into the high country led by Jim or one of the wranglers. Fishing is excellent on Cebolla Creek, which runs through the ranch, or in the stocked ponds. A guide will teach you beginning fly fishing and supply all equipment. There's also a square dance, campfire singalongs, supper in a mountain meadow, a raft trip, old western movies, and, if you're brave, tubing on the Cebolla Creek.

As a climax to a perfect week, there's a steak fry on the picnic island, and awards are given for the gymkhana (games on horseback) held that morning. A video will remind you of what went on during the week. It's a bittersweet time, because everyone knows they'll be leaving after breakfast in the morning — leaving new friends that have become so close in such a short time, leaving the peaceful Powderhorn Valley and a way of life they won't see again, at least not for another year.

Address: 1525 County Road 27, Dept. BP, Powderhorn, CO 81243

Telephone: 970-641-0220, 800-786-1220

Location: 234 miles from Denver, 204 miles from Colorado Springs, 164 miles from Grand Junction

Guest Capacity: 30

Season: June 1 to mid-September

Policy on Children, Pets: Children are welcome. No pets.

Rates: $$ American plan. Children's rates.

Credit Cards: None. Personal and traveler's checks accepted.

Colorado
Powderhorn Guest Ranch

WATERING THE HORSES BEFORE A RIDE.

POWDERHORN'S SCENIC OVERLOOK.

Colorado

Rainbow Trout Ranch
Antonito, CO

Rainbow Trout Ranch truly offers everything you could want in a dude ranch vacation. Its facilities include 16 guest cabins of various sizes and some built as early as the 1920s, in addition to a huge and historic lodge that has to be seen to be believed. Built with local Engelmann spruce, it is said that not a single nail was used in the entire 18,000-square-foot structure — just pegs. The ranch has a pool and a hot tub as well as guest laundry, a recreation room, and plenty of other small but nice conveniences.

But the real heart of the ranch is in the fantastic riding and fishing that lies not far from your cabin porch. The Van Berkums are very proud of their horses. Each person is expertly teamed with a horse that meets his or her riding level. The guest keeps that horse for the entire week, which allows them to really get to know each other in time to participate in the guest rodeo on Saturday afternoon. The riding is geared towards every ability level, and the terrain lends itself to both pleasant river rides and to more challenging mountain rides, with plenty of opportunity for anything in between. Rides are scheduled twice a day, and guests can sign up for all-day rides and an overnight ride that climbs to over 11,000 feet.

And the fishing... over a mile and a half of private fishing on the Conejos River right on the ranch. Rainbow trout are, of course, most plentiful, but there are also brown trout to be caught. Although catch-and-release is encouraged, the ranch kitchen is always delighted to prepare your cleaned catch for one of your meals. Or even send that trophy fish to the local taxidermist! There is also plenty of excellent fishing in local streams.

Along with the riding and fishing, the hiking is also great all around the ranch, and there are interesting places to visit within an easy radius. The drive up the Conejos River Canyon to Platoro Reservoir offers great fishing opportunities and gorgeous scenery. The nearby Cumbres and Toltec Scenic Railroad is America's highest and longest narrow-gauge railroad and makes for a spectacular day trip. East of Alamosa lie the impressive Great Sand Dunes; and the close proximity to New Mexico provides an interesting contrast to the ranch's location at an alpine 9,000 feet. Within an hour and a half of the ranch is Taos, New Mexico, and whitewater rafting on the Rio Grande River; the fascinating Santa Fe, New Mexico, area is only 2.5 hours away.

Part of the ranch's family appeal is its two great kids' programs (for 3-5 year olds and 6-11 year olds) and a fun teen program. Children don't have to be part of the programs, but most love them. Activities include horseback riding (including instruction), fishing, hiking, nature study, swimming, and crafts. Parents have a chance to go on longer rides, or to fish, or to just totally relax knowing their kids are in good hands and that they're having safe, creative fun.

Guests look forward to the wonderful family-style meals, with plenty of fresh fruits and vegetables, homemade breads and deserts, and entrees that vary from fresh trout to steak to ribs to turkey. Taking advantage of the great weather, there are frequent barbecue cookouts.

And if the summer season with all of the fun activities such as Western dance night, singalongs, horse-drawn hayrides, and such doesn't suit you, the ranch offers special early and later seasons. These are times for those of you who want to ride and fish and enjoy the quiet peace of this beautiful, little-known area in a more adult atmosphere.

There truly is something for everyone at Rainbow Trout Ranch. Enjoy the history, warm hospitality, and scenery that will make you want to stay forever.

 Address: P.O. Box 458, Dept. BP, Antonito, CO 81120

Toll-free Telephone: 800-633-3397;
Telephone/Fax: 719-376-5659
Location: 3.5 hours from Albuquerque, 1 hour from Alamosa, Colorado

Guest Capacity: 60

Season: late May to September 21

Policy on Children, Pets: Supervised children's program for ages 3-5, 6-11, teens. No pets.

Rates: $$ American plan. Minimum stay summer season.

Credit Cards: None

Colorado

COLORADO

RAWAH RANCH
GLENDEVEY, CO

Are you looking to escape? Are you interested in moose... elk... deer... coyotes... pronghorn antelope... wildflowers... everywhere? How about snow-capped mountains that pierce your heart, waterfalls, a 76,000-acre Wilderness, and a stunning, secluded valley? Imagine riding your "own" fine Rawah horse your way — high (12,000-ft. snowfields), low (aspen groves, ghost ranches), fast (riverside meadows, rolling sage hills), slow (breathtaking overlooks, alpine lakes) — all day, any day. Imagine hiking these same wonderful trails. Imagine private-water, wild brown trout.

To the Native Americans, *Rawah* meant "abundance." That's still true today. Rawah Ranch is a special place 8,400 feet high in the Colorado Rockies, just south of the Wyoming border in the stunning, secluded Laramie River Valley. Named for the 76,000-acre Rawah Wilderness adjoining it and the snowcaps above, Rawah blends the "best" from the numerous ranches where its owner/operators were themselves guests. The result is a unique combination of riding, fishing, hiking, and warm Western hospitality.

A maximum of 32 guests enjoy the ranch's cozy log cabins, all warmed by fireplaces. From the cabin porches, enjoy the peace and quiet or watch for the many deer and moose in the area. Also, in the ranch's beautiful log lodge, there are three guest rooms, each immaculate and with a private bath.

Meals are served family-style in the lodge dining room. Steaks and chicken grilled over aspen, plenty of fresh vegetables and fruit, homemade breads and mouth-watering desserts are just part of the wonderful country cookin'. Guests have even complained that the food is *too* good!

Mornings begin with coffee delivered to your cabin or lodge room, followed by a hearty ranch breakfast in the dining room. Most guests then head for a day of riding, fishing, or hiking — the core of the kind of ranch experience provided at Rawah.

Riding is *the* favorite Rawah activity. The riding possibilities are nearly limitless. Trails and open areas cover 100,000+ acres of wilderness, national forest, and private land. The experienced and seasoned wranglers carefully pair you with your horse for the week and provide plenty of hints to make time in the saddle enjoyable and comfortable. Take advantage of the complimentary individual or group instruction.

Riding destinations include a 12,000-foot pass near never-melting snow, breathtaking valley overlooks, cascading waterfalls, abandoned miners' cabins. There are half-day rides every morning and afternoon and all-day rides. There's even a riverside breakfast ride. High, low, fast, slow, short, long... it's your choice.

The quality of the horses, the number of wranglers, the unmatched scenic variety, and the flexibility of the riding program assure guests' satisfaction. Rawah offers easy access to trails in the Rawah Wilderness and the National Forest. It's a short car trip to hiking trails at Rocky Mountain Park.

Just steps from the lodge, the Laramie River meanders through the ranch. Experts say the fly fishing for native brown trout here may be Colorado's finest. Numerous nearby mountain lakes are home to rainbow, cutthroat, and brook trout. The ranch includes a professional fly-fishing clinic and a pond stocked with rainbow trout. They'll even let you borrow equipment if you've forgotten your own.

As the day winds down, the relaxing hot tub always beckons. Crackling fires in the lodge's stone fireplaces are the backdrop for evening activities. The ranch's rec room offers table tennis, pool, and table shuffleboard, and a weekly evening of western dancin' has everyone kicking up their heels!

Gorgeous scenery, comfortable and cozy accommodations, terrific horses, fabulous fishing... what more could you ask for!

Address: Summer: 11447 North County Road 103, Dept. BP, Jelm, WY 82063; Winter: 1612 Adriel Circle, Dept. BP, Fort Collins, CO 80524

Telephone: 800-820-3152, 970-435-5715 (Summer), 970-484-8288 (Winter) **Web:** www.jimbalzotti.com

Location: 135 miles from Denver

Guest Capacity: 32

Season: mid-June to late September

Policy on Children, Pets: Children must be at least 6 years old. No pets.

Rates: $$ American plan.

Credit Cards: None

(SEE PAGE 228 — 229 FOR COLOR PHOTOS!)

Colorado
Rawah Ranch

RUSTIC LUXURY.

COMFORTABLE SURROUNDINGS.

COLORADO

RED FEATHER GUIDES AND OUTFITTERS
GOULD, CO

Arnie Schlottman and Todd Peterson, owners of Red Feather Guides and Outfitters, have been in the outfitting business for the past 15 years. They located the business in North Park, one of Colorado's high mountain parks, to be in the midst of exquisite scenery and wildlife - North Park is the "Moose Viewing Capitol" of Colorado. Arnie and Todd developed several programs that include horseback rides and summer trips offered June through September. After the summer season, the focus is on wilderness big game hunting trips that last through November. All trips originate at the lodge, which is easily accessible by flying into Denver or Steamboat, Colorado or Laramie, Wyoming. The lodge is a short 1-1/2 hour drive from Steamboat and Laramie and 3-1/2 hours from Denver.

Together, Arnie and Todd have perfected the horse lover's dream - the High Country Pack Trip. Ride a couple of hours into the most spectacular country you have ever seen and camp next to high mountain lakes where the trout fishing is excellent. During the day, part of the family can horseback ride while the others fly-fish. While you are playing, Red Feather will do all the work - setting up camp, cooking and wrangling. Throughout your vacation, a full-time camp cook is provided to ensure three savory, home-style meals each day. Red Feather provides everything except your bedroll and personal gear. Don't forget your camera! What kind of weather should you expect? Colorado is famous for amazing blue skies and starry nights. You can experience highs in the 70's during the summer months, with lows at night down to freezing. Expect early afternoon rain showers with clear skies in the evening.

A favorite of North Park visitors is the horseback rides. Rides into the Mount Zirkel Wilderness Area, Routt National Forest and the Never Summers, Colorado State Park for hourly, 1/2 day, all-day or overnight trips provide an experience you will never forget. Stop for lunch in beautiful mountain meadows overflowing with wildflowers and snap pictures of elk, deer, coyotes or moose. Steak dinner rides are a specialty, including favorites like toasted marshmallows and s'mores, ending with all the wranglers telling stories over a cozy campfire. The wranglers are great hosts and have a well-trained string of gentle horses that even the most inexperienced of horse-lovers can handle.

If you prefer to stay at the main lodge, Red Feather has a plethora of activities for their guests, including fully-guided fishing trips to a private lease on the North Platte River. This stream is plentiful with rainbows, browns and brook trout. Guests may also enjoy exploring the hiking trails found within a short distance from Red Feather's lodge. Accommodations at the lodge include a rustic main house, where meals are enjoyed, and five cozy log cabins, all electrically heated. There is a newly constructed centralized shower-house with private facilities for both men and women. At the end of the day, relax in Red Feather's hot tub and sauna! Or, kick up your feet and view the elk and moose right from the front porch.

Whatever your interests may be...photography, fishing, mountain climbing, camping or horseback riding on the Continental Divide, the entire family can share this outing in the serene wilderness of the Rocky Mountain High Country. Your family will cherish the memories for the rest of their lives.

Address: 49794 Highway 14, Gould, CO 80480;
Mailing: P.O. Box 16, Walden, CO 80480

Telephone/Fax:
from November 1 to June 1, 970-723-4204
from June to November, 970-524-5054

Location: 3-1/2 hours from Denver

Guest Capacity: 25

Season: June 1st through November 1st

Policy on Children, Pets: Children and pets are welcome.

Rates: Call

Credit Cards: Call

Colorado

Red Feather Guides and Outfitters

LANDING A BIG ONE.

COLORADO

ROUBIDEAU WESTERN ADVENTURES, INC.
DELTA, CO

Rod and Dale Hall and Larry and Wanda Boyd graze their 900 mother cows on a public land grazing allotment totaling 75,000 acres and ranging in elevation from 5,700 feet to 10,000 feet above sea level.

Our ranching partnership in the unique Roubideau Canyon area offers a variety of vacation opportunities. Our guests will be able to participate in the daily ranch activities and management of the cattle as they are moved to fresh pastures throughout the summer. In addition to the rounding-up and moving cattle, fences need to be checked and mended, some salt needs to be packed to new grazing areas, grazing utilization needs to be monitored, and other daily ranch chores that take you riding through sage, pines, aspen, and spruce. Enjoy the fresh air and wide open spaces, the beautiful western Colorado scenery and the wildlife such as deer, elk, wild turkeys, coyotes and an occasional bear. At days end, join in a game of horseshoes, try your hand at throwing a rope, relax by the fireplace, or enjoy a peaceful hike in the stillness of a mountain evening.

Our cow camp accommodations are clean, warm and comfortable. The cabins have running water, propane lights and refrigerators. Showers are available, but restroom facilities are outside. Sleeping quarters at the Hall camp, (the UCC camp), are a separate bunkhouse style camp trailer. Boyd's cabin, (the 7N camp), has an upstairs with bunks for the guests. All meals are served homestyle, and the guests eat with their host families. On longer days, lunches are packed on your saddle. Occasionally, weather permitting, we have an outside barbecue and guests experience the flavor of authentic Dutch oven cooking.

The horses you will be riding are our ranch horses and will be suited to each guest according to riding experience. All tack and horses are in good, safe condition. Days can be planned for your comfort level. Non-riders can enjoy hiking, wildlife watching or just relaxing.

The cow camps are approximately 40 miles (1 hour drive) from the nearest town or store so we ask that you please bring any items you may need. A few suggestions are a jacket, hat, warm clothes, comfortable riding boots and comfortable shoes for hiking if you wish, gloves, sunscreen, insect repellent, flashlight, sleeping bag, personal toiletries, and personal first aid supplies such as Band-Aids, aspirin and allergy medications. Our cabins are non-smoking facilities. This list may vary according to the time of year of your vacation. At the time of reservation, a more specific gear list will be provided. Due to the distance, non-scheduled trips to town will be made at an extra charge.

Airport pickup from Montrose is available. When making your travel connections, your travel agent can use the Montrose-Delta area to enjoy other scenic attractions, rafting, festivals and our historical sites.

We are not a dude ranch. We are not fancy and we are not flashy, but we are real. We would like to invite you to share the joys and excitement of our daily ranch life, enjoy the beauty and wonder of nature in the Roubideau Canyon and on the Uncompahgre Plateau, and come join our crew for a vacation of a lifetime!

Address: 3680 Cedar Road, Delta CO 81416

Telephone: 888-878-9378
Web: www.dci-press.com/roubideau/index.html
E-Mail: trawest@dci-press.com

Location: 40 miles SW Montrose, CO

Guest Capacity: 6 at each of 2 ranches

Season: April through October

Policy on Children, Pets: 8 years and older. No pets.

Rates: $$

Credit Cards: None

COLORADO

Colorado

San Juan Guest Ranch
Ridgway, CO

The San Juan Guest Ranch is surrounded by more 14,000-foot peaks than any other spot in the United States. The ranch itself sits at 7,200 feet in the Uncompahgre Valley in southwestern Colorado, which has been called "the Switzerland of America." The San Juan Mountains are rugged and unrestrained, full of astounding beauty. Lush cottonwoods and willows line the edges of rivers and streams. The vivid sunsets are a stark contrast to the red sandstone buttes. These mountains were once home to the Ute Indians, and to them the mountains were sacred.

Pat MacTiernan and her son Scott feel the same way about the area. When you visit the San Juan Guest Ranch, you will learn the lingo, the history, the ecology, and the philosophy of the southwest Colorado region. To survey the beauty of the mountains is a pleasure; to truly understand this place is a joy.

Horses are essential on a Colorado ranch, and there is no better way to experience the beauty of the Valley than on the back of a horse. The ranch has a comprehensive riding program for riders of all ages and superbly trained horses. Experienced wranglers will soon have you riding from valley floor to the timberline like a seasoned hand, even if you've never been in the saddle before. There are daily half-day and full-day rides, as well as an overnight high-country pack trip.

Picture yourself on a day-long ride that ends high in the mountains. The camp's been set up; supper is on the fire, and it sure smells good after the ride! There's plenty of time to fish, explore high mountain trails and meadows, or just relax by the campfire and watch the gorgeous sunset. Fall asleep under the stars and wake to the smell of a hearty breakfast before a leisurely ride back to the ranch.

In addition to the riding, San Juan Guest Ranch has a wide variety of activities from which to choose. You can fish for trout in the private pond or in nearby streams or try hiking and mountaineering in the San Juans. There are roping and riding lessons, a trip and rifle pistol range, hayrides, volleyball, softball, and horseshoes. You can take a jeep ride to explore mountain ghost towns and abandoned mines. In the evenings, there are bonfires, singalongs, dancing, and storytelling. For those tired muscles at the end of the day, there's the hot tub. The nearby town of Ovray offers hot springs, tennis, shopping, and museums.

In the winter, you can ride on the snowy trails. There are also ice skating, Nordic skiing at your doorstep, snowshoeing and tobogganing, back-country skiing, and ice climbing. The famous Telluride Ski Resort is only 45 minutes away. Christmas at the ranch is festive and traditional, with a turkey dinner and all the fixings, caroling, and sleigh rides.

The ranch house is where the hearty, family-style meals are served. The fireplace and large comfortable sofas in the den invite you to sit, stay awhile, and join fellow guests for good conversation. The two-story guest lodge has accommodations for singles, couples, and families. The rooms are comfortable and homey and have decks from which to view the magnificent scenery.

The ranch has a supervised program for children 6 and older, activities including horseback riding, caring for and feeding ranch animals, an overnight pack trip, fishing, swimming, hiking, crafts, and wagon rides.

As a special treat, Scott has created a 33mm photo adventure/workshop for adults that is offered in the fall. Capture the aspens' magnificent show with advice from a professional outdoor photographer.

The San Juan Guest Ranch maintains an atmosphere that promises to make your vacation — if only for a week — a memory that you will cherish for years to come.

Address: 2882 County Road 23, Dept. BP, Ridgway, CO 81432

Telephone: 970-626-5360, 800-331-3015; **Fax:** 970-626-5015

Location: 32 miles from Montrose, 360 miles from Denver

Guest Capacity: 32

Season: Year round

Policy on Children, Pets: Children must be 6 years old. No pets.

Rates: $$ American plan. Children's rates.

Credit Cards: MasterCard, Visa

(See Page 230 and 232 for Color Photos!)

Colorado
San Juan Guest Ranch

COOKING OUT OVER THE OPEN FIRE.

THE FISHING'S GREAT!

COLORADO

SAN JUAN OUTFITTING
DURANGO, CO

Pack Trip Description Our pack trips take you into the heart of the Weminuche Wilderness which is Colorado's largest, most rugged and scenic area enjoying wilderness designation. All travel is restricted to foot or horseback, insuring the unspoiled effect of the many high alpine basins, streams and lakes. Several peaks are over 14,000 feet with many being just under that. Our campsite is reached by a 14-mile horseback ride into a location at 10,300 feet, just below the Continental Divide.

Our accommodations are quite comfortable, including a 16'x20' dining tent and cabin style sleeping tents accommodating two guests each. They are equipped with full-sized cots and pads. We also enjoy the luxury of a hot shower which has proven to be quite popular through the years. Our meals are excellent and include gourmet foods and fresh baked desserts.

During the course of a typical 5-day trip, we will ride loop routes from camp over passes on the Continental Divide, to nearly 13,000 feet. We experience fishing in 3 streams and 5 lakes for Brook, Cutthroat and Rainbow trout with some weighing up to 5 pounds. The Pine River flows right in front of the camp offering very good fly fishing for the novice and accomplished angler alike. Local wildlife includes elk, moose, mule deer, bighorn sheep and black bear. Mule Deer & moose are often photographed in our camp as they are quite accustomed to our presence. Elk are much more shy but can offer excellent photographic opportunities in the higher basins.

Our horses have all been selected for character and physical stamina enabling them to negotiate precipitous terrain in a very dependable manner. Since we operate in a wilderness area, all of our supplies and camp equipment is packed in on horseback. Most equipment is packed utilizing a single or double diamond hitch which results in a secure load.

In order to provide each person with personalized, high quality service, we limit our group sizes. Five-day trips are the norm. However, longer stays can be arranged if made well in advance. We generally plan our first trip in mid-June and continue through the end of September.

Continental Divide Ride Description The 'dive ride' leaves from the top of Wolf Creek pass and traverses the spine of the Rocky Mountains to the mining town of Silverton, Colorado. Because of its wilderness designation, all travel is restricted to foot or horseback, insuring the unspoiled effect of the many high alpine basins, streams and lakes.

During the course of our 8-day, 100 mile excursion, we will move our camp six times. Every riding day will be spent at elevations at, or above, 11,000 feet. During the ride, it is not uncommon to watch large herds of elk, individual deer and an occasional group of bighorn sheep. Each camp site has been selected for its scenic beauty and quality fishing in either lakes or streams.

On our fifth day, we will ride into our summer base camp and spend a day and a half there. In this camp we have a 16'x20' cook tent for our dining and a hot shower, which is quite popular.

In order to provide each person with personalized, high quality service, we limit our group size to no more than six clients per trip. The cost of the continental divide ride is $1,800 per person. This trip will leave from Wolf Creek Pass and will arrive in Silverton the first week of August.

Address: 186 County Road 228, Durango, CO 81301
Telephone: 970-259-6259; Fax: 970-259-2652
Web: subee.com/sjo/sjohome.html or www.jimbalzotti.com
E-Mail: sjo@frontier.net
Location: 6 miles north of nearest airport.

Guest Capacity: 10

Season: Year round

Policy on Children, Pets: Children 7 years or older. No pets.

Rates: $$

Credit Cards:

Colorado

San Juan Outfitting

WE KNOW ALL THE GOOD SPOTS.

San Juan Outfitting

Colorado

7W Guest Ranch
Gypsum, CO

You've worked hard for this vacation so why not treat yourself to the best - a week of high adventure at the 7W Guest Ranch. Escape to a place where friendly folks care and your happiness really matters. We're smaller than most, no more than 18 guests at one time, and we may be hard to spot, secluded among the quakies and spruce, but once you've hoofed among our high country trails, plucked trout from our clear mountain lakes and gazed in awe at the brighest heavens you've ever seen, you may wonder what took you so long to find us.

The 7W is one of Colorado's oldest and most respected guest ranches. Perched on a high bench at an elevation of 9,000 feet, the air is clear and crisp. The spectacular peaks of the Flattops Wilderness frame the horizon, wildflowers abound and wildlife magically appears and disappears right before your very eyes. You're in the heart of the Rockies where mountains are larger than life and the romance, freedom and adventure of the Old West lives on forever.

The 7W Guest Ranch offers you a special opportunity to relax and enjoy the Colorado Rockies. Guests enjoy lingering around an evening campfire, watching starry skies, exchanging wild tales of the West, a quiet walk through the forest, or listening for the cry of a coyote.

A visit to our small, mountain home is a blend of immaculate, rustic cabin accommodations, scrumptious homemade Blue Ribbon cookin', and an enthusiastic, talented and courteous staff. You'll ride through forest of aspen and spruce, fish for trout in our private stocked lake and mountain ponds, and walk where miners, trappers and outlaws traveled. Breathe fresh, clean air and wonder at the striking panorama in the distance. The weather is wonderful, lots of sunshine, cool and crisp nights.

Our emphasis is horses with almost unlimited riding opportunities. We'll pair you with a horse that matches your riding ability. You will ride over a variety of terrain - thousands of acres in the White River National Forest that adjoins the ranch property. In addition to trail rides and riding instruction, enjoy a breakfast ride, dinner ride, and an overnight pack trip. For the more adventuresome, we specialize in three to five day pack trips.

The 7W's reasonable rates include daily maid service, all activities on the ranch (including great trout fishing), overnight pack trip planned for the end of the week, as well as all the clean air and sunshine you can soak up. Reduced rates are offered. It's not too early to make reservations - many of our summer weeks fill up quickly.

We preserve the atmosphere of the Old West at the 7W Guest Ranch with a friendly and relaxed pace - hospitality at its best. We hope that you will be able to join us for an unforgettable adventure in western vacationing. Looking forward to hearing from you!

Address: 3412 County Road 151, Gypsum, CO 81637

Telephone: 970-524-9328
Web: www.jimbalzotti.com

Location: 175 miles from Denver International Airport, 134 miles from Grand Junction.

Guest Capacity: 18

Season: June through September

Policy on Children, Pets: Children are welcome. No Pets.

Rates: $$

Credit Cards: MasterCard and Visa

Colorado

7W Guest Ranch

YOU WON'T WANT TO GO HOME.

IT TASTES BETTER THAN IT SMELLS!

Colorado

SKY CORRAL GUEST RANCH
BELLVUE, CO

Located in the beautiful Roosevelt National Forest, the ranch is remotely situated on 452 acres with a friendly, family atmosphere. Our smaller capacity allows guests to relax and meet people from around the world.

Horseback riding in the Colorado Rockies is the focal point of Sky Corral. The trails will lead you to scenic overlooks of the Continental Divide, Colorado, & Wyoming. Aspen groves and open meadows are breathtaking while challenging your abilities as a rider through the forest. We own our own horses and offer beginning, intermediate and advanced rides. Guests love the way we take them "bushwhacking" through the tall timber, getting off the beaten path and being able to explore new places and experience the chance to see wildlife in their natural habitat. It is a great opportunity to see the country the best way possible - on the back of one of our fine horses. We also offer English and Western lessons for those who want to become more comfortable in the saddle or just want to learn to ride for the first time. Because of our smaller size, you won't get lost in the crowd and we will be able to spend time with you until you are comfortable with your horse. Our experienced wranglers will have you riding like a cowboy!

Activities are abundant with a tennis court, swimming pool, sauna, Jacuzzi & gym available to all guests. The ranch staff loves to challenge anyone to a volleyball game or to shoot some hoops on the basketball court. Evenings are usually filled with square dancing, hay rides, sing-a-longs around the campfire, or visiting with other guests.

Whitewater river rafting on the Cache la Poudre River is always one of the highlights of your vacation along with our overnight campout.

The Sky Corral Guest Ranch offers a wonderful family atmosphere with our children's program for 3-7 and 8-12 years of age. Our children's counselors will provide fun activities for all ages. Parents can be assured of the care and attention their children will receive while they horseback ride, hike, fish and do nature crafts. The program allows families with children of all ages to enjoy their ranch vacation. Your children will be given the opportunity to experience all the ranch activities we have to offer. Along with the enjoyment of these activities, your child will walk away with a better understanding of horses and a greater appreciation of the outdoors. Our goal is to make the week a learning experience as well as to have a wonderful time.

During your free time enjoy fishing in the stocked lake (our motto is "We'll cook'em if you catch'em."), a tennis game, volleyball game, a game of horseshoes, swimming in the heated pool or jacuzzi, hiking, working out in the gym, or just relaxing on your deck and taking in all the beauty that surrounds you. You are always welcome to come up to the barn and learn more about the horses or visit our petting zoo with the kids.

We invite you to come and enjoy the beauty of the Rockies and our Western hospitality at the Sky Corral Guest Ranch.

Address: 8233 Old Flowers Road, Bellvue, CO 80512

Telephone: 970-484-1362; **Toll Free:** 888-323-2531
Fax: 970-484-0331
Web: www.coloradovacation.com/duderanch/skyranch
E-Mail: jocon72553@aol.com

Location: 23 miles northwest of Fort Collins

Guest Capacity: 30

Season: May through October

Policy on Children, Pets: Children of all ages welcome. No pets.

Rates: $$ American Plan.

Credit Cards: MasterCard and Visa

COLORADO

SKYLINE GUEST RANCH
TELLURIDE, CO

Skyline Guest Ranch is a family owned and operated ranch offering an unequaled outdoor experience for approximately 35 guests each week. We feel the following information is important to consider when making your holiday plans.

A Skyline holiday centers around interaction with other guests. Daily activities are designed to accommodate small groups where adventure and camaraderie can be shared. We encourage guest to come to Skyline with enthusiasm and a desire to explore new ideas and experiences.

The ranch is located at an elevation of 9,600 feet. Our accommodations are small, yet comfortable. Phones and televisions in rooms, and locks on doors will not be found. We feel this promotes the feeling of being at home and is in harmony with our mountain setting. There are ten rooms in the main lodge and six cabins that surround the reflection pond. Since we are far away from fire protection, smoking is not allowed in any of our buildings.

Dave and Sherry Farny purchased Skyline Ranch in 1968 and ran the Telluride Mountaineering School, a summer mountaineering school for young adults, until 1980. Since 1983, Dave and Sherry have operated Skyline as a guest ranch. Throughout the years, the Farny children have joined the business, bringing with them new ideas and enthusiasm. They hope to continue the celebration of their parents' love and apprecia-tion for the mountains, as well as their joy in sharing the beauty of Skyline in both summer and winter.

Skyline is proud to have developed a riding program that takes an active role in sharing natural horsemanship

in the mountains. Skyline's horsemen share their knowledge in a very open and hands-on environment, encouraging guests to get as involved as they wish. The riding program begins on Monday with an orientation covering horse philosophy, riding skills, tack handling and safety. Tuesday through Saturday, with the exception of Thursday afternoon, guests can choose between all-day or half-day rides. All of our rides are led by qualified horsemen who skillfully match rides to meet riders' abilities. The riding pace is dictated by the terrain.

For cowboys and cowgirls six years and older, riding privileges are determined after assessing each one's capabilities. In the late afternoons, younger children are given the opportunity to ride their own horse under the supervision of a Skyline staff member. The overnight pack trip is offered on Friday. A sense of adventure and a few personal items are all a guest needs. Everything else is provided by the ranch.

This adventure involves a full day of riding through remote country, setting up camp and returning the following day after an equally unforgettable ride. You can expect the trip rain or shine. Watching the sun set while cooking out over an open fire, laying your sleeping bag under a blanket of stars or in a comfortable tent and awakening to the smell of cowboy coffee, make Skyline overnights memorable. The camaraderie and adventures shared on these overnights are the perfect ending to your Skyline holiday.

Address: P.O. Box 67, Telluride, CO 81435

Telephone: 970-728-3757;
Fax: 970-728-6728

Location: 8 miles from Telluride's airport.

Guest Capacity: 35

Season: Year round

Policy on Children, Pets: Children of all ages welcome.

Rates: $$

Credit Cards: American Express, MasterCard, and Visa.

Colorado

Sylvan Dale Guest Ranch
Loveland, CO

An established family operation based on the sincere love for people and the western lifestyle, Sylvan Dale Guest Ranch is easy to find but hard to leave! An easy hour's drive from Denver, the Ranch is nestled in a peaceful river valley of the Rocky Mountain foothills.

The 6-night dude vacation is designed with the entire family in mind. With four generations of Jessups living on-ranch and taking part in daily operations, it's easy to see why family fun comes naturally here. Sylvan Dale was purchased by Maurice and Mayme "Tillie" Jessup in 1946 and continues to prosper under the enthusiastic direction of their daughter, Susan, and her "stablemate", David.

Sylvan Dale is an authentic working ranch, raising cattle, horses and hay. Children and adults can take part in ranch chores or just kick back and relax. Take up a heated game of tennis followed by a refreshing dip in the outdoor heated pool. Challenge some new friends to horseshoes or sand volleyball. Go on a leisurely nature walk or an exhilarating whitewater rafting trip! Ages 5-12 can take part in our excellent children's program from noon till three daily. Evenings are family affairs - cook-outs and campfire sing-alongs, live western entertainment and line dancing, family softball in the "Field of Dreams," Native American program and a ranch party with skits, awards and jean branding!

The horseback riding package is optional at Sylvan Dale and includes a breakfast ride, five group riding lessons, horse care, an overnight pack trip, 10 hours of additional riding, and gymkhana games on horseback! Folks can choose from a variety of available "a la carte" rides, the pack trip, team-penning and more. Guests joining us the first week of July can take part in our annual Cattle Drive!

Sylvan Dale boasts first class protected fishing waters, fly-fishing instruction and licensed guides. The waters of the Big Valley lakes promise the excitement of trophy trout fishing and the Big Thompson River will challenge even the best of anglers.

Wholesome, family-style meals feature fresh vegetables from the garden and organic breads, not to mention home-baked favorites by owner "Tillie" Jessup - cherry pie, cinnamon rolls, and hot buttered biscuits. A traditional "Thanksgiving" feast with all the trimmings and fresh rainbow trout - your catch - are dinner-time favorites. If you'd like to ring the dinner bell, just check with the cook!

Accommodations are Comfortable Country: individual and family units in nine cabins and the two-story Wagon Wheel bunkhouse. No TV or telephones in the rooms. Cabins are carpeted and accented with antique furnishings. With its large gathering room, wood-burning fireplace, raised stage, and kitchen for midnight snacks, the Wagon Wheel is ideal for family reunions.

The Ranch is open year 'round for travelers, vacationers, and business groups. You can call and ask for brochures to be sent on: bunk & breakfast, holiday events, weddings, group picnics, retreats and conferences. Or, visit our website for all details. Sylvan Dale Guest Ranch is a wonderful place for people of all ages. A comfortable blend of authentic working-ranch activities and resort recreation for families who want to vacation together. Discover your inner cowboy at Sylvan Dale Ranch! Great fun! Great food! Great folks!

Address: 2939 N. County Road 31D, Loveland, CO 80538

Telephone: 970-667-3915; **Fax:** 970-635-9336
Web: www.sylvandale.com or www.jimbalzotti.com
E-Mail: ranch@sylvandale.com

Location: 7 miles west of Loveland

Guest Capacity: 60

Season: Year round

Policy on Children, Pets: Children of all ages welcome. No pets.

Rates: $$ American Plan.

Credit Cards: None

(See Page 233 for Color Photos!)

Colorado
Sylvan Dale Guest Ranch

OUR LODGE.

Colorado

T Lazy 7 Ranch
Aspen, CO

The T-Lazy-Ranch in Winter 1998/1999 marks the 60th/61st anniversary of the T-Lazy-7 Ranch located near the base of the world famous Maroon Bells. Well before skiing was popular, the Deane Family was hosting friends and guests on their working horse ranch. Today, the Deane Family is still at it and the T-Lazy-7 offers more outdoor activity and unique facilities than any other similar location in the West. Operating under a special use permit #766 from the U.S. Forest Service in the White River National Forest, the ranch offers rides into some of the most beautiful areas in the world. We are ready, willing and able to meet the western hospitality needs of groups, 2 to 200!

The T-Lazy-7 spreads over 400 private acres whose seclusion is enhanced by a quarter of a million acres of surrounding National Forest. 17 quality rustic cabins for rent encircle our large western lodge that is complete with heated pool and hot tub. Three miles of private fishing, streams, stocked beaver ponds, rolling pastures, thick sands of aspen and pine, waterfalls, and horses and animals galore grace our exquisite valley. And boy, can we throw parties...

If you want an evening of "organized chaos and wholesome hell-raisin fun" try our Private Western Party Night. Two fourteen-person, horse-drawn sleighs allow all interested folks to enjoy the beautiful scenery of the T-Lazy-7 on winter evenings. The sleighs make 20-minute loops throughout the evening. Dinner begins around 7:15. Each guest personally selects their own choice cut of steak and proceeds though the "chuck" line. The menu includes: a hearty Rib-Eye steak (we cut our own), our famous cowboy baked beans, fresh garden salad with special T-Lazy-7 dressing, corn-on-the-cob, fresh baked rolls and butter, with coffee and cheesecake for dessert. Your steak will be cooked to your satisfaction because you cook it yourself over one of our two large indoor grills. Our live Country/Western band plays from 7:00-10:00... you'll be sure to hit the dance floor for some foot-stompin fun! Learn some dance favorites such as the Electric Slide and the Cotton-Eyed-Joe. Western Party Nights are reserved for private parties only. We seat up to 140 people. Our dramatic log lodge is western all the way with animal trophies on the walls, authentic memorabilia, a real wagon hanging from the ceiling and two huge open fireplaces in each of the adjoining seating sections.

The T-Lazy-7 outdoor Chuckwagon Grounds are perfect for any group. Towering Colorado Blue Spruce canopy the massive open-air bonfire ring. The stage, dance floor and table/seating areas for 150 people are sheltered from the weather by rustic log beams and wooden roofing laced high up in the trees. Silent radiant heaters located in the roofs can be turned on if the evening becomes chilly. Old wagons, stagecoaches and surreys dot the grounds, while giant elk and deer antlers adorn the trees. Maroon Creek rushes by next to beaver ponds stocked with trout. Volleyball courts, horseshoe tossing pits, and hayrides add to the outdoor western flair. We've had parties for over 500! Kids and families love it. Wear your hats and boots and plan on an authentic Western hoe-down. Our dinner menu includes: BBQ ribs and chicken, chips & dip, fresh veggie & dip platter, fresh fruit salad boats, fresh garden salad, corn-on-the-cob, cowboy beans, fresh baked rolls and butter, cold soda pop, coffee, hot cocoa and tea. There are hot apple pies for dessert, and of course, marshmallows for roasting. The price for the party is $60 per person, which includes our Country/Western dance band and hayride.

Address: 3129 Maroon Creek Road, Aspen, CO 81611

Telephone: 970-925-7254; **Toll Free:** 888-T7 LODGE
Fax: 970-925-5616 ;   **Web:** tlazy7.com
E-Mail: tlazy7@rof.net

Location: 4 miles from Aspen

Guest Capacity: 85

Season: November 21 through April 4, Mid-June through October

Policy on Children, Pets: Children encouraged to come. No pets.

Rates: $ and $$

Credit Cards: Discover, MasterCard and Visa.

(See Page 234 for Color Photos!)

Colorado

Weminuche Wilderness Adventures
Bayfield, CO

Charles M. Russell, the famous Western artist once said "you can see what man made from the seat of an automobile, but the best way to see what God made is from the back of a horse." It is with this spirit that Weminuche Wilderness Adventures runs two separate horseback riding adventures — one into the Weminuche Wilderness area of Colorado (June - September) and one into the Superstition Wilderness area of Arizona (February - April). These are trips of a lifetime.

The Weminuche Wilderness is the largest wilderness area in Colorado, encompassing 433,745 acres. Here you'll be able to see and appreciate what the western mountains were like before their occupation by man. You'll return to the joys and exhilarations of a lost way of life on a pack trip by horseback into the wilderness area of the spectacular San Juan Mountains. Known as the "Switzerland of America" for their alp-like terrain, the San Juan Mountains of southwestern Colorado have more 14,000-foot peaks than any other range in the Rockies. On this exciting adventure, you'll reach deep into majestic mountain back country, comfortably mounted on well-trained horses, being outdoors for every part of every day where land and nature are still peaceful and free.

Starting above Vellecito Lake, 35 miles northeast of Durango, you'll mark a leisurely pace along meandering streams and through vast alpine meadows. The terrain will never cease to fascinate you throughout these six days — the forest changes from ponderosa pine to Colorado blue spruce and Douglas fir to the delicate high-altitude tundra on the Continental Divide.

If you like to hike and want to take a try at mountain climbing, there are a number of peaks, such as the 13,100-foot Buffalo Peak, that are climbable without technical mountain gear.

Many species of birds, deer, elk, coyotes, bighorn sheep, and an array of smaller wildlife are at home in these forests and towering mountain peaks. The cool mountain waters yield an abundance of trout, some of which you'll sample when they're cooked over the campfire.

The Superstitions of Arizona are rugged mountains of alluring beauty, as mysterious and colorful as the history and legends that surround them. The warm, dry desert air, giant saguaro cacti, and countless deep rock canyons provide welcome relief from winter's gloom. Though the trails are twisty and the riding adventuresome, ample rewards lie in the glory of the Sonoran Desert in bloom. By day, the sight of wild quail, javelina, and eagle reward a quick eye. At night, you'll listen to tales of the Lost Dutchman's gold, Spanish conquistadores, and marauding Apaches.

Troy and Michelle Shupe's "moving pack trip" starts at the Peralta Trail Head with camps at the old J. F. Ranch Headquarters, Angel Springs, and the historic Reavis Ranch. You'll take a "ride through time" — entering the thirteenth-century Salado Indian cliff dwellings. Explore Circle Stone, a Superstition Mountain mystery. Trails traverse from the desert floor to the pines of the 6,000-foot level of the Superstition Mountains. Everywhere, the renewed life and colors of spring abound.

Tasty cowboy specials are cooked each meal from scratch. Relaxing evenings are spent in camp, enjoying the company of the outfitters, the wranglers, and other new friends. You can talk over the day's adventures with the guide, sing along with the guitar while sitting around a glowing campfire, or gaze at the spectacle of the night sky where the stars seem so near through the clear air.

Join Weminuche Wilderness Adventures for the kind of horseback riding adventure that will make you feel like you've gone back in time. See some of the prettiest country the way it was meant to be seen — on horseback!

Address: Summer: 17754 CR 501, Dept. BP, Bayfield, CO 81122; Winter: P.O. Box 5744, Dept. BP, Apache Junction, AZ 85278

Telephone: Colorado: 970-884-2555; Arizona: 602-671-3372, 602-710-7431

Location: Colorado: 320 miles from Denver; Arizona: 45 minutes from Phoenix airport

Guest Capacity: Colorado: 8; Arizona: 11

Season: Colorado: June - September; Arizona: February - April

Policy on Children, Pets: Children should be 10 years of age and older. No pets.

Rates: $$ American plan.

(See Page 235 for Color Photos!)

Colorado

Weminuche Wilderness Adventures

OUR FOUR WHEEL DRIVE TRANSPORTATION.

RIDING TO HUNCHBACK PASS.

Colorado

Wilderness Trails Ranch
Bayfield, CO

A soaring eagle, the song of coyotes, a deer bounding across the meadow, horses grazing in the front pasture, and millions of stars in the clear mountain skies — these are just part of the sights and sounds of the Los Pinos Valley. Capture the spirit of the Southern Ute Indians who lived, rode, and hunted these meadows, forests, and mountains a century ago.

Since its beginning in 1950, the rustic, yet comfortable charm of the Wilderness Trails Ranch has been reflected in the authentic log cabins and lodge, which were built from trees on the property. The ranch blends aesthetically into the surrounding secluded mountain valley, providing guests with a vacation in harmony with a quality wilderness experience. Intimate, personalized, and family-oriented best describe the ranch philosophy. Gene and Jan Roberts and their family have been opening their doors and welcoming guests since 1970. They love to share meals, ride, and take part in many activities with their guests, who frequently become long-time friends.

Although the ranch has myriad activities, the program revolves around its high-quality horse operation. The Roberts raise and train Morgan and Quarter Horses on the ranch, although their fine string also includes Arabs, Appaloosas, Thoroughbreds, and Tennessee Walkers. The Roberts are both certified riding instructors, and riding instruction for every level is given daily. Rides are divided into small groups according to ability, thus providing an opportunity for experienced riders to jog and lope through the lovely meadows. An overnight pack trip, riding seven miles into the mountains is offered weekly. The month of September, when fall colors are at their peak, is for adults only. In late September, the Wilderness Trails Ranch cattle round-up is offered to adventuresome, skilled riders, who want to experience something very unique.

The 72-foot heated lap pool is popular with guests of all ages, and water skiing at nearby Vallecito Lake is available for adults and teens to enjoy. A soak in the spacious hot tub is sure to be soothing and revitalizing. During the week, there are also four-wheel drive trips, tours to Mesa Verde cliff dwellings, river rafting, hayrides, hiking, and fishing. Arrangements can also be made for trips to the historic Durango Narrow Gauge Steam Railroad.

The ranch has exceptional children's programs for 3-to-5 year olds, kids 6-11, and a special program for teens 12-17. Well-trained counselors ride with the 6-11 year olds, and teens have their own wrangler. Riding instruction, trail rides, hayrides, and a variety of other activities keep the kids entertained.

Accommodations are comfortable, two and three bedroom log cabins with porches. The cabins, nestled among pines, spruce, and aspen, all have lovely country furnishings, modern private baths with shower/tub combinations, and individually controlled heat. Three-bedroom, three-bath cabin suites feature wood-burning stoves in living rooms, coffee bar and refrigerator.

Meals are a special event at the Wilderness Trails Ranch. Guests love the varied menu: the hamburger and steak-and-seafood cookouts, scrumptious breakfasts, salad bars, terrific Mexican buffet, and especially those homemade desserts and breads. A weekly gourmet candlelight dinner is a very special evening for adults.

In the evenings, experience a variety of fun — a hilarious staff show; two-steppin', country-swingin' western dance; horse-drawn hayrides; singalongs beside a blazing fire, complete with hot chocolate and marshmallows to toast.

Whether you visit in the summer or in September, you can be certain at the end of your stay, you won't want to leave!

Address: Summer: 23486 County Road 501, Dept. BP, Bayfield, CO 81122; Winter: 1766 County Road 302, Dept. BP, Durango, CO 81301

Telephone: 970-247-0722, 800-527-2624; **Fax:** 970-247-1006

Location: 35 miles northeast of Durango, 190 miles from Albuquerque

Guest Capacity: 48

Season: June through September

Policy on Children, Pets: Family-oriented resort with exceptional children's programs; no children under age 3. No pets.

Rates: $$ American plan. Family rates. Discounts in early June, September. Overnight pack trip extra. Airport transfers extra.

Credit Cards: Discover, MasterCard, Visa

(See Page 236 — 238 for Color Photos!)

Colorado
Wilderness Trails Ranch

WAIT UNTIL YOU TASTE OUR FOOD!

BEAUTIFUL COUNTRY TO RIDE IN.

Colorado

Wilderness Trails Ranch

GUESTS GOING FOR A WAGON RIDE.

INSIDE OR OUTSIDE, YOU ARE GOING TO LOVE
WILDERNESS TRAILS RANCH.

Colorado

Colorado

Wind River Ranch
Estes Park, CO

Wind River Ranch is a family, Christian guest ranch. There is something for everyone at Wind River. Located at 9,200 feet in the magnificent Tahosa Valley, our guests experience spectacular views of Long's Peak and the Rocky Mountain National Park. Wind River, now 110 acres, was originally homesteaded by Elkanah Lamb in the late 1800's. Reverend Lamb was an itinerant Bible teacher and mountain climber. The fireplace from the original homestead is now part of our recreational center. On Mrs. Lamb's 70th birthday, as a present to her husband, she climbed Long's Peak which is over 14,200 feet. What a gift that must have been!

Our horse program is one of the best in the country. Unlimited lessons and riding are all part of the price. We can accommodate riders of all skills. Wind River has the use of thousands of acres in the Rocky Mountain National Park and Roosevelt National Forest. Our rides vary from one hour to half-day rides. We have a wranglers' breakfast one morning and a steak cook-out in the evening. The activities include a volleyball court with authentic beach sand for our court. You will really love the square dance on Monday night on our basketball court under the stars. On Thursday, we have a hoe-down in the barn with songs and skits from the staff. You sit on bales of hay in the 1889 barn and your family will create memories that will last a lifetime as you laugh with one another. Add golf, whitewater rafting, hiking, fishing, rodeos and a hayride and you have the ingredients for a vacation with a purpose.

As a Christian guest ranch, we have nationally known Christian speakers, authors and ministers who lead the optional studies each morning in the Ranch House. We also have an inspirational fireside every evening in our upper meadow. Our staff leads us in some songs around the campfire, and, again, our speaker delivers a short message. Fireside every evening is at dusk and each sunset is awesome. Wind River Ranch gives our guests the time to reflect and renew their relationship with the Creator of all this beauty that surrounds us. Time is a valuable commodity that enables all of us to pause and consider all that is around us. We do not get opportunities like this in the city.

The choices are endless at a Christian guest ranch. You can fill your time with activities or just sit back in one of our cypress rockers and enjoy the cool mountain air. If a rocker is not your style, try our hammocks in the aspen grove by the upper meadow. As you sway in the hammock, you can hear the cool, gentle breeze drift through the aspens as the stream babbles under you.

To learn more about this opportunity for an investment in your family, visit our web site. We really hope you will be our guest this summer. We will treat you like you are our family.

Address: P.O. Box 3410, Estes Park, CO 80517
Telephone: 800-523-4212
Web: windriverranch.com
Location: 7 miles south of Estes Park
Guest Capacity: 50

Season: End of May to early August
Policy on Children, Pets: No pets
Rates: $$
Credit Cards: Visa, MC, Discover

(See Pages 239 for Color Photos!)

Colorado

Florida

ACE OF HEARTS RANCH
COCOA, FL

Not far from all the excitement of Central Florida, you will travel down a country road and find, tucked away from crowds and lines, a beautiful, peaceful ranch. The ACE OF HEARTS RANCH is proud of the various breeds of horses, numbering 30, that enjoy participating in the many activities of the ranch such as, trail rides, day trips, beach rides, lessons, pony parties, horse-drawn hayrides and weddings.

Dennis and Sandra Bressler started the ranch in 1996 and are proud to be such an important part of the vacation of thousands of travelers. Fulfill your fantasy while riding on one of the most pristine beaches in the world with the Kennedy Space Center and the Space Shuttle launch pad in the backdrop. You will enjoy a beautiful ride on a beach with no hotels or buildings. The Canaveral National Seashore is home to many creatures such as dolphins, bald eagles, manatee, roseate spoonbills, seatrout, redfish, snook, deer, bobcat, armadillo, and many, many more. Right Whales calve offshore in the early spring. Birds such as peregrine falcons, bald eagles, ducks, warblers, and more rest and feed in the fall, winter and spring while on their long flights. The beauty of this beach has been enjoyed since 1513 when it was viewed by the crews of Ponce de Leon and has remained almost unchanged - pristine, natural and beautiful.

The wooded trails on the ranch will take you through live oak hammocks, past flowering meadows - many times graced with butterflies. The trails also provide for half-day and full-day trips of horseback riding where they will lead you to one of the many state and national forests Florida has to offer. You can choose from a wide variety of scenery at the Ocala National Forest, Wekiva Springs, San Sebastian Buffer Reserve, Seminole Forest, and others.

The ACE OF HEARTS RANCH can help you with a country outdoor wedding or a wedding with a special theme. For your company party, birthday feast, or any occasion, the fun begins before sunset at the ranch with a horse-drawn hay ride. Tour through the ranch before or after a wonderful southern barbecue. After a great meal, it's on to the campfire for story-telling and songs under the stars. The ranch offers the finest state-of-the-art stable in central Florida. Your horses will enjoy the special attention they receive at our ranch, and you will enjoy the beautiful trails you can share with your horse.

The Bresslers enjoy being a part of family celebrations, and watching the joy of riders beginning lessons, friends and families going on trail rides together, couples planning their weddings at the ranch, and making holiday times special with group gatherings, such as campfires, hayrides, and barbecues. The biggest hit when school is out is summer day camp, where young people ages 9 and up can learn first hand how to care for, ride and have fun around horses all week. So as you can see, there is something for everyone at the ACE OF HEARTS RANCH.

Address: 7400 Bridal Path Lane, Cocoa, FL 32927

Telephone: 407-638-0104;   **Fax:** 407-638-0104
Web: aceofheartsranch.com or
www.jimbalzotti.com
E-Mail: s1vann@worldnet.att.net

Location: Within an hour of 3 international airports.

Guest Capacity: Wooded trails - 10, beach rides - 6, daytrips - 5, parties - 200

Season: Beachrides are November 1 through April 30. Everything else is year round.

Policy on Children, Pets: Children need to be 8-years old to ride trails alone. No pets.

Rates: $

Credit Cards: Visa, MC, Amex

(SEE PAGE 240 FOR COLOR PHOTOS!)

Florida
Ace of Hearts Ranch

WHAT A GREAT PLACE TO CHAT WITH DAD.

OK, OK, JUST THREE MORE MINUTES AND
WE'LL GO RIDING.

Georgia

Victoria's Carriages & Trail Rides
Jekyll Island, GA

Join us for a trail ride through history on Jekyll Island. Give rein to your imagination. Seek out the majestic ancient moss-laden live oak where Blackbeard buried his treasure - the one with the hook deeply embedded in the trunk! Balmy ocean breezes beckon the romantically inclined to the silvery moon-drenched beach where lapping waves whisper secrets of a bygone era, "Who goes there?" Are those the ghostly forms of young lovers escaping from prudish Victorian chaperones?

Follow the trails used by J.P. Morgan and William Rockefeller as they pursued wild boar, deer and wild turkeys through the maritime forest. Live out your fantasy of being a millionaire in the peaceful tranquility of this small barrier island off the southern coast of Georgia.

Or just take a trail ride for the fun and exhilaration of communing with nature. Here is your opportunity to photograph the beautiful marshes of Glynn immortalized by Georgia's illustrious and beloved poet, Sidney Lanier. 'Shoot' a wild turkey or deer as it crosses the trail. Capture a rainbow bending over the golden isle. Bird watchers will zoom in on the amusing pelicans catching a midday lunch, or egrets gracefully descending to moss-covered branches for the night. Dolphins playfully leap in the gentle surf and an occasional loggerhead sea turtle may make a rare daylight appearance on the beach during nesting season.

A golden era existed for over half a century on Jekyll Island beginning in 1886. And before that, the moccasins of peace loving Indian tribes roamed the island followed by Spanish, English and French. Each group left a smattering of history for us to investigate and enjoy. Old tabby ruins, a crumbling family cemetery and the Millionaires' Village where one fifth of our nation's wealth wintered until the mid-forties in their secluded private haven. The original clubhouse was the centerpiece of the compound with over a dozen of the millionaires' cottages, nestled among magnificent live oaks and magnolias surrounding it. The clubhouse has been restored to the four-star Jekyll Island Club Hotel. Your carriage ride will take you on a slow leisurely tour through the National Landmark Historic District and along the edge of the Jekyll Creek where luxurious yachts once dropped anchor. Immerse yourself in the luxurious past.

Make plans now to visit this blending of past and present. Enjoy Jekyll Island as the millionaires did at the turn of the century: in the saddle or perhaps on the seat of a horse-drawn carriage. Narrated carriage tours of Jekyll's Historic District are offered Monday through Saturday.

Trail rides leave from the Clam Creek picnic area on the north end of Jekyll Island. Victoria's Carriages & Trail Rides offer morning, afternoon and sunset beach rides. Rates are $30 per person for a one hour ride; $40 per person for an hour-and-a- half ride. Wedding carriage service, group hay rides, birthday parties and gift certificates are available.

Address: 100 Stable Road, Jekyll Island, GA 31527

Telephone: 912-635-9500; Fax: 912-635-2818

Location: on Jekyll Island, off the southern coast of Georgia

Guest Capacity: 1000's

Season: Year round

Policy on Children, Pets: Children welcome. No pets.

Rates: $

Credit Cards: Discover, MasterCard and Visa

HAWAII

PRINCEVILLE RANCH STABLES
HANALEI, HI

Princeville Ranch dates back to before the great western cattle drives, before the annexation of California and the southwestern territories, and before the settling of the West. Princeville Ranch Stables was established in 1978 by Donn and Gale Carswell, who have built the ranch and stables into what it is today. Originally, the Carswells intended to offer trail-riding activities to visitors to the island of Kauai, the Garden Island. Today, Princeville Ranch Stables provides both visitors and residents of Kauai with an activity that preserves and promotes the natural beauty, history, and culture of Kauai.

If you are vacationing on the island of Kauai, E Komo Mai (come) experience Princeville Ranch Stables' spectacular trail-ride excursions. All of the stables' guided outings are as safe for the beginner as for the expert rider, and there is a wide range of horseback adventures.

The culmination of the waterfall picnic ride is the perfect Hawaiian setting — at a fern-lined waterfall plunging into a mountain pool. This four-hour excursion starts with a ride across ranch land. Then you'll have a short (thank goodness!), steep hike to reach majestic Kalihiwai Falls. Enjoy a picnic lunch followed by a cool, brisk swim and meditate over the breathtaking beauty of the Falls.

For the more adventurous, you might want to experience a little of the life of a paniolo (Hawaiian cowboy) by joining the Princeville Ranch cattle drive ride. Help the cowboys round up a herd of cattle at sunrise and drive them across ranch lands to their final resting pen. The ranch's border collies — Buster, Lani, and Jumbo — will also help move the herd. An unforgettable experience!

You might want to take part in the bluff-to-beach ride along the extraordinary scenic vistas overlooking Anini Beach. You'll ride across lush green pasture land and meander along the bluffs of the spectacular North Shore. Here you'll follow the trail toward the beach and tie-off the horses nearby. (Hawaiian law forbids riding horses on the beach.) A short stroll takes you past the polo fields and to the beach for a light snack and a relaxing swim.

Enjoy a wonderful hour and a half of horseback riding across the Princeville Ranch range land and take in the incredible panoramic views of the magnificent Hanalei Mountains. Below is the splendid vista of the Hanalei Valley and its mosaic patches of taro fields. In the distance, you can see the famous Hanalei Bay and, even farther off, you can catch a glimpse of the legendary Bali Hai.

Princeville Ranch Stables will also arrange an exclusive, personalized, private ride! On the ride you'll learn about the history of Princeville Ranch and hear fascinating stories about the legends of Kauai.

Princeville also offers a complete horsemanship program of horse training, riding lessons, and seminars. The Carswells believe in the philosophy that rider and horse are partners and that it's important to understand how a horse behaves and thinks.

When you are vacationing on Kauai, contact Princeville Ranch Stables for a truly unique riding adventure.

Address: P.O. Box 888, Dept. BP, Hanalei, HI 96714

Telephone: 808-826-6777; **Fax:** 808-826-7210

Location: 45 minutes from Lihue Airport

Guest Capacity: 6 per ride

Season: Year round

Policy on Children, Pets: Children must be 8 years of age and accompanied by an adult. No pets.

Rates: $-$$. Riding only. Accommodations available at nearby bed & breakfast facilities.

Credit Cards: MasterCard, Visa

IDAHO

BAR H BAR RANCH
SODA SPRINGS, ID

If you are looking for a real hands-on working vacation ranch with beautiful scenery, plenty of good food, great horses, and a friendly atmosphere, the Bar H Bar Ranch is the place for you.

The 9,000-acre cattle ranch is located in the Bear River Range of the Wassatch Mountains and borders the Caribou Cache National Forest in southeastern Idaho. Here, you'll see Idaho at its best — a splendid variety of stately pines, shimmering aspen, cold mountain streams, the breathtaking panorama of beautiful wildflowers, verdant green meadows, and an abundance of wildlife. You can expect to see deer, elk, and moose and smaller animals like red fox, porcupine, skunks, and rabbits. For those who like bird-watching, you'll see plenty of sand hill cranes, ducks, Canadian geese, and a variety of smaller birds.

The Bar H Bar Ranch is steeped in rich pioneer history. It is one of the oldest ranches in southeastern Idaho and was originally homesteaded by the Mormon Church. Their unique irrigation system is still used by the present owners, McGee and Janet Harris. Take the time to ride out and see the deep ruts of the Oregon Trail, which are still visible on the ranch. You'll have an undescribable feeling as you try to visualize the many wagon trains that passed through here on their way to a new life. You'll feel the heartache and grief of those who had to leave a loved one along the way. It is an awesome experience.

The Bar H Bar Ranch runs 2,000 head of beef cattle and pastures most of them on private land. As a result, there's always plenty to do making sure the cattle have enough food and the fences are mended. Guests can participate in all ranch activities, although there are some who just like to sit in a big comfortable chair, relax, and drink in the views.

Those who do get involved will ride daily to check the cattle, doctoring some and making sure they all have plenty of salt and water. Other activities will vary according to the season. In May and June, you can expect calving, branding, fence mending, and moving cattle to the spring and summer range. During July and August, you'll mostly check for sick cattle and round up the escapees who have located the only hole in the fence for miles! In September and October there's the fall roundup — gathering the cattle down off the mountain range, doing a little branding, and preparing the cattle for market and for the winter range.

For your "off" time, you might try nature hikes or fishing in the great little fish pond stocked with trout. Local attractions include Lava Hot Springs, Minnetonka Cave, the historic Pioneer Tabernacle, and nearby natural soda-water springs. In the winter, the ranch has both spectacular snowmobiling and cross-country skiing.

Guest accommodations are in the old bunkhouse, which was renovated and refurnished with native lodge pole furniture. The beds are clean and comfortable and the showers are good and hot. Each of the four guest rooms has a private entrance. Most of our guests appreciate the fact that there are no televisions or telephones in the rooms.

Most meals are cooked and served family style in the old ranch cookhouse. You'll enjoy three healthy, home-cooked meals a day in the tradition of the West, which may include homemade bread, pies and other goodies, Dutch oven cookouts, and steak frys. The cookhouse is open 24 hours a day, just in case someone gets the midnight munchies!

The Bar H Bar Ranch promises a vacation you won't soon forget. You'll come as a guest, but leave as a friend!

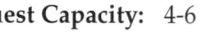

Address: 1501 Eight Mile Creek Road, Dept. BP, Soda Springs, ID 83216

Telephone: 208-542-3082, 800-743-9505; **Fax:** 208-547-0203
Web: www.jimbalzotti.com

Location: 100 miles from Jackson, Wyoming; 120 miles from Idaho Falls; 184 miles from Salt Lake City

Guest Capacity: 4-6

Season: May through September

Policy on Children, Pets: Children are welcome but are the responsibility of parents. No pets.

Rates: $$ American plan.

Credit Cards: None

Idaho
Bar H Bar Ranch

A CALF PLAYING HIDE AND SEEK IN THE GRASS.

RUSTIC FENCING SURROUNDS THE RANCH.

IDAHO

DIAMOND D RANCH
STANLEY, ID

The Diamond D sits in a beautiful valley surrounded by rugged and remote snow-capped mountains. It is located in the Frank Church Wilderness which is the largest wilderness in the lower 48 states. The Demorest family has owned and operated the ranch since the 1950's. The ranch has a colorful history of gold prospectors, settlers, Indian skirmishes and wagoneers. The ranch was the center of mining activity before the area was federally designated wilderness.

The winding road into the valley was built for the miners in 1903. Heavy ore wagons drawn by teams of eight, rumbled back and forth across the ranch on their four-day journey to the railroad spur 125 miles away. Our barn today was a store and saloon along with the ranch house being an inn for those thirsty wagoneers. For the past fifty years, the valley has been quiet and peaceful, to be enjoyed by the rancher and now the many guests who join us each summer. Mining is still a pastime, but mostly just for guests who sift through stream beds in search of souvenir gold nuggets.

Open June through mid-October, the ranch offers a variety of summer activities. Each guest is matched to a horse according to their riding experience, and will ride each day except for Saturday which is our check-in and out day. We offer two rides a day and our saddle-savvy folks really enjoy our all-day ride to Pinyon Peak that features a view of four encompassing states: Montana, Wyoming, Washington and Oregon.

Fishing is another popular activity at the Diamond D. The area has trout-rich lakes and streams perfect for hooking an eighteen-incher worth bragging about. There is a private lake where "You catch it, we'll cook it up" applies. Twice a week there are organized hikes but anytime is the right time for smaller, casual excursions. Wild company abounds, including elk, deer, moose, mountain lion, sheep and goats. A five-mile hike along Loon Creek will be rewarded with the soothing warmth of hot springs.

Today the Diamond D accommodates about three dozen guests at a time. Guests have a choice between individual cabins with private baths and rooms in the main lodge with shared baths. One large cabin is available for families or bigger groups and sleeps ten comfortably. There are also private suites and a honeymoon suite, decorated with hearts, lace and flowers. When the evenings cool off or on rainy days, guests often play Scrabble, Dominos or card games around the stone fireplace in the lodge. At dusk, it is not uncommon to hear the clank of a horseshoe game or the crackle of a bonfire with songs and stories. Barbecue smells will lure hungry guests away from a volleyball match. The scents of home cooking and baking are as much a part of the lodge as its old mining lore. Three meals a day, made with fresh produce and including homemade baked desserts and homemade bread, keep the kitchen staff busy and the guest satiated.

Address: P.O. Box 35, Stanley, ID 83278,
Mailing Address: P.O. Box 1555, Boise, ID 83701

Telephone: 800-222-1269; **Fax:** 208-336-9772
E-Mail: diadlld@aol.com

Location: 4 hours from Boise Airport, or land on our landing strip.

Guest Capacity: 35 - 40

Season: June 10 through October 1

Policy on Children, Pets: Children 6 and over to ride on trails. Pets okay, if housebroken.

Rates: $$ American Plan

Credit Cards: None

(SEE PAGES 242 FOR COLOR PHOTOS!)

IDAHO

DIAMOND D RANCH

OH, DON'T STOP!

Idaho

Hidden Creek Ranch
Harrison, ID

Dawn, mist, eighty horses, thundering hooves, dust, and the adventure of the Old West.... As the sun rises, the herd is wrangled up, grained, cleaned, and saddled — a horse for every type of rider. Welcome to Hidden Creek Ranch, a dude ranch in Idaho's panhandle that has a social conscience. Owners Iris Behr and John Muir are committed to living with nature and helping their guests to, first, become become more aware of their own connection with nature and then to understand how to respect and not abuse nature or its resources.

The horse program is excellent and includes all-day rides, morning or afternoon rides, a sunset ride, champagne brunch ride, hayrides, Centered Riding© instruction, and a barrel-racing rodeo. With daily scenic, challenging, and fast rides, will find just what they're looking for! Novice riders are given particular attention, with orientation, instruction, and steadily increasing ride time to instill confidence.

The ranch itself sits on 570 acres, surrounded by national forest — adding over 200,000 acres to the riding area. Trail rides deep into timbered forest or across grassy meadows are experiences long remembered. From mountain plateaus, guests are awestruck by the beauty of the Coeur d'Alene "chain of lakes" river system.

More experienced riders may want to sign on for the overnight pack trip into the back country. The wilderness camp is set up with comfort in mind — everything is provided. Evenings promise great food, good company and "stretched" campfire stories. After two days in the saddle, riders can return to their luxurious log cabins, eat gourmet meals in the big lodge, relax in hot tubs, or have a soothing massage.

Whether as exhilarating as horseback riding through the mountains or as relaxing as sitting back and reading a book, the choice of activities is exceptional: fishing, trap shooting, mountain biking, boat tours, archery, nature awareness programs or exercise, gold mine and medicine trail hikes. Evening activities differ nightly and include a sunset ride, western dance, Native American pipe and sweat lodge ceremony, campfire activities, awards night, and quick draw contest. Golfing, rafting, boating, shopping, and sightseeing are available in the area.

A full children's program is in place June through August. Counselors provide special activities for kids including nature awareness, beading and leatherwork, picnics, hiking, biking, campfires with storytelling, a cookout, an overnight in Native American teepees, and, of course, daily horseback riding. All of the ranch activities help guests, whether adults or children, develop an appreciation of actually being a part of nature.

Accommodations are available for large or small families, couples, singles, corporate retreats, and family reunions. Authentic log cabins, each with two or four units equipped with modern log furniture are decorated in Native American motif and appointed with private baths. The 7,000-square-foot Lodge features a volcanic rock fireplace, library, and deck overlooking our valley and timber-ridged mountains.

The Lodge also contains a spacious dining facility. Classically trained, the chef offers an eclectic gourmet dining experience including buffet lunches, four-course family-style dinners, and a seven-course candlelight dinner. Evening cocktail hours offer complimentary hors d'oeuvres with wine, beer, or soft drinks.

Return to a simpler time with rides, rest, recreation, and, above all, respect for nature. Experience a horseback riding vacation that celebrates life and generates joy and excitement. With European attention to detail, first-class accommodations, and a beautiful Idaho landscape, Hidden Creek Ranch will leave memories of a great vacation that will last a lifetime!

Address: Dept. BP, 7600 East Blue Lake Road, Harrison, ID 83833

Telephone: 800-446-DUDE, 208-689-3209; **Fax:** 208-689-9115; **E-mail:** hiddencreek@hiddencreek.com; **Web:** www.nidlink.com/~hiddencreek or www.jimbalzotti.com

Location: 5 miles east of Harrison, 78 miles southeast of Spokane

Guest Capacity: 40

Season: May through October

Policy on Children, Pets: Excellent children's program. No pets.

Rates: $$ - $$$ Six-day stay. American plan. Overnight pack trip extra. Children's rates. Group discount.

Credit Cards American Express, MasterCard, Visa

(SEE PAGES 244-252 FOR COLOR PHOTOS!)

IDAHO

HIDDEN CREEK RANCH

RIDING THROUGH THE MAGNIFICENT COUNTRY OF IDAHO.

IDAHO

IDAHO ROCKY MOUNTAIN RANCH
STANLEY, ID

Summer Program Idaho Rocky Mountain Ranch, a unique 1000-acre property, is located in the scenic Sawtooth National Recreation Area of central Idaho and surrounded by more wilderness than any location in the continental United States. Originally built in 1930 as a private hunting club and currently on the National Register of Historic Places, the charming handhewn log lodge and cabins are handsomely decorated with handcrafted furnishings. All 21 accommodations - four lodge rooms and 17 cabin units - have private baths. Each cabin, as well as the lodge living and dining rooms, have gorgeous stone fireplaces. The absence of telephones, televisions and radios in the lodge and cabins helps our 42-50 guests to relax and enjoy the native beauty and tranquillity of their spectacular surroundings.

Dining in the historic lodge, with its large central fireplace and hand quilted tablecloths, is a relaxing and congenial experience. Each morning, guests enjoy a full continental buffet as well as a choice of hot breakfasts. A five-course dinner is served "country fine dining" style each night and on Monday, Thursday and Saturday evenings the ranch features a specialty guest dining experience: Monday, a horse drawn wagon transports guests to our Dutch-oven cookout site; Thursday we have the IRMR BBQ buffet, featuring salmon, chicken, beef, ranch beans and other hearty Idaho fare; Saturday is Idaho Night, when all five courses, including three entrees, are served family-style. We have musical entertainment 3-4 nights a week. Breakfast and dinner are included in the cost of guest stays.

Deb and Tom Proctor, owners of Pioneer Mountain Outfitters, operate the ranch's horse program. Scenic half- and all-day rides, specialty rides and overnight pack trips are available for riders of all ability levels. (Riding is not included in the basic cost of stay at the ranch.) Other favorite on-ranch activities for guests are: our natural hot springs-fed swimming pool which is delightful for daytime swimming as well as nighttime soaking and star gazing; fishing in the stocked catch-and-release pond or the Salmon River; hiking; photography; biking; and watching the sun set over the Sawtooth Mountains from the lodge's front porch. Additionally, there are countless opportunities for hiking, biking, fishing, horseback riding, climbing, scenic drives (two and four wheel), photography, ghost town and museum visits and whitewater rafting and kayaking in the surrounding public lands and area. The 1999 Summer Season is June 11 through September, 1999.

Winter From Thanksgiving through March each year, the ranch offers two accommodations for guests: The Homestead Cabin, a log cabin studio suitable for two people; and The Winter Home, a three bedroom house for a maximum of 6 people. Both accommodations have private baths, full kitchens with stove, oven, refrigerator and microwave, wood burning stoves and electric heat, and are stocked with breakfast items such as oatmeal, homemade granola, coffee and teas.

Guests ski tour the ranch's 1000 acres by day, and at night snuggle up to the wood burning stoves or stargaze from the hot springs pool. Elk and coyote are abundant during the winter months and their bugles and howls provide nighttime serenades. Skate, cross country and telemark skiing can be done on the ranch or on groomed trails nearby.

Address: HC 64 Box 9934, Stanley, ID 83278

Telephone: 208-774-3544; **Fax:** 208-774-3477
E-mail: idrocky@cyberbighway.net

Location: 120 miles from Boise

Guest Capacity: 42-50 (summer), 8-10 (winter)

Season: Summer: June 11 through September 18, Winter: Thanksgiving through March

Policy on Children, Pets: Children allowed with some restrictions. No pets.

Rates: $ and $$

Credit Cards: Visa, MC, Discover (summer only)

(SEE PAGES 243 FOR COLOR PHOTOS!)

IDAHO

KINGSTON 5 RANCH BED & BREAKFAST
COEUR D'ALENE, ID

With the Coeur d'Alene Mountains as its backdrop, this picturesque 1930s country farmhouse is a wonderful place to escape to and relax. The 4,500-square-foot farmhouse, which was completely renovated in 1993, is the perfect place for that "quiet, relaxing retreat" or "wonderful romantic getaway." The house offers spectacular views of the valley and surrounding mountains from every room and every window. Enjoy the huge family room, with its wall of windows. Read, curl up in front of a crackling fire, listen to your favorite music. Watch the horses grazing in the pasture - and if you're early enough, catch a glimpse of the deer grazing in the valley.

The ranch caters expecially to guests who bring their own horses on their riding vacation, although owners Walter and Pat Gentry can make arrangements for horses and riding with some advance notice. Walter and Pat do have some horses on the ranch for more experienced riders who want to improve their skills on performance animals. Horses are accommodated in four separately fenced pastures and four indoor/outdoor stalls with individual runs, and two holding pens adjacent to the round pen. There's also a quarter-mile exercise track and a 200-ft. x 100-ft. arena.

The Ranch offers a wide variety of outdoor activities and miles of mountain horse trails in some of the most scenic areas you'll ever encounter; mountain lakes, streams, and waterfalls, lush green valleys surrounded by towering stands of pines and a blue sky that seems to go on forever!

Riders have access to the widely publicized "Taft Trail System," the largest continually linked trail system in the United States. Built on the old Northern Pacific rail bed, the trails are wide and flat, with dirt and some gravel footing. The system stretches east from Coeur d'Alene to St. Regis, Montana, and then south to Avery, Idaho, and provides hundreds of miles of gently sloping trails through beautiful mountain scenery. In the spring, summer, and fall, you can use the trail for horseback riding, mountain biking, or hiking; in the winter, it's popular for snowmobiling and cross-country skiing.

The private suites in the farmhouse offer in-room fireplaces and private baths with Jacuzzi tub and private outdoor spas. Lazy mornings begin with the delightful tastes and smells of homemade specialties, including a generous full country breakfast. In addition, there are always fruit, cookies, juices, coffee, and tea available for those evening snacks. Many of the fruits and vegetables are grown right here at the ranch, so you know they're fresh!

Other activities include canoeing and rafting on the Coeur d'Alene River, mountain biking, tennis and golf nearby, fishing and hunting along the river. Just ten minutes to the east of the ranch is the Silver Mountain Ski area, with the world's longest gondola. This area also has mountain biking and summer mountain concerts.

There's sightseeing aplenty in the area! The ranch is five minutes east of historic Cataldo Mission in Old Mission State Park. It is the oldest building in Idaho still in use today. Fifteen minutes east is the historic town of Wallace, Idaho. The entire town is on the Historic Register. In Wallace, you'll find an underground silver mine tour and the Northern Pacific Depot Railroad Museum. West of Coeur d'Alene is a fun amusement park and great shops at the Post Falls Factory Outlet Malls.

The Kingston 5 Ranch Bed & Breakfast offers guests the best of both worlds — comfortable, peaceful accommodations and the chance to do some riding through scenic mountain valleys.

Address: P.O. Box 130, Dept. BP, Kingston, ID 83839

Telephone: 208-682-4862, 800-254-1852; **Fax:** 208-682-4862

Location: 65 miles east of Spokane, Washington

Guest Capacity: 4

Season: Year round; horseback riding May through October

Policy on Children, Pets: Children must be 10 years old. Outdoor boarding for pets in barn; indoor boarding within 1 mile.

Rates: $ Special ranch breakfast included. Horseback riding extra.

Credit Cards: MasterCard, Visa

IDAHO

MOOSE CREEK RANCH
VICTOR, ID

The Van Ordens own and manage a traditional dude ranch with an emphasis on families. The ranch is located in the Teton Mountains, just 30 minutes from Jackson Hole, Wyoming, in a beautiful and secluded valley. Moose Creek runs through the property, which is surrounded by the Targhee National Forest.

In the mid-1960s, Kelly Van Orden's father purchased Moose Creek Ranch and operated it as a boys' ranch. It was here at Moose Creek that Kelly discovered that he loved riding and loved the mountains. He worked and lived with the boys in the summers, until his parents decided to close the ranch to the public. Kelly married Roxann in 1971 and they made Moose Creek Ranch their home and are raising five children here. It was always Kelly's dream to live on a dude ranch again and give his own children a childhood like his. In the spring of 1990, the family decided to reopen the ranch to the public. After much remodeling and re-education, they greeted their first guests in the summer of 1991. "There have been experiences and advice from other ranch owners that have really helped us to grow as a family and as a business." The boys, Travis and Chase, are wranglers and guides at Moose Creek. Holly (Travis's wife), Kresta, Kathryn, and Karlie are also very active in the ranch operation, working in the kitchen and in the children's program.

Moose Creek's unique riding program is designed to teach guests to ride properly and safely in high mountain terrain. Guests are also taught to groom and saddle their own horses, which are expertly matched to each rider's ability. Horses from the "Adopt a Wild Mustang Program" are trained and integrated into the ranch's remuda. These horses were born in the wild mountain regions of Oregon and Nevada, so they're very surefooted on the mountain trails of Idaho and Wyoming.

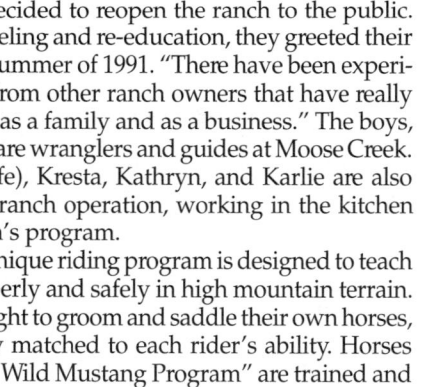

The comfortable modern cabins near the Lodge are very spacious and have private baths. The ranch house itself has five bedrooms and three baths. The main Lodge is a beautiful cedar building and houses the dining room, sitting areas, fireplace, a covered porch with large wooden swings, the trading post, and the B.Y.O.B. Wolf Trap Saloon, complete with a pool table. Just out the front door is the pool house with a large indoor heated swimming pool, a hot tub, and sauna. Table tennis, board games, volleyball, a horseshoe pit, and other yard games round out the recreational fun.

An Indian Teepee houses the children's program. While parents are enjoying their activities, the little ones are supervised by the staff counselor according to the theme of the day: mountain men, Indians, animals, etc. The games, hikes, pony rides, and crafts both entertain and educate.

An antique school bell hung in the yard rings at mealtime and anytime there is a special activity to attend. Square dancing, campfire entertainment, a night at the Western show, or an afternoon on the Snake River are some of the included activities. Climb aboard the ranch's authentic 1800s stagecoach and ride along as you head into a little Idaho town. You'll really get a taste of old-tyme cross country travel.

Three wholesome, family-style, meals, are served daily in the main lodge dining room. Meals are hearty Western fare, with homemade breads, tempting desserts, and lots of cookies and lemonade to refresh you after a long trail ride. There are also several cookouts around an open campfire. The Van Ordens can even help you plan that special occasion.

Family fun, personal challenge, enjoyment of the beautiful outdoors, and lasting memories of the Western lifestyle are the hallmark of Moose Creek Ranch. The whole family is looking forward to your visit!

Address: P.O. Box 3108, Dept. BP, Jackson, WY 83001

Telephone: 800-676-0075, 208-787-2784; **Fax:** 208-787-2284

Location: 25 miles from Jackson Hole, 65 miles from Idaho Falls

Guest Capacity: 35

Season: June 1 to October 1

Policy on Children, Pets: All ages welcome; one-on-one program for those under age 8. No pets allowed in cabins.

Rates: $$ - $$$ American plan.

Credit Cards: Amex, MasterCard, Visa

Idaho

Moose Creek Ranch

SNACK TIME.

ALL SMILES.

Idaho

Shepp Ranch
Boise, ID

Shepp Ranch is a haven of comfort, healthy meals, good fellowship and western traditions. Located in the roadless wilderness of the Nez Perce National Forest, the ranch lies on the north bank of the Main Salmon River at its confluence with Crooked Creek. The only way to Shepp Ranch is by jetboat or charter plane from Boise. The nearest town, Riggins, is 45 miles down river.

Spring is a beautiful season at the ranch. Historical tours explain the beginnings of Salmon River settlements. Hiking, boat rides to see deer, elk, eagles, bear and Rocky Mountain Bighorn Sheep are always a thrill - a great getaway from busy city living with the opportunity to just relax and enjoy the serenity. Don't forget to bring your camera! Our summer vacation schedule starts the in middle of June and ends after the Labor Day weekend. The five day/five night package starts on Monday mornings and ends on Saturday mornings. If you're not able to spend five days with us, you're welcome to stay as many days as your plans allow.

Shepp Ranch is without televisions and there are no telephones to interrupt the tranquility. Guests will find the living casual and comfortable. Mornings lend themselves to hiking, trout fishing and riding. We start the week with a peaceful ride along Crooked Creek while your body adjusts to the feel of the saddle and your mind succumbs to the serenity of your surroundings. Later, climb mountains to scenic overlooks or ride along the river to a sandy beach. For those wanting an all-day ride, we'll pack a lunch and check out an old gold mine or spend the day fishing for trout. Afternoons are spent rafting, playing at our favorite swimming hole or relaxing in the shade of a tree with a good book.

The ranch is famous for its great home cooked meals served family style. Barbecued pork chops, blueberry and sourdough pancakes, and handcranked ice cream are just a few of the favorites you will enjoy. Let us know if you have any special dietary needs and we'll provide them.

We have accommodations for sixteen guests: six rustic log cabins with showers and wood-burning stoves. Each sleeps four comfortably. Two of these, Filer & Shepp cabins, have separate bedrooms.

A hot tub overlooks the Salmon River - right on the river's edge. A great place to end your day under a blanket of stars.

The Shepp Ranch offers a Western welcome to anyone eager to encounter the wilderness in its unspoiled splendor— horseback riders, fishermen, hunters, photographers, naturalists or anyone just looking to get away from it all. This is the best place to just relax and enjoy the tranquility of your surroundings. We are known for providing swimming, riding, hiking or complete relaxation for the individual or the family. This is the perfect destination for those who want to experience one of the most popular wilderness areas. Come see nature at its best!

Address: P.O. Box 5446, Boise, ID 83705
Telephone: 208-343-7729
Web: www.ioga.org/shepp/.
Location: 45 miles east of Riggins
Guest Capacity: 16

Season: March 1 - November 21
Policy on Children, Pets:
Rates: $$ - $$$
Credit Cards: None

Shepp Ranch

A TRAIL RIDE THROUGH THE TIMBER.

Idaho

Western Pleasure Guest Ranch
Sandpoint, ID

Located on 960 acres of summer pastures with tall pines and breathtaking views of the Selkirk and Cabinet mountain ranges is the Western Pleasure Guest Ranch, a third-generation cattle ranch that has been operating on the same land since 1940. The ranch is located in Idaho's scenic Panhandle region, about 16 miles northeast of Sandpoint and has been a recreational ranch since 1990. The lodging was uniquely designed and built in 1993 by the Schoonover family for the comfort of their guests.

Guests at the ranch will discover the peaceful serenity of secluded, modern log cabins. Each cabin has one bedroom, a loft, a furnished kitchen, and a classic wood stove and can accommodate up to eight yet is cozy enough for two. The 10,000-square-foot log lodge has six guest rooms with private baths; a large great room featuring a river rock fireplace; a recreation room with pool table, air hockey, and television; a loft area that's great for reading a good book; and a 2,000-square-foot deck where you can just put your boots up and do nothing but relax. Outside there are horseshoe pits, a volleyball court, a playground for children, Indian teepees, and a fire ring for those cool summer evenings.

Enjoy some good ol' fashioned fun horseback riding in some of the prettiest country you've ever seen! The Schoonovers raise a lot of their own horses and use mostly Appaloosas and Quarter Horses. You'll ride through grazing cattle or perhaps catch a peek at deer, elk, bear, and other wildlife in their native surroundings. They can accommodate all levels of riding experience. The minimum riding age is six, and guests may participate in morning and afternoon rides, all-day rides, and overnight rides. For extended fun, take a two-day trail ride into the Panhandle National Forest. Enjoy an overnight campout, singing songs around the campfire, and sleeping in a teepee. It'll be an adventure you and your family will never forget.

Top off one of your trail rides with a hearty feast. Perhaps it will be a mouth-waterering cowboy steak with all the fix'ns. Or sit down to a delicious home-cooked meal served family style in the lodge's great room. Evening activities include group wagon rides, family buggy rides, line dancing, and singalongs around the fire.

The winter season brings its own special activities. See the beauty of northern Idaho as it's transformed by winter. You'll enjoy a horse-drawn sleigh ride through beautiful meadows blanketed white by the snow and enjoy a peace and tranquility like no other. After the ride, the hot coffee and hot chocolate will warm you inside and out. A variety of trails are groomed for skiers throughout the winter.

There are a number of things to do off the ranch and in and around the Sandpoint area. Sandpoint sits on the shores of one of the largest freshwater lakes in the United States. Lake Pend Oreille has many opportunities for outdoor enthusiasts. Schweitzer Ski Resort is a short 45-minute drive from the ranch and is beautiful in summer or winter. An 18-hole golf course is only 10 miles from the ranch. Sandpoint also offers shopping for the entire family.

The Schoonovers invite you to share in the history of Western Pleasure Guest Ranch — relax a little, have lots of fun, participate in the ranch activities, sing along like the cowboys used to, and return home with a renewed interest in your own future.

Address: 4675 Upper Gold Creek Road, Dept. BP, Sandpoint, ID 83864

Telephone/Fax: 208-263-9066

Location: 90 miles southeast of Spokane

Guest Capacity: 34

Season: Year round

Policy on Children, Pets: Children are welcome; must be 6 to ride. No pets.

Rates: $$ American plan. Children's rates.

Credit cards: MasterCard, Visa

IDAHO
WESTERN PLEASURE GUEST RANCH

GREAT RIDING. WINTER...

...OR SUMMER.

IOWA

GARST FARM RESORTS
COON RAPIDS, IA

Located on a commercial Iowa farm on the edge of a 4,500-acre private park, Garst Farm Resorts offers a truly unique experience and a wide variety of activities and lodging choices. In addition to great horseback riding, we present a commercial agriculture tour, stargazing, canoe and mountain bike rental, bobsled rides, fishing and hiking. We offer history too...Khruschev visited this farm in 1959 and you can learn about that visit during your stay. Coon Rapids, the classy and friendly small town nearby, offers several dining and shopping options and a great golf course, bowling alley, tennis courts, racquetball court, and an outdoor aquatic center. The resort lies one hour west of Des Moines and two hours east of Omaha.

Garst Farm Resorts is an advocate of eco-tourism, and there is a lot to learn if you are interested in land conservation and bio-diversity. The horseback and hiking trails meander along 8 miles of the Middle Fork of the Raccoon River and feature gorgeous woodlands, native prairies, several prairie reconstruction projects, natural and reconstructed wetlands, game food plots and tree stands for wildlife viewing. Riding trails cross the river and pass high ridge lines, fishing ponds, 100 million year-old rock faces, and an old log cabin. Woodland song birds, waterfowl, raptors, deer, wild turkey, and coyote abound and catfish, bass and bluegills frequent the 35 fishing ponds and the river. Our astronomer believes the valley to be among the darkest places in Iowa...the night skies are fabulous. You can learn much about soil conservation which is a primary goal of the corn, soybean, cattle and buffalo farm that we operate.

The resort rents horses for all types of riders or you can bring your own. On guided rides, our fun and helpful trail boss will mount you in the riding ring and ensure that you can cope with the fact that the trees do not move. Then, off you go into the beautiful valley for a few hours or for the whole day. Generally, we ride side by side, rather than nose to tail. Guests enjoy the freedom they are offered. If you bring your own horses, an open pen or stalls are available or you can keep your horses with you at our campsites. Guides are available.

If you are in the mood to be pampered, try our Bed and Breakfast, originally the home of Roswell and Elizabeth Garst, fascinating and innovative farmers. The large farmhouse has the feel of a warm and prosperous farmhouse from the 1940's, full of family antiques and historical photos of farming and the Khruschev visit. We have both private and shared bath rooms. For a more private experience, or for larger groups, rent one of our cottages or our five bedroom farmhouse, each with a fully-equipped kitchen. Camp sites offer a picnic table, fire ring and firewood, a cold water shower, privacy, beautiful locations and plenty of grass for your horses. 805 River, a primitive camping cabin is miles off the road, with the river at your front door and a spring-fed pond at your back.

 Address: 1390 Highway 141, Coon Rapids, IA 50058

Telephone: 712-684-2964; Fax: 712-684-2887
Web: www.farmresort.com or www.jimbalzotti.com
E-Mail: gresort@pionet.net

Location: 1-1/4 hour NW of Des Moines airport, 2 hours east of Omaha airport

 Guest Capacity: 28

Season: Year round

Policy on Children, Pets: Children welcome, minimum age of 6 for horseback riding. Pets welcome in some facilities, kennels available.

Rates: $

Credit Cards: Amex, MasterCard and Visa

IDAHO
GARST FARM RESORTS

MY NEW BEST FRIEND.

KENTUCKY

FIRST FARM INN
BURLINGTON, KY

We have a 120-year old house and barn. The property includes 20 acres with woods, pasture and pond. Middle Creek Park offers several hundred acres of hills for riding about 15 miles away.

Horses and a pony are available for guests and their children to ride. Riders must demonstrate their ability before being allowed to trail ride. The Kentucky farm animal liability act ensures that you ride at your own risk. Other activities include an outdoor hot tub, hiking, biking, rowing or enjoying the rockers and swings around the house and in our 200+ year old trees.

First Farm Inn offers two guest rooms furnished with family antiques fitting the period. Beds are converted to queen-size. Baths are accessible from inside the rooms. Kentucky allows us only to serve breakfast, but a wide variety of excellent restaurants are available within a 20-minute drive. Feel free to bring your own picnic. We're 6 miles from Burlington, 8 miles from Florence, 20 minutes from downtown Cincinnati, Ohio and 3 miles from Lawrenceburg, Indiana.

Entertainment: Dana's original music and arrangements accompany breakfast or guests may play the keyboard. Free passes are available for Turfway Park horse racing. Rabbit Hash General Store, operating continuously since 1830, is open year-round. Antique shops are in Burlington and Lawrenceburg. Right across the river are the Argosy and Grand Victoria casinos. First Farm Inn offers discounted lift tickets for Perfect North Ski Resort, one of the best maintained and largest downhill ski areas in the Midwest. Discounted tickets are also available for Kings Island amusement park. Downtown Cincinnati offers a world-class zoo, arboretum, art and history museums, ethnic restaurants, professional sports, etc.

Well-behaved children are accepted, portacrib available - arrange in advance. Well-behaved and house-trained pets are accepted - check in advance.

Bring your horses and ride or try a trail on one of First Farm Inn's happily retired herdmates. Watch the horses graze in front of the setting sun as you recline in the luxurious outdoor spa. Our updated 1870's farm home is set on 20 acres of Kentucky's rolling hills above the Ohio River where Indiana, Kentucky and Ohio join. Listen to rain on the tin roof, paddle around the pond, rock or swing on the verandah, watch the birds, warm yourself by the fireplace, create your own grapevine wreath, play the piano or sing along, check out the playhouse or explore the barns or woods.

First Farm Inn's spacious rooms have queen-sized beds and private baths. Business travelers will appreciate our work space, dual phone lines, FAX receiving and Internet access. Your room is decorated with century-old oak furniture.

Wake to the smell of bread baking. Geared to guests' comfort and convenience, a full hot breakfast is presented when you like. Try something new like Pasta Carbonara or request your old favorites. Let us know if you prefer vegetarian, low fat, low sodium, etc. Enjoy Dana's original music and arrangements while you eat.

We're 10 minutes away from the Greater Cincinnati-Northern Kentucky International Airport. Spend a night with us before your plane takes off or when you return.

Address: 2510 Stevens Road, Burlington, KY 41005

Telephone: 606-586-0199; **Toll Free:** 800-277-9527; **Fax:** 606-586-0299
Web: www.bbonlinr.com/ky/firstfarm
E-Mail: firstfarm@goodnews.net

Location: 10 minutes from Cincinnati-Northern Kentucky International Airport

Guest Capacity: 4 to 7

Season: Year round

Policy on Children, Pets: Children welcome. Pets welcome, check in advance.

Rates: $

Credit Cards: None

MAINE

FOOTHILLS FARM
BROWNFIELD, ME

Although horse lovers often dream of riding their mounts through the open spaces of the American West, those planning to visit the New England region should be aware of the unique opportunity for trail riding that the three northern New England states offer. The forest-covered Appalachian Mountain range, which dominates the landscape of the area, is traversed by an extensive network of back roads, logging trails, and snowmobile routes that local equestrians enjoy year round.

Panoramic views, cascading mountain streams, and rolling meadows abound. An added bonus of the region are the abundant signs of the former communities that were abandoned after the Civil War when young soldiers realized that easier agrarian lifestyles were available to them in southern regions. Farmsteads that had been worked for centuries were abandoned and quickly reclaimed by the forests. Others, of course, remained — their sturdy post-and-beam and clapboard construction still standing as testament of that bygone era.

Situated in a small valley and surrounded by the forest-covered countryside, the Foothills Farm is one such homestead. Constructed around 1860, the main farmhouse has been lovingly restored by its current owners to retain the charm, craftsmanship, and beauty of the house. Indeed, guests have often remarked that visiting the Foothills Farm is reminiscent of visiting their grandmother's home — the ultimate compliment for restorers of old houses! For nearly fifteen years, Kevin and Theresa Early have been hosting visitors who relish quiet, unspoiled New England countryside and clean, comfortable accommodations at a reasonable cost.

At the urging of some of their guests who are horse owners, the Earlys constructed a four-stall barn and fenced in paddocks for guests' horses. They've also mapped out and marked a system of trails, from 8 to 20 miles in length, that can be explored on your own or with a guide. Some trails simply follow the routes of town-maintained dirt roads, which make them excellent for driving teams. Others follow routes that were once major thoroughfares when horseback was the main mode of transportation between the farmsteads and communities that once dotted the region. Today, stone walls, forlorn graveyards, and granite-lined cellar holes are all that remain of what were once thriving communities. Yet many of these old roadways are still maintained and provide equestrians a glimpse of New England that most visitors never see. Best of all these trails are accessible from the Foothills Farm's barnyard — there's no need to traverse highways or follow busy roadways. You'll get a sense of riding through the New England of one hundred years ago.

Accommodations for two-legged visitors are provided in the farmhouse. There are four guest rooms, all with shared baths, furnished in tasteful period antiques. A favorite socializing area is the den/library where guests can enjoy the warmth and cheeriness of cocktails in front of a blazing fire. In the summer months, a screened-in porch overlooking the gardens and fields provides a restful place to end the day. The Bunkhouse, an old carriage shed, is now a cottage complete with private bath, kitchenette, fireplace, and bar and can accommodate up to six guests comfortably.

In addition to the natural amenities at hand, there are also those offered by nearby Conway, New Hampshire, and the bustling Mount Washington Valley, where visitors will find scores of restaurants and tax-free factory outlet shops. Seven major ski areas lie within a 40-mile radius, and there is canoeing available on the nearby Saco River. For those who enjoy shopping for older things, there are numerous out-of-the-way antique shops.

So, if you're a horse owner who loves to explore new trails in a peaceful setting, visit the Foothills Farm.

Mailing Address: P.O. Box 1368, Dept. BP, Conway, NH 03818

Telephone: 207-935-3799

Location: 10 miles from Conway, New Hampshire; 40 miles west of Portland, Maine

Guest Capacity: 12

Season: Year round

Policy on Children, Pets: Children and pets are welcome.

Rates: $ Full country breakfast included. Horse boarding extra.

Credit Cards: American Express, MasterCard, Visa. Personal checks accepted.

MAINE

SPECKLED MOUNTAIN RANCH
BETHEL, ME

Nestled in the foothills of Maine's White Mountains is a unique getaway for horse enthusiasts and nature lovers alike. Open year round, Speckled Mountain Ranch is a working farm and riding stable where you can connect with nature and take part in a rural way of life. Individuals, couples, families, and small groups enjoy ample opportunities for hiking and outdoor adventure - on foot as well as on horseback.

From the moment you arrive, we'd like you to feel at home. Have a look around - we'll introduce you to our horses and friendly dogs and cats. Take a stroll through the woods, hike up a mountain trail to a wilderness pond, watch a golden sunset on a cool summer evening from a horse-drawn carriage, or just sit back and enjoy the beauty of the area. The farmhouse is sunny, spacious and smoke free, with 3 rooms (queen and double beds) and 2-1/2 shared bathrooms.

After a good night's sleep, you'll wake to the smell of fresh coffee, raspberry pancakes, homemade bread or other delicious goodies (many fresh from our organic garden). While you are enjoying breakfast, you'll also enjoy sweeping views of the countryside. Keep your eyes open and you might see a moose and her calf or other abundant wildlife wandering across the hayfields.

Our intention is to create challenging and rewarding riding experiences. Our horses are responsive, well-trained and energetic. Come and get to know the exceptional animals we are privileged to share our farm with. Haflingers, Morgans, Warmbloods, Quarterhorses, a Mustang, and others all call Speckled Mountain Ranch home.

May through October we offer 2 to 5-day English trail riding vacations. Groups are small (2 to 5 riders with 1 to 2 guides), and trail rides vary according to riding levels. Advanced riders who can comfortably handle a horse at all gaits over varied terrain will enjoy galloping up the Miles Notch Trail, crossing the Pleasant River, and jumping a log or two along the way. We also offer instruction for less experienced riders who are serious about learning riding and horse care. (Our weight limit for riders is 180lbs.) Plan on riding 2 to 3-1/2 hours per day.

Grooming, tacking up, and care after riding are all part of your vacation experience. Riding, lodging, breakfast and lunch, and taxes are included in package prices. During the month of July we offer an overnight camp for girls ages 12 to 15. Our camp is for intermediate to advanced level riders who are interested in riding out on the trail. We ride English, and our emphasis is on handling a horse safely and successfully outside of the ring, over varied terrain. Camp sessions run for 6 days with 2 to 4 hours of riding each day. Groups are limited to 5, so there is a lot of individualized attention. For more information, ask for our summer camp brochure and application.

You're welcome to join us taking care of the horses and other farm 'chores', but some of the other activities you may enjoy here are: horseback riding, carriage rides, bird-watching, hiking, photography, orienteering, rock hounding, mountain biking, swimming, fishing, canoeing, snowshoeing (guides available), cross-country skiing (guides available), relaxing, stargazing, and moonlight strolls and gardening.

Address: RR 2 Box 717, Bethel, ME 04217

Telephone: 207-836-2908

Location: Approx. 70 miles NW of Portland, Me

Guest Capacity: 9

Season: May through October for riding and driving, December 24 through mid April for B & B

Policy on Children, Pets: Children welcome. No pets.

Rates: $ and $$

Credit Cards: None

Maine

MICHIGAN

DOUBLE JJ RESORT RANCH
ROTHBURY, MI

Double JJ Resort Ranch is located 20 miles north of Muskegon in west Michigan on Big Wildcat and Carpenter Lakes and only 10 miles east of the Great Lakes. For over 60 years, the Double JJ has entertained thousands of guests in its friendly and relaxing atmosphere. Located on 1200 wooded acres, the ranch has miles of beautifully wooded trails and nationally acclaimed Thoroughbred Golf Club, an 18-hole championship course designed by Arthur Hills. Guests can participate in activities from dawn 'til dusk each day offering everyone as much or as little as they want.

In the summer, resort activities include: daily riding with weekly breakfast and dinner rides. Private lessons and corral sessions are available. Rides are divided into small groups by experience levels. Rides vary from 1 to 2 hours, and go out several times each day. Outdoor swimming pools, hot tubs, water slide, tennis courts, target sports, volleyball, horseshoes, hayrides, campfires, evening entertainment, championship golf and more. In the winter, activities include: horseback riding, sleigh rides, snow tubing, dog sledding, cross-country skiing and snowmobiling.

All-inclusive packages are available May through November. Daily, weekend, midweek and weeklong horseback riding or golf packages available. Weekend packages available year-round. Ask about special holiday packages. Extensive children's programs. The Back Forty Ranch offers kids-exclusive packages with full-time counselors. Packages include accommodations, three meals daily, activities and entertainment. Full-service bars, snack bars, general store, trading post, and pro shop are located throughout.

Summertime evening entertainment includes: live bands, staff shows, talent shows, staff vs. guest volleyball, professional rodeo each week with guest competitions and games, hayrides and campfires, game and video room, theme weeks and costume parties. Separate adult and children's programming.

Accommodations include the Double JJ Ranch - adult exclusive bunkhouses and lodge rooms, the Back Forty Ranch - family log cabins and kids-exclusive teepees, Conestoga wagons and bunkhouses, and the Thoroughbred Golf Club - hotel and condominiums.

Address: P.O. Box 94, Rothbury, MI 49452

Telephone: 616-894-4444;
Toll Free: 800-368-2535
Fax: 616-893-5355
Web: www.doublejj.com or www.jimbalzotti.com
E-Mail: info@doublejj.com

Location: Just 3 hours from Detroit or Chicago

Guest Capacity: 300

Season: Year round

Policy on Children, Pets: Children welcome.

Rates: $$

MICHIGAN

DOUBLE JJ RESORT RANCH

A ROUND OF GOLF ON A SUNNY AFTERNOON.

YOUR HOME AWAY FROM HOME.

MISSOURI

MERAMEC FARM B&B
BOURBON, MO

Meramec Farm is situated smack dab in the middle of natural beauty. River bends, open fields, woods, and towering bluffs grace its vistas. The wild setting provides a lovely backdrop for frolicking calves, grazing cattle, horses, and mules. The yard is adorned with llamas, chickens, a goose, and sundry other creatures. This seventh-generation working farm offers visitors a chance to experience family farming, a variety of outdoor adventures, and a quiet relaxing atmosphere.

Family history is significant on the farm. In 1811, William Harrison came to this land from Kentucky. He swam his cattle across the Mississippi and proceeded to higher ground here in the Ozarks. In 1883, his grandson Benjamin built the house in which your hosts, Carol and David, now live. Today, cattle and hay are the main farming enterprises. Guests are invited to hop onto a bale of hay in the old truck and "tour the herd."

In 1982, Carol and David opened the farm to bed-and-breakfast guests, including those traveling with their own horses. Families enjoy the educational aspects of a visit to this working farm, as well as the setting and the good food. Many guests enjoy bird-watching, fishing, and walking. Miles of private, wide trails on the land feature a variety of terrain and ecosystems. The farm's "pampered trips" travel at the pace the guests prefer and include delicious meals before, during, and, on Saturdays, after the adventure. The high panoramas and river views leave memorable images!

The Meramec River adjoins the farm for 1.5 miles. Great blue herons nest here and these magnificent birds lend an exotic touch to the sycamores they favor. Wild turkey and white-tailed deer abound in the surrounding woods. Bass and trout are in the river and nearby streams. Stocked ponds on the farm may be fished without a license. Float trips are recommended. Canoe outfitters are nearby and rubber rafts are also available. For the less ambitious, a trip to the five-acre gravel bar with an inner tube is a great way to relax after a long ride!

Meramec's horse tours are a nice mix of wide wooded trails with challenging hills and vistas, open fields with rolling hills, and long flat river bottoms. River crossings are an option depending on water levels. In addition to approximately 10 miles of riding on the farm, the 24-mile Berryman Trail is an easy 20-minute drive away. A hilly, wooded trail on National Forest Service land, the Berryman can be significantly extended by using logging trails.

For guests who have several days, Carol arranges rides on other farms in the area. Most have a balance of open pasture and timbered trails, some are along bluffs. The experience of meeting local farmers is an opportunity to truly know this area of the Ozarks. Groups are limited to 10 riders (special arrangements can be made for larger groups). Breakfast is included with your lodging at Meramec Farm. Home-baked breads, farm fresh eggs, fruits, juices, meats, cereals, and a yummy array of specialties are featured. Lunch is provided on the full-day rides. On Saturdays, a big family style dinner is served. Much of the food is raised locally or on the farm.

This region of the Ozarks has many fine wineries, which offer tours as well as free samples. Trips to local caves are dandy rainy or hot-day options. Antique shops, flea markets, square dancing, rodeos, and local bluegrass music can keep riders and nonriders alike busy with little driving.

At Meramec Farm B&B, guests are invited to design their own tour of the region. Carol and David will take care of the details. The atmosphere is relaxed and friendly, the rides have great variety, and the food is fresh farm cooking. What more could you ask for!

Address: HCR 1, Box 50, Dept. BP, Bourbon, MO 65441

Telephone/Fax: 573-732-4765;
E-mail: mfarmbnb@fidnet.com

Location: 9 miles south of Bourbon, 70 miles southwest of St. Louis

Guest Capacity: 15

Season: Year round

Policy on Children, Pets: Children are welcome. Horses only.

Rates: $ Breakfast included, dinner on Saturday.

Credit Cards: None. Personal checks accepted.

Missouri

Meramec Farm B&B

OUR WELCOMING COMMITTEE.

JUST LYING AROUND.

MISSOURI

TURKEY CREEK RANCH
THEODOSIA, MO

Built in 1902, the once fine home near the Little North Fork of the White River was the 1953 honeymoon cottage that Dick and Elda Edwards would remodel for a fishing lodge.

The newly completed Bull Shoals Lake was nearly full in 1953, but went dry in 1954, then flooded in 1955 as Dick and Elda finished school and started a family. Two cottages were completed in 1959, a dock and another cottage followed. But the lake flooded again in 1961 and forced them to rethink catering only to the fishing trade. Stretching their credit to the limit, they bought four riding horses and the materials to build a smimming pool. The long hours and hard work began to pay off as family vacationers came to relax and play. In 1968 adversity struck again. Fire leveled the twice remodeled farm home and two attached units. The Dick and Elda, now with five children, began again. A home was built, a recreation building added and an aggressive building program begun. Throughout the changes, Turkey Creek has remained the same, a great family vacation place.

Horses and cattle are part of a ranch, and that's what Turkey Creek, the Swimming T, is...a working 700-acre ranch. Depending upon the season, you can see the various operations of a cattle ranch - attending to newborn calves, vaccinating and ear tagging, shoeing the horses or bringing in the hay for winter feed. Ours is a wonderful place with its hills and hollows, woodlands and pastures. Add to that a beautiful lake and you have Turkey Creek Ranch, an interesting, fun place for a great vacation.

Horseback riding is a favorite activity of guests. There are a string of Western pleasure saddle horses and free instruction for beginners. Wrangler escorted rides are scheduled mornings and afternoons, except Saturdays. Scenic trails wind through the area as bridle paths for the horses or nature trails for hikers. The dogwood and redbud in spring, the wild flowers of summer and the fall Ozark Colorama are magnificent. A sharp eye may see turkey, quail, deer and other wildlife.

There are so many fun things to do, that Turkey Creek Ranch is sure to appeal to just about everyone. Our recreation building has a whirlpool spa, video games, three regulation pool tables, bumper pool, shuffleboard, air hockey, table tennis and a piano for sing-alongs. Two pools, one outdoor and one heated indoor, a fountain wading pool and playground are for our guests' enjoyment. Tennis anyone? There are two tournament Cosmicoat surfaced courts to test your skill.

Then there are the fun activities for those who love the water. The sparkling, clear water of Bull Shoals Lake is not just for fishing - pontoon boats, paddle boats, canoes, sail boats, and ski boats are all at your disposal. So, for a memorable vacation, come to Turkey Creek Ranch, the lake resort with a dude ranch flavor...a fun place, a family place, make it YOUR place.

Address: H.C. 3, Box 3180, Theodosia, MO 65761

Telephone: 417-273-4362
Web: www.jimbalzotti.com

Location: 5 hours from Kansas City and St. Louis

Guest Capacity: 139

Season: Year round

Policy on Children, Pets: Children welcome, but no special children's programs. No pets.

Rates: $ and $$

Credit Cards: None

MISSOURI

Turkey Creek Ranch

MAIN OFFICE OVERLOOKING POOL AREA.

ALL SORTS OF ACTIVITIES ARE
AVAILABLE AT TURKEY CREEK.

MONTANA

63 RANCH
LIVINGSTON, MONTANA

If you wake before the morning breakfast bell, lie still and listen. From the hills echo the distant shouts of wranglers herding horses in streams of morning sunlight. From Mission Creek drifts the murmur of rushing water. Overhead, you'll hear the rustle of aspen leaves in the morning breeze.

The rich aroma of bacon and eggs, pancakes and hot coffee, borne on the brightest breath of air in the world, drifts from the main lodge. But no need to hurry. You take your own sweet time. Once you are in the log dining room, dig in! Get ready for an invigorating morning ride through some of the West's most awesome mountain scenery. You will be famished by noon if you don't eat hearty!

The horses are saddled and waiting when you and other guests congregate at the corral, snapping pictures, playing with the dogs, kidding the cowboys, ready to mount up.

Now you're off through sunlight and sagebrush and wildflowers. You've chosen a lazy ride up a timbered canyon to a waterfall. Or a spectacular climb to a mountain pass to round up a few stray dogies. Or a thrilling gallop with the wind in your face, on grassy hills rolling down to the regal Yellowstone River.

After a healthy lunch, how about another stirring ride on a breathtaking trail? A hike in the woods? Fishig in Mission Creek? Swimming in the pond? Taking photos of flower-filled meadows? A refreshing nap in a quiet cabin? You do what you like at The 63.

Later, after an unforgettable dinner, its a round of billiards in the game room, or checkers in the corner, or tall tales in front of a roaring fire while the teenagers and the kids dance and play off in the recreation hall.

Or perhaps something quieter...a romantic walk in the moonlight...a bit of reading...constellation counting in the crystalline air of the Big Sky. Then...the best night's sleep you ever had.

Tomorrow? Well, one thing's for sure...it will be a western adventure to remember at Montana's National Historic Site, The 63 Ranch.

Address: Dept. AAA, Box 979, Livingston, MT 59047

Telephone: 406-222-0570; **Fax:** 406-222-9446 (winter only)

Internet: www.ranchweb.com/63ranch

E-Mail: sixty3ranch@mcn.net

Location: 12 miles Southeast of Livingston

Guest Capacity: 30

Season: Mid-June through mid-September

Policy on Children, Pets: Children welcome. No pets.

Rates: $$ All inclusive.

Credit Cards: None

MONTANA
63 Ranch

FOLLOW US INTO SOME OF THE PRETTIEST
COUNTRY YOU'VE EVER SEEN.

MONTANA

B BAR GUEST RANCH
EMIGRANT, MT

Whether you seek the serenity of the Basin or the countless recreational outings beyond our gates, you will find the B Bar to be the perfect destination for your vacation. Here at the ranch, you can explore several hiking trails to hidden lakes, ridges and peaks, or set out cross-country to discover remnants of the petrified forest that once hemmed the Basin. Within a short drive of the ranch are spring creeks and blue-ribbon trout streams for anglers. For white-knuckle enthusiasts, there is white-water rafting through Yankee Jim Canyon. Or spend a day with your own naturalist getting better acquainted with Yellowstone Park.

The Black Canyon of the Yellowstone is a perfect back drop for a horseback ride. For longer horse packing trips, there are a variety of routes through the vast back country that encircles the ranch. Groomed trails invite you to explore the winter quiet and beauty of the ranch. Electric and Emigrant peaks are just two of many popular day hikes that reward weary explorers with a soak in nearby hot springs at the end of the trail.

The B Bar offers its guests many opportunities to observe the cycle of life on a western ranch. We live by a seasonal calendar marked with spring calving and plantings, summer haying and vegetable harvests, fall cattle drives, and winter feeding and grooming with young draft horses in training. Throughout the summer we work on horseback, monitoring the condition of both cattle and range, and moving the herd from one area to another as often as needed to preserve the health of the ecosystem. In the spring, there are ditches to be pulled, headgates built, fences mended, fields irrigated, and garden beds double dug. In the fall, there are cattle to be weaned and shipped, firewood to be gathered and cut. The long days we put in are rewarded with a deeply satisfying way of life and the opportunity to practice the stewardship we preach in a place called "paradise."

Today, operating in both the Basin and Big Timber, the B Bar owns and manages a conservation herd of Ancient White Park cattle, raises and works Suffolk Punch draft horses, and runs a commercial herd of natural beef cattle. The B Bar Guest Ranch has evolved from our commitment to the Rockies, our concern for stewardship of the land and its resources, and our desire to share the unique experience of living, working and playing in Tom Miner Basin. Here, we have established a high-altitude intensive, biodynamic vegetable garden as well as perennial borders, shrubs, and trees that circle the main lodge and extend to the cabins. Our guest cabins sit at the edge of an aspen grove. From your porch, you are likely to see elk moving quietly along the far sidehilll into the forest at dawn or watch the moon rise over Sheep Mountain as moose graze nearby at dusk.

Address: 818 Tom Miner Creek Road, Emigrant, MT 59027

Telephone: 406-848-7523; **Fax:** 406-848-7793

Internet: www.bbar.com

E-Mail: guestranch@bbar.com

Location: 45 miles south of Livingston, 40 minutes from south entrance of Yellowstone National Park

Guest Capacity: 34

Season: Summer: late May through mid-September; Winter: mid-December through end of February

Policy on Children, Pets: Children welcome. No pets.

Rates: $$ and $$$

Credit Cards: None

Montana

B Bar Guest Ranch

SLEIGH BELLS, CROSS COUNTRY SKIING AND SNOWSHOEING AT THE B BAR RANCH.

MONTANA

BAR N RANCH
YELLOWSTONE, MT

The Bar N Ranch, located in West Yellowstone, Montana, is located just a stones throw away from Yellowstone National Park. Orginally built by the Kephart family back in 1908, it can claim to be one of the original dude ranches that catered to guests coming from the east to enjoy life on a working ranch. The Bar N came to be known as having some of the best fishing to be found anywhere and is still considered a fly fisherman's paradise. Guests could ride our gentle saddle horses into the unspoiled beauty of Montana, see wildlife such as elk, antelope, deer, numerous birds and birds of prey in their natural habitat. Today, guests are able to enjoy the same wonderful vacation at the Bar N Ranch. Still owned and operated by the original Kephart family, they delight in offering their guests the same unique vacation but with more creature comforts.

At the Bar N Ranch, guests may do as little or as much as they like. The owners will arrange for guides so you can take advantage of the still fabulous flyfishing, whitewater rafting, helicopter rides, day trips to Yellowstone National Park, trail rides of various lengths of time to suit whatever level of rider you are, hay rides, buggy rides and much more. In wintertime, the fun doesn't stop. Guests can go on a thrilling dog sled ride, ride in an antique snowcoach or a horse drawn sleigh, snowmobile, snow shoe, cross country ski, and much, much more.

For accomodations, guests will be pampered in a variety of different lodge rooms and cabins. The newly built "Lodge Room" features all the amenities you would expect at a first rate guest ranch: double or queen bed; sitting area with sleepsofa or a deluxe futon; wood burning fireplace; private balcony with French doors; custom tile and slate bath with extra deep jacuzzi tub and heat lamp; color tv and vcr; phone/fax-modem connection; stereo; and mini-fridge. There are some deluxe lodge rooms that are larger and offer jumbo jacuzzi tubs. The cabins are the same as the rooms except they offer separate living room and bedroom, with a shower bath, and the hot tubs and jacuzzis are outside!

The Bar N specializes in offering the best in conference and seminar facilities with fax/modem data ports and audio or video format. It is the perfect place to not only conduct a successful business meeting but enjoy all that the Bar N offers when the business day ends. We would be happy to customize a package that suits your specific needs.

Food at guest ranches is always a special treat and at the Bar N this is one area where we excel! Guests enjoy a complimentary Happy Hour in the Great Room with a piano cabaret in high season. "Montana Country Gourmet" dinners served with wine delight our hungry guests. We offer wonderful ranch breakfasts and hearty picnic lunches. All of this is included with your stay. Guests who stay for three days or more enjoy free horseback riding.

So come and recharge your batteries, enjoy quality time with your family and come to see life the way it was meant to be lived.

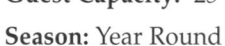

Address: 3111 Targhee Pass Highway, West Yellowstone, MT, 59758

Telephone: 1-800-BIGSKYS (244-7597)
Fax: 406-646-9682; **Voice Mail:** 406-646-7121
Web: www.wyellowstone.com/bar-n-ranch or www.bar-n-ranch.com
E-Mail: barnranch@wyellowstone.com

Location: Right in W. Yellowstone, MT

Guest Capacity: 25

Season: Year Round

Policy on Children, Pets: Children welcome with parent supervision. Pets—call for information

Rates: $-$$

Credit Cards: Mastercard and Visa, Personal Checks and Cash

MONTANA

BEAR CREEK LODGE
VICTOR, MT

Welcome to Bear Creek Lodge, a mountain hideaway nestled in a blanket of pine trees. Bear Creek Lodge is a civilized approach to a wilderness experience. Our guests are drawn to the serene, mountain setting and come to us primarily for fly fishing, hiking, and horseback riding, all of which are accessed directly from our property. We are located on 115 acres adjacent to the Selway - Bitterroot Wilderness, the largest wilderness in the continental United States. The Bear Creek Trailhead is a half-mile from our barn and there are more than 30 additional trails within an hour's drive from the lodge. A person could quite literally ride all summer and never be on the same trail twice. The horse facilities include three deluxe indoor stalls, pens, and pasture that are available to our guests traveling with their own livestock. We have no horses ourselves but a neighboring ranch provides trail rides from an hour to all-day.

The lodge was built in 1991 from logs harvested from the Yellowstone fire and is both tasteful and exquisite. With only eight guest rooms, each with private bath, the lodge creates a warm, intimate atmosphere. Each of its rooms are appointed with comfortable beds, antique furnishings, plush carpeting, a private bathroom, and a door that opens onto the forest. Three meals, lovingly prepared by Elizabeth, are included - her cooking has been featured in national publications on fine dining. They are lovingly prepared with fresh ingredients and designed to satisfy the palates of both discerning and hearty appetites. The guests dine, family-style on entrees such as chicken breasts served on a bed of marinated wild mushrooms or shrimp sauteed with apples and snow peas accompanied by homemade bread and followed with a scrumptious dessert. A fine wine from the well-stocked cellar rounds out the dinner meal. After dinner, guests can unwind in the pine-scented night air with a soak in the redwood hot tub on the front deck. Stargazing into the night skies is a favorite way to end the day. The lodge amenities include a living room, dining room, spacious deck, library and pool table, and exercise room/sauna.

We have a mile of private access to Bear Creek for our fishing guests and the Bitterroot River is just minutes away, filled with wild, native trout. We are on the Lewis & Clark and the Nez Perce Trails and have several historical sites in the area. Missoula is 45 miles north and is serviced by the major airlines.

Those who don't fish have plenty to do. They can capture the Old West with a horse trip into the Bitterroot Mountains. An experienced backcountry guide takes trips up Bear Creek Canyon ranging from an hour to several days. Mountain biking, canoeing, and whitewater rafting are also available at or near the lodge. For those seeking a daily fitness program, the lodge has an exercise room complete with sauna. The valley is a hiker's paradise with 29 trails throughout the Bitterroot Mountains for the beginner to experienced hiker. The trails lend themselves to bird watching, photography, backpacking, nature walks, and cross-country skiing in the winter months.

At Bear Creek Lodge, you'll feel as if you are visiting old friends. Come be part of our wilderness family.

Address: 1184 Bear Creek Trail, Victor, MT 59875

Telephone: 406-642-3750; **Fax:** 406-642-6847
Web: www.bear-creek-lodge.com
E-Mail: info@bear-creek-lodge.com

Location: 45 miles south of Missoula

Guest Capacity: 16

Season: In season: June through September, off season: October through May

Policy on Children, Pets: Children welcome with whole lodge booking. No pets.

Rates: $$ and $$ American Plan only in season, American Plan or B & B off season.

Credit Cards: Mastercard and Visa

Montana

Bear Tooth Ranch/JLX Outfitters
Nye, MT

First of all, we want to invite you to be with us during the summer. This is a spot that combines what is good about many things and that attracts a very special kind of person...one who enjoys horses or fishing, or both, or can't get enough of the informal ranch life and scenery that surrounds you every day of a vacation here. If you are that kind of person, you can't help but be an ideal guest, entertained by an enthusiastic ranch crew. We know that your vacation means alot to you. On a Montana ranch, a vacation is getting away from the routine of everyday life and putting your thoughts and plans in order for the months ahead.

Dude ranching at Beartooth Ranch and JLX Outfitters brings visitors and crew together to keep up with all that goes into running a mountain ranch. We welcome your help with trail cleaning, wrangling in the horses early in the morning, grooming and feeding them, fixing some fences, irrigating, hauling and stacking hay for winter use and cutting wood for the fireplaces. A ranch vacation here is a challenge to your sense of adventure. It is an invitation to saddle a horse and ride the old trails of the Stillwater canyon, rich in the lore of Indians. Only 100 years ago, it was still being explored by fur trappers and prospectors and settled for the first time by cattlemen and sheepmen. When we stop to think about the millions of people streaming toward their vacationlands each year, we realize that ranches are virtually unknown as destinations: "undiscovered paradises" that grow more attractive each year to those few who have found the romance of our way of life.

Beartooth Ranch adjoins the Absaroka-Beartooth Wilderness Area of nearly one million acres. Backpackers enjoy it, but we think the best in a wilderness experience comes with a ride over a mountain trail on a horse that walks along at a good pace...not at a trot or gallop. This way, our guests can view the spectacular scenery that lies beyond each ridge in a safe manner and save the fast riding for morning horse round-ups or by helping wranglers take the herd out to pasture each afternoon.

Work and recreation gravitate mainly to the horses. Basic riding instruction starts the day at 9:00 a.m. on Monday and a short trail ride follows. From then on, the riders' capabilities and wishes determine the scheduling. In general we have: morning rides at 9:00 and afternoon rides at 2:00 on Tuesday. Wednesday through Saturday, we feature pre-arranged or all-day rides. No horseback riding is available on Sunday. We raise and train trail horses and pack stock, so we reserve some hours during the day or evening to spend in the corrals to work with horses. Other evening events include volleyball, softball, touch football, badminton, soccer, beaver watching, trout fishing, ping pong and horseshoe pitching.

 Address: HC 54, Box 350, Nye, MT 59061

Telephone: 406-328-6194

Location: 90 miles Southwest of Billings

Guest Capacity: 20

Season: June 1st — September 1st

Policy on Children, Pets: Children welcome.

Rates: $

Credit Cards: American Express, Mastercard and Visa

MONTANA
BEAR TOOTH RANCH/JLX OUTFITTERS

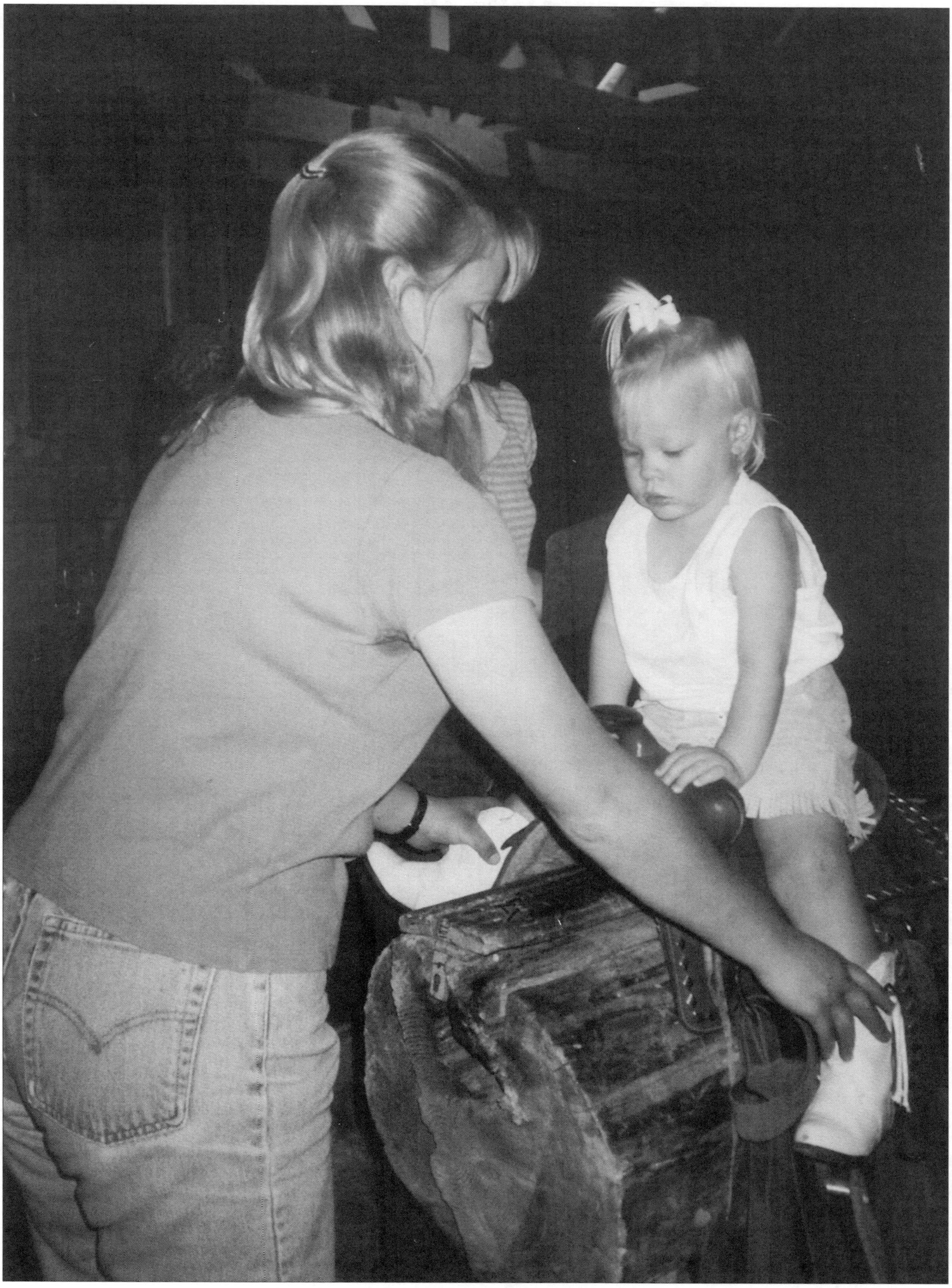

HEY, WHAT HAPPENED TO HIS HEAD?

BEAR TOOTH RANCH/JLX OUTFITTERS

MONTANA

BEAVER CREEK TRAIL RANCH
MALTA, MT

Our 10,000-acre farm and ranch includes a 300 crossbred cow/calf operation, 30 sheep and a variety of other animals: goats, rabbits, donkeys, ducks, calves, lambs, miniature pony, mule, elk, buffalo, llama, miniature potbellied pigs and a small herd of longhorns and all types of horses and more! We will have 4 horned Jacob sheep coming soon. Wheat fields, wide open spaces and beautiful prairies with coulees are common sights. This is a rugged area for riding and exploring. There is also great hiking and rock hunting in this area.

Our ranch vacations are very relaxing and a great chance to see how Montana ranch life really is. Guests have a lot of free time to explore or catch up on a good book. While staying on the ranch, we may have local people from our area come to the ranch to see animals or to shop the country store. You are invited to help out or watch the morning and evening chores.

The meals are home cooked and are served with the family and hired help. Evenings are spent visiting on the front porch. Campfires can also be started to enjoy a roasted marshmallow or s'mores. (We may even sing you a cowboy song!) Beautiful sunsets are enjoyed.

Cabins on the main ranch are a bunkhouse which has 2 full beds and 4 single bunks. A shower cabin with toilet is shared by your group. (We will be doing some changes to our shower cabin in 1999.)

Our place is located 30 miles from the closest town, Malta. We have very few neighbors so you are able to enjoy alot of peace and quiet!! Waking up to a big blue Montana sky and fresh clean air is something you won't want to miss. Most mornings, our wake-up call is a rooster and a donkey!! Our wildlife consists of white deer, mule deer, antelope, coyotes, foxes, prairie dogs and all kinds of birds. Wild flowers are found on the wide open prairies while riding horses. Daily rides range from one to two hours. A limited time is spent chasing cattle or just looking through the herd to see if everything is okay. If timing is right, guests can help with some of the round-ups.

Another adventure on the ranch, is what we call a **Back Country Trip.** We travel about 14 miles (by Suburban) to a remote coulee, with cabins that have NO power. You will feel like you are stepping back in time! Here you can see Indian teepee rings and graves. A sack lunch will be taken along for the day, then we come back to the main ranch to retire.

We are not a dude ranch but an authentic working family ranch. we do not hire extra help for recreation. Your stay would be like visiting friends or family. We try to plan our day according to your interests! Our springs are busy with calving and lambing. Fall brings hunters. We offer pheasants, ducks, antelope, mule deer, white tail deer and coyote.

We do not allow alcohol on the ranch.

Address: HC 65, Box 6180, Malta, MT 59538
Telephone: 406-658-2111
Web: www.jimbalzotti.com
Location: 1-1/2 hours from nearest airport
Guest Capacity: 8

Season: May through September
Policy on Children, Pets: Children welcome, adults are responsible at all times. No pets in cabins.
Rates: $$
Credit Cards: None

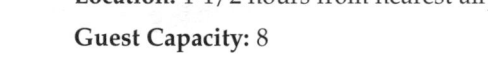

MONTANA
BEAVER CREEK TRAIL RANCH

WHAT DO YOU MEAN I'M NOT A HORSE?

MONTANA

BONANZA CREEK COUNTRY
MARTINSDALE, MT

Since the 1870s, Montana ranch hands have roamed the range and worked cattle along the banks of Bonanza Creek. A century later, very little has changed . . . the only difference is that now there is a place at the table set for *you*.

Bonanza Creek Country welcomes guests who are seeking seclusion, comfort, and a chance to do what they want. If you're looking for a vacaton with a relaxed atmosphere, Bonanza Creek has it! An escape from pre-set schedules. And because there are only eight to twelve guests at any given time, you're sure to get more personal attention than at a resort that can accommodate more guests.

This is beautiful country and the views are breathtaking. Bonanza Creek Country lies between the legendary Crazy Mountains and the picturesque Castle Range. On a clear day, you can see up to seven different mountain ranges. You'll want to bring along your camera and plenty of film. There are plenty of photo opportunities in this land and you'll want to save them for posterity.

The ranch's horses are a pride and joy. There's a horse to match each rider's size, age, and riding ability. There are two group rides every day, and the groups are limited to eight riders. The rides themselves are varied enough for any ability. There are new horseback rides every day, most half a day. One day you might ride through open meadows covered in wildflowers; another day explore the canyons and the secrets they hold. Wend your way through the wooded mountainsides. An all-day ride is offered once a week. You'll head out to see the Indian pictographs, high in the Castle Mountains, stopping to eat lunch on the trail.

Bonanza Creek Country is a family-operated working ranch. The ranch is home to about 1,200 head of cattle, and that means lots of work. Intermediate and better riders are invited to join in doing whatever might be required — putting out salt, checking for sick or injured cattle, moving the cattle from pasture to pasture, gathering bulls. It's your chance to create some experiences that will fit into your "best times" treasure chest.

When you're not on a horse, you might try fishing for trout in the ranch's private pond, mountain biking, watching for the abundant wildlife in the area, or taking a nature hike.

The accommodations at Bonanza Creek Country are the finest. The lodge and the three private cabins are crafted of sweet-smelling cedar logs. The lodge also contains the dining room, where you'll also enjoy the best home cooking this side of the Crazies. Your appetite may surprise you (it's the mountain air), and the meals served family style in the lodge are prepared to satisfy and please — from hearty country breakfasts to homemade desserts daily to sizzling steaks off the grill.

The cabins are situated so as to ensure privacy and provide the best views possible. The Cowboy cabin, decorated in (what else?) cowboy style, sleeps eight and has a bedroom, bathroom, living area, kitchen area, and loft. The Indian cabin has a Native American decor and sleeps four and has a bedroom, bathroom, living area, and porch. In the duplex, which has a separate entrance to a bedroom/bathroom unit with a connecting porch, you can choose either the Mountain Man or Antique side. For the more adventurous, there are a sheepwagon and a teepee.

Bonanza Creek Country provides an Old West adventure in the Big Sky country. It's a place to relax and get away from it all — a place to dream under starlit skies.

 Address: Lennep Route, Dept. BP, Martinsdale, MT 59053

Toll-free Telephone: 800-476-6045; **Telephone/Fax:** 406-572-3366

Location: 100 miles northeast of Bozeman, 135 miles northwest of Billings

Guest Capacity: 16

 Season: June through August

Policy on Children, Pets: Children welcome. No pets.

Rates: $$ American plan. Children's rates.

Credit Cards: MasterCard, Visa

MONTANA

C-B CATTLE AND GUEST RANCH
CAMERON, MT

MT-C-B Cattle and Guest Ranch This family-owned cattle ranch is in the famous Madison River Valley. Fly fishers know this river for its trophy brown and rainbow trout. C-B was established in 1971 by Mrs. Cynthia Boomhower of Palm Beach, Florida who, as a young girl, fell in love with the West and dude ranching. Today, her daughter, Sandy, and son-in-law, Chris, are the hosts. The ranch encompasses 21,000 acres and raises 200 head of Charolais crosses and a small herd of longhorn cattle.

At 5,000 feet elevation, the ranch lodge and cabins are situated where Indian Creek comes out of the Madison Range which rises to 11,000 feet behind the property. You'll have a great feeling of Western nostalgia as you pass under the C-B Ranch gate and experience the grandeur of Montana Big Sky Country.

There are three large double log cabins each with two double beds and a fold-out couch with private entrances. All cabins have private baths and open fireplaces or one Franklin stove with firestarter. Furnished with real Navajo rugs, clip-on reading lights, candles, and fresh flowers, C-B is very private, quiet and relaxing - the perfect place to escape from the fast-pace life. Additional electric heat and air-conditioning (seldom needed) is in all cabins. Just come put your feet up and relax.

There are no planned activities at the C-B Ranch. Traditionally, the men fish in the morning, noon and night, and the ladies enjoy horseback riding, walking and reading. Today, more women are fly-fishing, and so they should - fishing in this part of the country is superb. If you want to fish the famous Madison River, Sandy will put you in touch with one of the local guides (extra). You may also fish the waters of the Henry's Fork.

Horseback riders are assigned their own horses for the length of their stay. Guided rides go out daily. Because the ranch is small, riding is flexible but always guided. Our wranglers, with their Montana cowboy stories, will make your visit more than just an ordinary vacation. They can guide you into some of the most beautiful Montana backcountry that you will ever see and, up until now, have only heard about. Ask about the Indian Creek Waterfall ride. There is a weekly lunch ride with barbecue, weather permitting.

White-water rafting can be arranged. Your entertainment at night is the howls of coyotes and you may occasionally see deer, antelope, coyotes, elk, moose and bear. Most of all, families usually do things together. A kiddie wrangler is available but the ranch is better for older children, 10 years of age and up.

All meals are prepared fresh. There's always plenty of food and Sandy is proud of the fare - meat, potatoes and fresh salads. Also, fresh berries, melons, vegetables, cakes, pies and Sunday turkey dinner. Occasionally, wine is served with dinner but BYOB.

The horses, the blue Montana sky, the delicious food, the surperb fishing and the peace and quiet will make your vacation at the CB Cattle and Guest Ranch an experience you and your family will never forget.

Address: Box 148, 1106 Bear Creek Loop, Cameron, MT 59720. Winter: 4321 Orange Hill, Falbrook, Ca 92028

Telephone: 406-682-4608; **Toll Free:** 888-723-FISH. Winter: 760-723-1932, **Fax:** 760-723-1932

Internet: www.cbranch.com

Location: 23 miles from Ennis

Guest Capacity: 14

Season: Mid-June through end of August

Policy on Children, Pets: Older children welcome. No pets.

Rates: $$

Credit Cards: None

(SEE PAGES 262 FOR COLOR PHOTOS!)

MONTANA

CIRCLE BAR GUEST RANCH
UTICA, MT

The Circle Bar is a guest ranch located in the heart of Montana. In operation as a cattle ranch since 1890, the Circle Bar still provides an authentic Western experience complete with Spring and Fall cattle drives as well as routine checking and moving of the beef herds. In 1938, the ranch began accepting guests and building cabins along the clear waters of the Judith River. Today, the ranch is a pleasant blend of the Old West and modern conveniences. Nestled in the foothills of the Little Belt Mountains, the ranch and the bordering Lewis & Clark National Forest provide ideal terrain for horseback riding which varies from the cattle country's vast rolling hills to rugged mountain trails. The quality-bred horses are suited to each individual's ability and are chosen for their responsiveness and temperament. The Judith River provides excellent fishing.

The ambiance is definitely western. Our main lodge is graced by over-sized picture windows, a two-sided native stone fireplace, and high ceilings which reflect the spaciousness of the Big Sky Country. There are nine individual, one to four bedroom, log cabins with private baths. Most cabins have fireplaces or wood stoves and living rooms with wet bars. There are also four large, well-appointed suites in the main lodge with daily maid service. Laundry facilities are available.

Your stay can be as relaxed and unhurried or as structured and adventure-packed as you'd like. The Circle Bar tailor-makes each vacation in order to fulfill individual needs and exceed every expectation. The riding program is flexible. The accomplished rider will appreciate the quality of the fine string of horses while those who have never ridden will be paired with a trail-wise horse and taught the fundamentals. Four-wheel drive sightseeing, wildlife viewing and wildflower trips are available and a Middle Fork pack trip with advance notice. Ask about Blackfeet Indian Caves, Hole-in-the-Ground, and Elk Refuge rides as well as our exclusive Charlie Russell Ride and the Spring and Fall Cattle Drives. Fishing on the Judith River is a favorite. Back on the ranch, there is hiking, photography, nature walks, horseshoes, outdoor heated pool and indoor hot tub, and basketball and volleyball in the barn loft.

Children of all ages are welcome and invited to participate in unstructured children's activities. Family participation in all ranch happenings is encouraged. Team-drawn wagon rides, square dances, campfires sing-a-longs, ping-pong/pool table in the recreation room, mountain biking (B.Y.O.bike), extensive library, and, of course, solitude

You are served as part of the family in the dining room: hearty breakfasts, scrumptious lunches and delightful dinners. Homemade cookies, pies, breads as well as fresh vegetables and herbs from our garden, along with steak fries, picnics and gourmet meals bring rave reviews from our guests. You are ensured exceptional service and a memorable vacation at the Circle Bar Guest Ranch. Satisfaction guaranteed!

Address: HCR 81, Box 61, Utica, MT 59452

Telephone: 406-423-5454; **Toll Free:** 888-570-0227; **Fax:** 406-423-5686; **Web** www.jimbalzotti.com

E-Mail: cbr@circlebarranch.com

Location: 85 miles east of Great Falls airport

Guest Capacity: 35

Season: June 1 through September 15

Policy on Children, Pets: Children welcome. No pets.

Rates: $$

Credit Cards: None

MONTANA
CIRCLE BAR GUEST RANCH

FOR A UNIQUE ADVENTURE — COME VISIT OUR BUFFALO.

FALL ASLEEP TO THE SOUND OF OUR BABBLING BROOK.

MONTANA

COVERED WAGON RANCH
GALLATIN GATEWAY, MT

The Covered Wagon Ranch has been accepting guests since 1925. The ranch is operated as a western mountain ranch, accepting only a limited number of guests - 24 at maximum. Over the years, the ranch has become a tradition as experienced by the fourth generations that now enjoy the hospitality of the uniquely American vacation - the Dude Ranch. We tailor activities geared to each individual's desire just as you would when entertaining personal houseguests.

Located just three miles from the Northeast boundary of Yellowstone National Park, wildlife abounds. Elk, mule deer, moose, black bears and grizzlies are seen periodically on our trail rides. From Paradise Valley, through the Park, over the Gallatin Mountain Range and through our backyard, to Madison Mountain range and beyond, migrating mammals have traditionally called this Gallatin country, "home". There are many photo opportunities to capture the animals in their natural surroundings. We hold an outfitter permit to take day rides into the Park. This is a great way to see the backcountry of Yellowstone National Park.

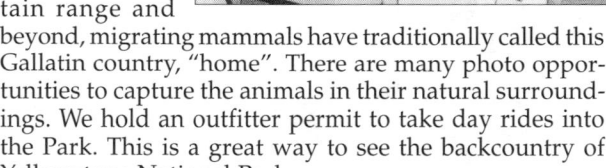

We are very proud of our string of horses. The breeding is mainly Quarter horse mixed with Morgan, Appaloosa, Percheron and Thoroughbred. Will King, the owner of the ranch, has several brood mares that he breeds. Every spring there are a few more horses to begin working with. Everyone has their favorite horse and looks forward each year creating new memories and adventures with their trusty mount.

An average day at the Covered Wagon ranch begins with starting the wood stoves in each cabin an hour before breakfast is served. The day's activities and plans are hashed out over breakfast. We know that plans change overnight so rides are not set in stone until the morning. Usually, one all-day ride will go out. Some folks like to ride in the morning and then, in the afternoon, try their hand at fishing the blue ribbon streams that surround the ranch in the Gallatin canyon. Afternoon rides are planned at lunch time. After a short snooze or a game of horseshoes, it is back down to the corrals to begin another adventure on horseback. When the day's rides come in and the horses are unsaddled, fed and looked after, the wranglers and kids (of all ages) meet at the swimming hole that is right on the ranch. The Taylor Fork River provides the best afternoon relief to a day in the saddle.

The dinner table is always abuzz with conversations of the day's activities, follies and adventures. Kids will migrate outside after dinner for a scavenger hunt or a game of horseshoes. When the campfire lights up and the guitar picking starts, the adults soon gather outside and sing-along with kids and wranglers. With the campfire embers glowing, the stargazing begins. After a few good attempts at trying to name all of the constellations, the guests and crew say goodnight and head to bed for a good night's rest to prepare for another day of enjoying Montana on horseback.

Come on out and experience a week full of adventures at the Covered Wagon Ranch and be our guests.

Address: 34035 Gallatin Road, Gallatin Gateway, MT 59730

Telephone: 800-995-4237; **Fax:** 406-995-2061
Web: www.gomontana.com/coveredwagon or www.jimbalzotti.com

E-Mail: coveredwagonranch.com

Location: 34 miles north of West Yellowstone

Guest Capacity: 24

Season: Mid-May through mid-October, for horseback riding.

Policy on Children, Pets: Children welcome. Must be at least 6 years old to ride. No pets.

Rates: American Plan

Credit Cards: Mastercard and Visa

(SEE PAGE 263 FOR COLOR PHOTOS!)

Montana
Covered Wagon Ranch

THE TAYLOR PEAKS ON A CLEAR FALL DAY.

A MOOSE ENJOYING THE LUSH VEGETATION IN
YELLOWSTONE NATIONAL PARK.

MONTANA

DJ Bar Ranch
Belgrade, MT

Come visit Montana and enjoy the Big Sky country! At the DJ Bar Ranch, you'll stay in a log home in a picturesque country setting. Explore the flora and fauna of the largest continuous stretch of wilderness in the continental United States. Savor the fabulous sunrises and sunsets of the Big Sky country where the horizon seems to stretch forever in every direction, interrupted only by the mountains.

Your horses will be in horse heaven in their own 20-acre pasture (fenced in high tension smooth wire), with an additional corral and loafing shed. If needed, a box stall and run are available at the main ranch about a mile away. If ol' Dobbin throws a shoe, owner Jehnet Carlson, a certified AFA farrier, can replace it.

There are many trail heads located within an easy day-trip distance to the surrounding mountain ranges. The nearby Gallatin National Forest has hundreds of miles of picturesque trails. There are a number of spectacular destinations, including high alpine lakes.

Spring rides in the area include sightings of baby elk, deer, coyotes, and foxes. In June, nearby Bozeman hosts the College National Finals Rodeo. On summer weekends, the Backcountry Horsemen of Gallatin Valley gather for trail rides in the wilderness. "Poker Rides" are put on by many area horse clubs. Fall is greeted with the sounds of bull elk bugling in the mountains. The Spanish Peaks are a pristine wilderness area, with majestic views of craggy peaks inhabited by bighorn sheep and mountain goats. You can travel the opposite direction for a high desert ride.

You might plan a day of trout fishing on your own private piece of the East Gallatin River. If you bring your fishing rod along, you'll have your choice of rainbow, brook, brown, and even golden trout. Experience the magnificence of the rare Arctic grayling on your line.

Relax under the stars in the private hot tub while planning to enjoy the endless recreation the area has to offer, including golf, hiking, mountain biking, hunting, water sports, and exploring the area's history. Thrill to the springtime sight of bald eagles soaring overhead and the dances of the sand hill cranes. Fall asleep in the evening to the coyotes' lullaby. Plan a float trip down one of the many rivers in the area. Yellowstone National Park, Lewis and Clark Caverns, the Museum of the Rockies, and historic Virginia City are within easy driving distance. The Madison Buffalo Jump, a historic Native American archaeological site, is nearby.

In winter, you will be conveniently close to two excellent ski areas, Bridger Bowl and Big Sky. Yellowstone National Park offers snowmobiling opportunities galore.

As you treasure the panoramic views of the Gallatin Valley and surrounding mountains from this hillside retreat, you'll feel like you've captured your own little bit of paradise. Rich Western decor accents this luxury home with its three bedrooms, two baths, satellite TV, hot tub, and laundry facilities. The house is a combination of quaint and homey, filled with exquisite art and antiques. The kitchen is complete with everything you'll need to prepare your own meals, including microwave, dishwasher, and a barbecue on the deck. The area also offers many fine restaurants for gourmet dining, and a selection of menus is provided at the guest house.

Experience the most relaxing vacation you'll ever have, away from the stresses of city life and along the many happy trails of Montana's Big Sky country.

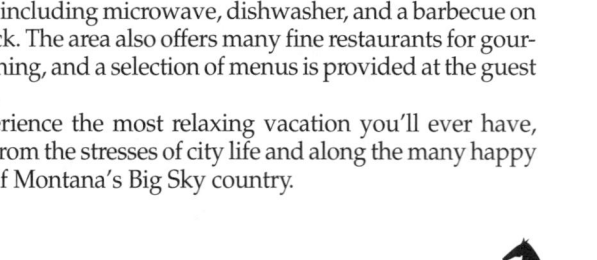

Address: 5155 Round Mountain Road, Dept. BP, Belgrade, MT 59714

Telephone: 406-388-7463; **Fax:** 406-388-7241
Web: www.jimbalzotti.com

Location: 9 miles from Belgrade

Guest Capacity: 12

Season: Year round

Policy on Children, Pets: Children are welcome. No pets.

Rates: $1,200 per week for house. No meals included. Horse stabling extra.

Credit Cards: None.

MONTANA

JJJ WILDERNESS RANCH
AUGUSTA, MT

The JJJ Wilderness Ranch is a family-owned and operated guest ranch about 100 miles south of Glacier National Park on the eastern front of the Rocky Mountains. Located above the beautiful Sun River Canyon and Gibson Lake, the "Triple J" specializes in family vacations emphasizing horseback riding, trout fishing and hiking.

Hospitality is a tradition of the West we come by naturally. The Barker family has shared this western tradition for over 2 decades with a limited number of guests. Not being too big allows us to get to know our guests on first name basis.

With over 60 head of mountain quarter horses suited to all riding experience levels, there are unlimited riding opportunities including trail and meadow riding plus an optional overnight packtrip. The famous Bob Marshal Wilderness, a million acres of pristine, roadless wilderness, is just a short horseback ride away. The week begins with horsemanship and riding lessons assisting and instructing as needed or desired. Daily activities are planned with you each morning, allowing flexibility for personal interests.

For the youngsters, our Kiddie Wrangler supervises fun rides and games and teaches basic horsemanship. Exploring, interpretive nature hikes, swimming in the heated pool, and playing in Fort Mortimer offer many activities for the day to fly by.

Out of the saddle, you can enjoy fly fishing instruction at the ranch trout ponds to wet your appetite for fishing rainbow, cutthroat or brown trout on the Sun River or Gibson Lake. Guided interpretive tours, non-technical mountain climbing of Mortimer Peak, endless hiking opportunities and wildlife watching for elk, deer, bear, bighorn sheep and golden eagles, provide an ultimate outdoor experience. Western cook-outs and evening campfires with authentic cowboy poetry and singing provide a relaxing end to a busy day at the JJJ.

The ranch has six separate cabins that sleep from two to six and are nestled among the aspen and spruce. At nighttime, the crisp mountain air will coax you into a deep, comfortable sleep.

All meals are home-cooked and served family-style in the main lodge, featuring homemade goodies and vegetables from the garden. Traditional steak barbecues and outdoor cookouts highlight your stay.

For those who enjoy "roughing it", there's high adventure on 5 to 8-day horse pack-trips into the heart of the remote Bob Marshal Wilderness. This is an experience like no other. Fully-outfitted, your trip into this magnificent, unspoiled country by sure-footed packmules and mountain savvy horses will reveal awesome beauty of rugged landscape, pristine lakes and streams, elk, deer, bighorn sheep, mountain goats and black bear. It's a fisherman's paradise with native cutthroats and rainbow trout. You'll experience the challenge of adventure, the fellowship of guests and crew, and appreciation of the horse.

Providing an ultimate outdoor vacation experience is a goal the Barker family is happily committed to. Our family invites you and your family to a vacation adventure of a lifetime in Montana's Rockies.

Address: Box 310, Augusta, MT 59410

Telephone: 406/562-3653 or 406/562-3651

Web: www.jimbalzotti.com

Location: 80 miles West of Great Falls, 25 miles Northwest of Agusta

Guest Capacity: 20

Season: June — September

Policy on Children, Pets: Children welcome

Rates: $$

Credit Cards: None

(SEE PAGES 265 FOR COLOR PHOTOS!)

MONTANA

HARGRAVE CATTLE AND GUEST RANCH
MARION, MT

Live the legend on a historic 87,000-acre cattle ranch! While you'll be quite comfortable as a Hargrave guest, with down comforters on your bed and a cat curled up on the antique rocker, you'll not be visiting a dude ranch. Hargrave is a working cattle ranch in the tall timber country of the northwestern Montana Rocky Mountains. You'll become part of a small group of special people who learn what this way of living offers to harried lives. A ranch guest list of 15 means individualized attention, whether at the pool table, the campfire, or the dinner table. All year round, people join the ranch crew and play or work, depending on their individual fantasies.

Spring calving in March-April has guests out in the calving pastures helping the cowboys ear-tag and doctor the newly birthed calves and sort them to fresh pasture. Joining the 2 a.m. herd check can mean seeing the Northern Lights in all their splendor. May's spring days bring green grass, fence building, and neighborhood branding parties. Early June is our "turn out" time, when cattle are pushed to their summer ranges. These are true cow punchin', long-saddle days.

Summer means checking and controlling the cattle on horseback, but this is also a time when there's time for plain ole fun. Like rides to the ridge tops, overnight camp-outs, outdoor dinners, and plain ole star-gazing on the front porch. There are five lakes within 5 miles to settle in on with the canoe or rod and reel, or to cool off in their crystal waters after a day in the saddle. Try your fly rod in the 2 miles of the Thompson River that run through the ranch. You might want to try a river float on a nearby blue ribbon stream or tingling whitewater rafting.

Rainy day? No problem. It's a great opportunity to take some time to see Kalispell's Western stores, cowboy outfitters, galleries, custom jewelry, authentic Blackfoot artwork, or the bronze foundry.

Autumn sees guests joining the ranch to herd the bunches of cattle down to the home valley. The wranglers locate the cattle out in the timbered mountain valleys; everyone trailers out and then pushes the cattle towards the main ranch. Miles of green pine trees, meandering streams, and sharp ole cows quite knowledgeable about their territory make this a challenging season. Fall in the Rocky Mountains means less folks about, hungry trout, and glorious colors.

When a blanket of white covers the valley, the ranch welcomes bed and breakfast guests. The cows are close in the meadow, and you can watch or join the cowboys in daily feeding. The meadows and rolling upland pastures make for fine cross-country skiing, or challenge yourself at the Big Mountain Ski Resort just 1.5 hours away in Whitefish. You can even "take a day off" to settle around the fireplace with book from the library.

Whatever the season, a day trip to spectacular Glacier National Park is a must. For decades, this alpine wonderland has thrilled millions of people from around the world. Whether it's the heart-stopping motor drive up the "Going to the Sun Road" during the summer or the hush of the cedar forest alongside an ice-rimmed stream in winter, you'll not soon forget the sights.

Hargrave offers a unique opportunity to really live the legend that runs through the core of American history and myth. Leo and Ellen were drovers and outfitters on the Great Montana Cattle Drive of 1969 and are committed to sharing that way of life with their guests. Come be a partner in the real West — you'll always carry the memories.

 Address: 300 Thompson River Road, Dept. BP, Marion, MT 59925

Telephone/Fax: 406-858-2284
Web: www.jimbalzotti.com

Location: 40 miles west of Kalispell

Guest Capacity: 15

Season: Year round

Policy on Children, Pets: Special riding program for kids in July-August. No pets.

Rates: $$. American plan.

Credit Cards: MasterCard, Visa

(SEE PAGES 264 FOR COLOR PHOTOS!)

MONTANA
HARGRAVE CATTLE AND GUEST RANCH

COME TO A REAL WORKING CATTLE RANCH
AND LIVE YOUR DREAMS.

WE LIVE AND RIDE IN UNSPOILED, BEAUTIFUL COUNTRY.

HARGRAVE CATTLE AND GUEST RANCH

MONTANA

HORSESHOE HILLS GUEST RANCH
SEELEY LAKE, MT

Located in a secluded meadow surrounded by foothills and mountains, you will find Horseshoe Hills Guest Ranch. Mount Morrill to our north, the Horseshoe Hills to our west and a range of foothills to our east surround and protect the 220-acre ranch. The Blackfoot Clearwater Game Range to our south is a haven for wintering elk and other wildlife. It is also the elk's spring calving area - cows and calves can be seen 'til mid summer. Bordered by state and national forest, the Bob Marshall Wilderness is accessible directly from the ranch. Located between Highway 200 and scenic Highway 83, our location is 15 minutes from Seeley Lake where most services can be found.

The ranch has premier accommodations for horse and rider. The ranch rooms, with private entrances, can accommodate 8 guests. Appropriately decorated, the bear room sleeps 2 while the cowboy room hosts 6 guests. Our 4 campsites located in a wooded area have full hookups. While folks are settling in, our horse guests can enjoy a lush pasture with a mountain creek flowing, corrals and covered outdoor stalls, or our indoor heated stalls in the barn.

The riding opportunities are endless! The miles of trails behind the ranch can be explored on your own or one of our ranch staff would be glad to guide riders on half or full-day trips. The Bob Marshall Wilderness has many well-marked trails taking you into beautiful unspoiled areas. Trailering to a trailhead expands the range of riding possibilities allowing riders to view more of our pristine wilderness.

Give your horse the day off and take in some area attractions in the Swan Valley. Drive scenic Hwy. 83 between the Mission & Swan Mountain Ranges to visit Glacier Park or the Flathead Lake area. The area ghost town, bison range, Fort Missoula and Missoula offer a variety of activities to suit everyone's tastes. Summer highlights are local rodeos, pow wows, cowboy poetry and local artisans.

The ranch raises Quarter Horses, Paints and Scottish Highland Cattle. We are a breeding, boarding and training facility. Occasionally, our outdoor arena has horse clinics, 4-H events or shows scheduled. Our area also lends itself to endurance riding.

Horses are required to have current vaccines and health papers. Obedient dogs are welcome but not in the guest rooms. Kennels are available. A vet and farrier are on call. Hay and grain are available at the ranch.

We are open year-round for winter activities too! Snowmobiling, (bring your own or rent), is next to none on the mountain powder or groomed trails. Snowshoeing or cross-country skiing are also popular. Local dog sled teams can also show you the area. Ice fishing on area lakes appeal to some guests.

Visit us at the ranch. Kick up your heels or prop up your boots. Ride or just relax in Big Sky Country. The Last Best Place - Montana.

 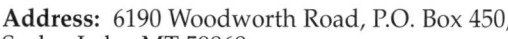

Address: 6190 Woodworth Road, P.O. Box 450, Seeley Lake, MT 59868

Telephone: 406-677-2276; **Fax:** 406-677-2266
Web: www.horseshoehills.com

E-Mail: slk3833@montana.com

Location: 15 miles Southeast of Seeley Lake

Guest Capacity: 8, plus camping

Season: Year-round

Policy on Children, Pets: Children welcome. Pets are allowed, but not in rooms. Kennels available.

Rates: $

Credit Cards: Mastercard, Visa and Discover

MONTANA
HORSESHOE HILLS GUEST RANCH

SNOW-CAPPED MOUNTAINS SURROUND OUR RANCH.

A VIEW FROM THE TIMBER.

Montana

The Leffingwell's G bar M Ranch
Clyde Park, MT

If you are thinking about sampling life on a Montana cattle ranch, the G bar M may be just what you are looking for.

A family-owned and -operated cattle and horse ranch, the G bar M is a living history of Montana ranch life in the sage-covered foothills of the Bridger Mountains of southwestern Montana. The ranch still maintains the standards and quality of life passed down from generations of their family, who homesteaded the land nearly one hundred years ago. The Leffingwells began taking guests in 1934 and found the cattle and guest businesses worked so well together that they have been at it ever since. They are proud to share their Western heritage and hospitality with guests.

The G bar M accommodates only about 10 to 15 people at a time. Families, singles, and couples are all made to feel welcome. You're only a stranger for the first few minutes; then you are among friends. New guests as well as those who come back year after year are often at a loss to choose what they enjoy most about their stay at the ranch. The home-cooked meals, oven-fresh breads, and delightful desserts add to the relaxed atmosphere and camaraderie around the kitchen table. The riding can include regular ranch riding chores such as moving the cattle to mountain pasture in the early part of the summer; checking the condition of fences, water holes, or grass; and hauling salt with a pack horse to the cattle on the high mountain pastures.

Although riding is the main activity, there are many other things to do including fishing, hiking, photography, and best of all just taking it easy in a beautiful, relaxed atmosphere. The Leffingwells are always happy to give lessons as well as pointers in riding and horsemanship, or you might want to learn some basic roping skills. Depending on the time of year you go, you might be interested in watching the hay being put up, lumber being made at the sawmill, or horses being shod at the blacksmith shop.

The range land is the grazing area for the cattle herd. It is also the habitat for deer, elk, moose, a multitude of bird species ranging from golden eagles to hummingbirds, as well as a number of predators. But there are no grizzly bears, rattlesnakes, or poison ivy, and the ranch does just fine without them, thank you! For those who enjoy fishing, there are always trout to be caught out of the two creeks that wind their way through the ranch. The wildflowers provide a pleasing variety of color to the breathtaking mountain scenery that changes not only with each bend in the trail or ridge-top vista but also from hour to hour during the day and from week to week throughout summer.

If you are interested, the Leffingwells will tell you not just about the history of the valley but about the people they've known who made that history happen . . . or about the geology of the mountains and how and why the valley came to be . . . or about the local ecosystem that they fight hard to protect, having seen the effects of change swing from the buffalo to the cellular phone. You may be interested in seeing and photographing the over two hundred varieties of wildflowers, the many species of birds, and the wild game. The ranch is a private game preserve open to a limited amount of federal, state, and private wildlife research, but not to hunting.

The Leffingwells are justly proud of their beautiful country, good food, good horses, and returning guests.

Address: Box 29, Dept. BP, Clyde Park, MT 59018

Telephone: 406-686-4423

Location: 35 miles northeast of Bozeman

Guest Capacity: 15

Season: May 15 to September 15

Policy on Children, Pets: Children are the responsibility of parents. No pets.

Rates: $ - $$ American plan. Minimum stay one week (Sunday to Sunday).

Credit Cards: None. Personal checks accepted.

Montana
The Leffingwell's G bar M Ranch

WELCOME!

MONTANA

O SPEAR GUEST RANCH, LTD.
BROADUS, MT

If you like a warm family atmosphere and good home cooking, book a vacation with us on the Little Powder River in the Big Sky Country of the plains of southeastern Montana. Our family is the third generation to live on the original homestead of 1890. We are a working ranch and the family does all the work!

The accommodations are great - at least, we've been told that many times! The large cabin accommodates up to fifteen guests and has Primestar TV, telephone service, a pool table, and is tastefully decorated with many family antiques and heirlooms. Photos of yesteryear adorn the walls. The smaller cabin, The Bungalow, accommodates two to three people. It is very warm and cozy - a smaller replica of the large cabin.

All meals are homecooked and served family style. If you go away hungry, it's your own fault! Guests enjoy country fried chicken, ranch raised beef served many ways, homemade bread, rolls and cinnamon rolls, hearty stews, chili, etc. The list goes on. Many of the ladies request recipes to take home with them.

Most of the guests who visit want to spend most of the time horseback riding. The cowponies are matched to the riders for size and temperament. The guest wranglers give riding instructions and make sure each rider feels comfortable with their mount. Then they spend their time helping us move cattle to pasture and herding the sheep to new feed grounds and riding the beautiful countryside viewing all the wild game and pretty wildflowers.

Other activities, depending on the season or the weather for the day, include: fishing (catfish in June), hunting, branding time, assisting with general ranch chores, hay rides, hiking to see abundant wildlife (deer, antelope, coyotes, bobcats, fox, prairie dogs, rabbits, raccoons, badgers, porcupines, etc.) and cookouts.

In the spring, we brand, irrigate the hay meadows and get all the livestock moved to summer pasture. Summertime is busy checking fences and getting the hay put-up for winter use. Fall brings beautiful leaves on the trees and time to get all the livestock moved home again and the offspring shipped to market. Wintertime is spent feeding the hay we put up in the summer and enjoying wintertime activities such as ice skating, cross-country skiing, snowmobiling, sledding, and sitting in front of the warm fireplaces!

Guests from many countries and all across the United States have enjoyed our ranch in the heart of "Custer Country". We've hosted family reunions, Christmas vacationers, honeymooners, birthday parties and anniversary parties.

Located within a few hours of us are the Devils Tower National Monument, Historic Deadwood famous for its gambling casinos and rich history, Battle of Little Big Horn (Custer's Last Stand), and Mount Rushmore National Monument.

Local activities include both a historical and wildlife museum, the annual 4th of July Ranch Rodeo, and in August the Country Fair and the Powder River Let'r Buck Broncs and Bulls Rodeo.

After all the exciting events for each day, we return home to the porches to sit on the swings and enjoy beautiful sunsets and evening sounds.

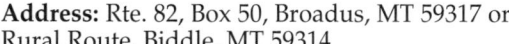

Address: Rte. 82, Box 50, Broadus, MT 59317 or Rural Route, Biddle, MT 59314

Telephone/Fax: 406-427-5388 or 406-427-5406

Location: 3-1/2 hours form Billings, MT or 3-1/2 hours from Rapid City, SD

Guest Capacity: 17

Season: Year-round

Policy on Children, Pets: Children welcome. Small children may not ride horses. No pets

Rates: $ - $$

Credit Cards: None

MONTANA

MONTURE FACE OUTFITTERS
GREENOUGH, MT

Monture Face Outfitters provides horse and mule pack-trips in the magnificent Bob Marshall Wilderness in northwestern Montana. This protected area encompasses over 1.5 million acres of lush rich meadow, high alpine passes and gorgeous rivers. The fishing is legendary, the scenery breathtaking. Wildlife, including elk, deer, black bear, Grizzlies, mountain goats and big horn sheep are commonly viewed during the trips. Wildflowers are abundant, carpeting hillside after hillside with a tapestry of color. This area is vast, remote, pristine and unspoiled. Here you will find the 'real' big sky country.

This is a small, quality-oriented business. Tom, Tim and Valerie are the knowledgeable owners of Monture Face Outfitters. By staying small, they do not need to hire outside help and so do all of the outfitting themselves. While the Forest Service guidelines under which they operate allow them to take groups up to 15, they rarely do. The ideal number of people is 6 to 8 and they will take as few as 4. Personalized attention is of utmost importance.

They offer a variety of different wilderness adventures. Choose a comfortable three-day trip into their complete, clean and already set-up base camp. Try a five or six-day half base camp and half roving/progressive trip, and sample our forefathers experiences. For the truly adventurous, sign up for the ultimate 10-day roving excursion deep into the 'Bob', riding over 100 miles and camping in new places each night. They can also customize a trip just for you but you will need to book early to insure your itinerary and dates. Monture Face Outfitters also provides big game hunts and strictly adheres to ethical, fair chase methods.

The heart of a good outfitters operation is, of course, their animals. Only seasoned trail horses and mules in absolutely top-notch condition are used on their trips. The equines are treated like members of the family. For a real treat, try riding a mule, the surest footed, smoothest ride in the Rockies.

The pack trips are all inclusive. You arrive at the trailhead with your personal gear and sleeping bag then let Monture Face Outfitters take care of you with no hidden charges or extra costs. They also provide free transportation to and from the Missoula, Montana airport, arrange first and last night lodging and take you to the Monture Trailhead.

The meals are notable, even fantastic. No expense is spared in providing you with the freshest, choicest foods ever packed-in. The baked goods and non-stop coffee will leave you fully satisfied as well as the grilled meats, lots of salads, whole grains and Valerie's famous raspberry pancakes.

Imagine riding through rich, waist deep, mountain grasses on a strong and conditioned horse or mule. Stop at a hidden, babbling spring for cool refreshment and to take a snapshot of the towering mountain pass you'll soon ride over. At night, lie back on the soft, green grass and gaze up at a sea of stars so clear and bright, you'll think you can reach up and touch them. Let's go 'packin'!

Address: Box 27 Clearwater Jct., Greenough, MT 59836

Telephone: 888-420-5768; **Fax:** 406-244-5763
Web: www.montanaoutfitter.com
E-Mail: info@montanaoutfitter.com

Location: 35 miles east of Missoula, Montana

Guest Capacity: 8

Season: Late May through September

Policy on Children, Pets: Children 14 years or older only. No pets.

Rates: $$

Credit Cards:

(SEE PAGE 266 FOR COLOR PHOTOS!)

MONTANA
MONTURE FACE OUTFITTERS

DEEP IN THE HEART OF MONTANA.

OUR LUXURIOUS WILDERNESS BASECAMP.

Montana

Parade Rest Ranch
Yellowstone, MT

Located ten miles north of West Yellowstone, the Parade Rest Ranch is sheltered by the arms of the giant cottonwoods while Grayling Creek, gurgling past, calms your spirit. Do what you dream about - anything or nothing. Kick back with a book, walk through the serene countryside, ride horseback on mountain trails, soak in the hot tub, or flyfish in the early morning mist. Not a dream, a getaway. Casual, quiet, comfortable.

Decidin' on Ridin'? Flower-filled meadows, towering mountain peaks and streams and blue lakes will greet your eyes and glimpses of elk, moose, coyotes, sandhill cranes, and even bald eagles only add to the adventure. The peace and tranquillity spread serenity through your soul.

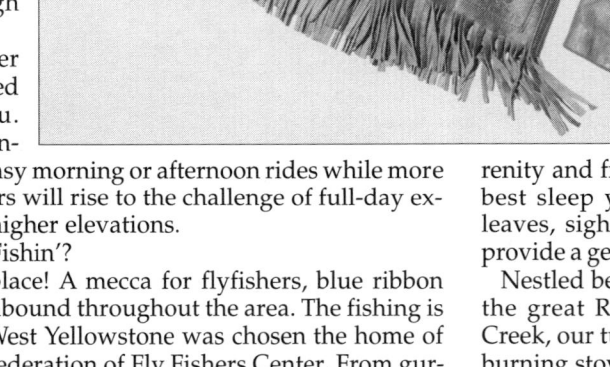

Our wrangler will be delighted to guide you. Novices can enjoy leisurely, easy morning or afternoon rides while more advanced riders will rise to the challenge of full-day excursions into higher elevations.

Wishin' for Fishin'?

This is the place! A mecca for flyfishers, blue ribbon trout streams abound throughout the area. The fishing is so great that West Yellowstone was chosen the home of the National Federation of Fly Fishers Center. From gurgling Grayling Creek right on the premises to the mighty Madison and famous Snake Rivers - there's always a place for you! Easily accessible from the Parade Rest, you can take them in by yourself, or we'll help you make arrangements with any one of a number of professional fly fishing guides in the area. Bring your favorite rod and other gear - nature will supply the rest!

In no time at all your mouth will start watering every time the dinner bell rings. The warm, friendly atmosphere of the dining room is well-matched by great home cooked, family-style meals. Forget the diet - it can't be done - not with so many treats to tempt your tummy!

All meals are included. Breakfast and dinner are generally served in the dining room. Lunch is packed to be taken with you for your daily activities. Outdoor steak barbecues are held mid-season among the aspens. You'll enjoy riding to and from your outdoor dinner via haywagon or horseback.

The quiet serenity and fresh mountain air are sure to encourage the best sleep you're likely to ever experience. Rustling leaves, sighing prairie grasses and the gurgling creek provide a gentle lullaby drawing you into restful dreams.

Nestled between the foothills of the Gallatin Range of the great Rocky Mountains and delightful Grayling Creek, our turn-of-the-century guest cabins boast wood-burning stoves, comfortable beds, cozy atmosphere and modern baths. You'll get so relaxed you may never want to leave!

Address: 7979 Grayling Creek Road, West Yellowstone, MT 59758

Telephone: 406-646-7217; **Fax:** 406-646-7202
Web: www.parade-rest-ranch.com
E-Mail: res@parade-rest-ranch.com

Location: 6 miles north and 1 mile west of West Yellowstone airport

Guest Capacity: 60

Season: May 10 through October 5

Policy on Children, Pets: Children welcome, there are no special programs. No pets

Rates: $$

Credit Cards: Visa and Mastercard

(See Page 267 for Color Photos!)

Montana
Parade Rest Ranch

YOU'LL LOVE OUR HOME COOKIN'.

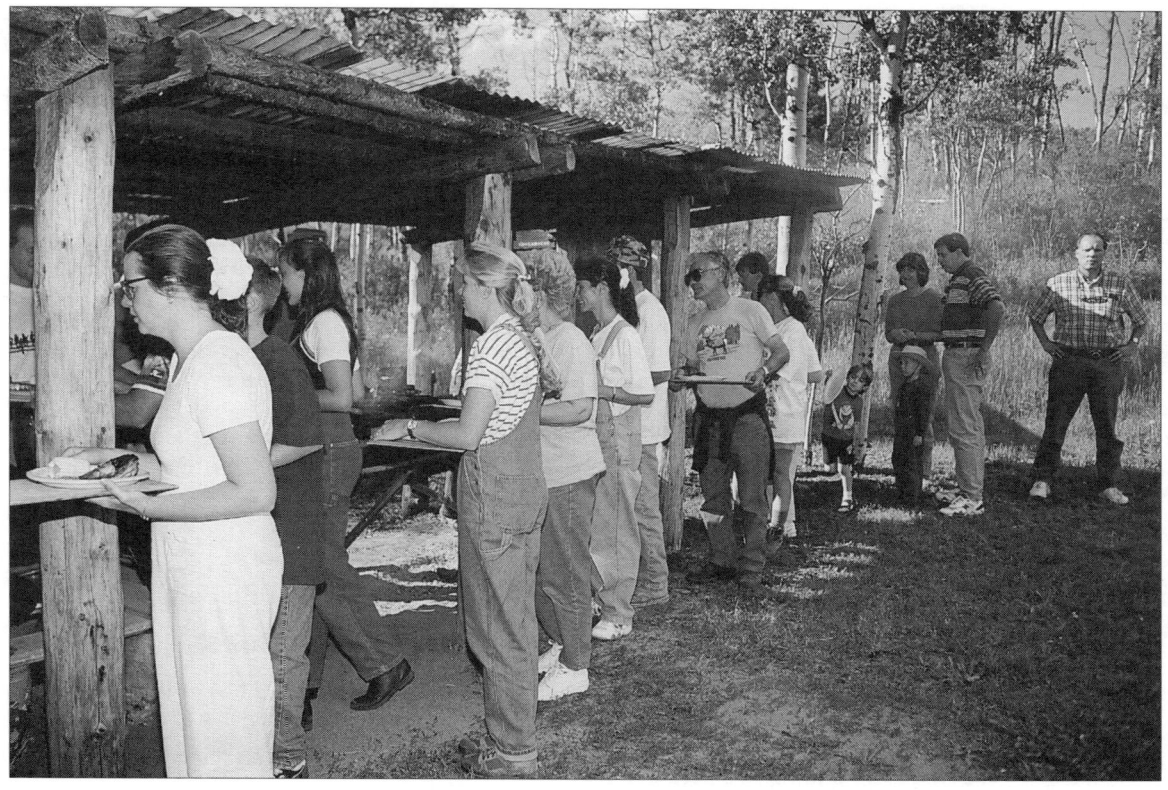

LEAVE ROOM FOR SECONDS!

MONTANA

POWDER RIVER WAGON TRAINS
BROADUS, MT

Powder River Wagon Trains & Cattle Drives and Ranch Vacations is located in extreme southeastern Montana. This is the heart of "Custer Country" — where many of the skirmishes leading up to the Battle of the Little Bighorn took place. The area is rich in the history of our Native American and pioneer forebears.

Powder River's "Premiere" wagon train and cattle drive takes place annually during the first full week of August. This 40-mile trip is an authentic Western adventure. You'll travel through high country plains, rolling hills, and pine-topped ridges — ending up six days later in beautiful downtown Broadus, Montana. Along the way, you're invited to ride horseback and help move a herd of longhorn cattle from summer pastures, like cowboys have been doing for a hundred years. You'll ride a cow-savvy ranch horse that has been carefully matched to your riding ability, whether novice or experienced. Or, choose to ride in one of the six or eight accompanying covered wagons.

On any given day in the "Land of the Big Sky," you are likely to see pronghorn antelope, whitetail or mule deer, elk, jackrabbits, eagles, and maybe even a herd of buffalo as well as ancient Indian teepee rings.

At night, you'll sleep in a cowboy teepee. Powder River's supply crew will set up your own personal teepee each night as well as deliver your luggage in advance from the previous night's camping spot. At the campsite, sanitary portable toilets and generator-powered showers are a welcome end to a long day in the saddle. Powder River believes in pampering its guests just a little!

Meals on the trail are prepared chuck-wagon style, and you can bet there is plenty of wholesome food. Montana is beef country, so be prepared to find meat in various forms — steaks, prime ribs, bacon, barbecued beef, sausages, and so on — at every meal. Still, for those who subscribe to vegetarianism, there are enough servings of salads, veggies, pastas, beans, breads, fresh fruits and desserts to keep malnourishment at bay!

Each evening after dinner (or "supper" as folks out West call it), some form of musical entertainment is provided. Guests become fast friends joining in to learn the basics of country western dancing or simply relaxing around the campfire.

As an alternative to this Premiere cattle drive, Powder River offers once each month from May through September a 5-day/5-night "Ranch Vacation." On this adventure, the horses, accommodations, meals, and entertainment are much the same as in the Premiere cattle drive. But you will take part in actual ranch activities. Depending upon the time of the year you come, you'll find yourself helping brand calves, gather cattle, move herds to fresh pastures, or simply ride horseback to your heart's content.

Regardless of which trip you choose, you'll have plenty of time to roam through this incredible countryside. You need not be an experienced horseperson to take part in either trip, but an open, adventuresome mind will enable you to truly enjoy this experience. The whole family can participate in these unforgettable vacations in the "Land of the Big Sky."

For those of you looking to live the life of a "cowpoke," join the others on this ride of a lifetime!

Address: P.O. Box 676, Dept. BP, Broadus, MT 59317

Telephone: 800-492-8835, 800-982-0710; **Fax:** 406-436-2388 **Web:** www.jimbalzotti.com

Location: 2.5 hours from Billings, 2.5 hours from Rapid City, South Dakota

Guest Capacity: 60 for Premiere vacation, 20 for Ranch vacation

Season: May to September

Policy on Children, Pets: Children are welcome. Pets are not encouraged.

Rates: $$$ American plan. Children's rates.

Credit Cards: All major credit cards accepted.

Montana
Powder River Wagon Trains

TEEPEES ON THE OPEN RANGE.

LONGHORN CATTLE.

MONTANA

RJR RANCH
EUREKA, MT

At RJR Ranch We care. We care about you and making your vacation the best possible. We chose our staff, our horses, our activities, our menu with one overriding concern - will this be what you will enjoy and value?

To enhance the horseback riding experiences, natural horsemanship is taught before any guest leaves for a trail ride. For those guests desiring more skill and knowledge in dealing with a horse, personal attention is provided by very accomplished riding instructors. Whatever your level of riding ability, whether you've never been on a horse before, or are pretty comfortable in the saddle, we will help you improve your skill level. Before you know it, we will have you riding and whooping it up like a cowboy.

The ranch is surrounded by 2 million acres of Kootenai National Forest and Glacier National Park. Some of the most spectacular mountain scenery, rivers and lakes found anywhere are enjoyed with the guided assistance of our naturalists. You will see for yourself why they call Montana "Big Sky Country". It is a perfect opportunity to explore this country the old fashioned way - from the saddle of one of our horses. Make sure you bring your camera and plenty of film. You never know what wildlife you will see in their own natural setting.

Fishing starts on the ranch pond with fly casting lessons. We'll have the novice fisherman comfortable enough to get out there to hook a big one. The experienced fishermen are challenged by the variety of streams, rivers, ponds and lakes that extend in every direction and include many trout species. Fishing in such a beautiful setting will be an experience you'll never forget.

Rafting is a highlight for everybody. The Middle Fork of the Flathead in Glacier Park provides a pristine river and setting. Splashing down the river is alot of fun but make sure you bring your bathing suit. This will be one vacation you and your family will never stop talking about!

Accommodations are spacious, comfortable, completely authentic log construction with wonderful mountain views. The resulting deep sleep and physical activity require healthy food and our chef emphasizes healthy and plentiful by providing variety and tasteful presentations. Each meal is a delight and always has you looking forward to the next meal. Platters of fresh vegetables, fruits, freshly prepared meats and of course, who can forget dessert! After dinner, you may want to step outside and gaze at the stars that will appear so close, you'll think you can reach out and touch them. When it's time for bed, the cool night air and the stillness of the night will have you sleeping like a baby. Three platform tennis courts, a hot tub, archery targets and equipment, canoeing and swimming help fill in those spare moments. Since this is your vacation, you can feel free to do as much or as little as possible. Hike on the many acres of beautiful country, go horseback riding, fish the many streams, go rafting, or just put your feet up and take a nap in the sun - it's your vacation!

For more information and testimonials, please call toll free, Dick and Gail Reilly (owners) at 888-563-9012.

Address: P.O. Box 117, Eureka, MT 59917

Telephone: 406-889-3395
Web: www.jimbalzotti.com

Location: 70 minutes north of Kalispell

Guest Capacity: *18*

Season: June 1-October 31

Policy on Children, Pets: Children over age 6. No pets.

Rates: $$

Credit Cards: None

MONTANA
RJR RANCH

A PICNIC BY THE LAKE.

YOU MAY NOT WANT TO GET OUT OF BED.

Montana

Roundup Cattle Drive, Inc.
Roundup, MT

Anyone can show you Montana, but Roundup Cattle Drive has the expertise to help you experience the Big Sky country to its fullest. Roundup Cattle Drive — a once-a-year, five-day horseback adventure in the grandeur of central Montana — gives you that chance. Relive the life of the cowboy. Lose yourself in the tens of thousands of acres of jackpine timber and sandstone cliffs. Listen to the creak of the wooden-wheeled wagons on the rock-strewn trail. Savor your campfire coffee as the remuda of 150 horses and mules are wrangled at the crack of dawn. Round up a herd of hundreds of cattle. Bring in the strays. Marvel at the unlimited vistas of the "Land of the Big Sky." Relax with a cold beverage after a hard day of riding "drag." Jangle your spurs on the horse-drawn dance floor doing the western swing. Inhale the aroma of the mouth-watering grub fresh from the old chuckwagon. Blissfully sleep under the stars to the serenade of the coyote. Sound like fun? Come and enjoy.

This event is special as it occurs only *once* a year in late August or early September. The town of Roundup is true to its name, as it was the trail head of the historical "Great Montana Centennial Cattle Drive" of 1989. It was so well received as the experience of a lifetime that it is repeated on a yearly basis.

You arrive in Roundup, which is 50 miles north of Billings, to spend five days and four nights on the trail enjoying the true Western life. This is the vacation of a lifetime for guests seven to seventy, from novice riders to the experienced, for families, singles, and honeymooners.

The expert wranglers will match you with a horse that meets with your experience and approval. (You can even bring your own horse and/or saddle.) First-time riders receive instruction.

The horses are pre-screened, "well broke," and gentle. You might catch a cow-chasing Quarter Horse, a placid lovable "gentleman," a peppy Arab, a colorful Indian pony or Appy. Most of these 100+ horses come from local area ranchers. Somewhere in the remuda is a horse you'll want to take home with you!

Should you become tired of riding or feel you've chased enough cows for a while, you can catch a ride in a covered wagon. There are side trail rides and a ranch rodeo where you can show off your riding skills. Saddles and all tack needed are provided.

You'll have the opportunity to work cattle just as cowboys have done for over one hundred years in the West. You will be one of fifty guests who join the experienced cowboys to drive the large herd of cattle on the trek over 30,000 acres in this incredible country. No smog, no haze, just clean and healthful mountain air in the most beautiful scenery you'll find anywhere.

The food is delicious and well satisfying — somehow better because it is served under the big blue sky. You may sleep in your bedroll by the campfire or in provided "cowboy teepees." Before sleep, however, there is entertainment — nightly live country music and western dancing. If you play a guitar, bring it along and join in. There is also a talent night and team competitions. On the last night of the cattle drive, there is an awards dinner and a motel at the end of the line.

You might want to extend your vacation a few days either before or after the cattle drive to enjoy scenic Montana. The folks will help you put together an exciting itinerary. Try fly fishing or rafting on the Big Horn or Madison rivers. Yellowstone and Glacier national parks are not too far away. Retrace the route of Lewis and Clark. See the Little Big Horn Battlefield or an Indian powwow. And for those die-hards, there's a golf course or shopping mall around somewhere.

Your hosts are all experienced, courteous, and exhibit the caring and friendliness that guests are sure to talk about long after the drive is over. All you have to do is come, relax, and have a good time.

Address: P.O. Box 205, Dept. BP, Roundup, MT 59072

Telephone: 800-257-9775, 406-358-2454; **Fax** 406-358-2454

Location: 50 miles north of Billings

Guest Capacity: 50

Season: Annual event, late August to early September

Policy on Children, Pets: Children must be 7 years old and with parent. No pets.

Rates: $$ - $$$ American plan.

Credit Cards: MasterCard, Visa

MONTANA
ROUNDUP CATTLE DRIVE, INC.

HERDIN' THE CATTLE.

LOOK AT ALL THESE COWBOYS!

MONTANA

SCHIVELY RANCH
BILLINGS, MT

Montana: "The Big Sky Country". It's a lifestyle you've heard about all your life...now experience it firsthand at a real, working cattle ranch at the foot of the Pryor Mountain Range in Montana. Established in 1902 and nestled on the Pryor Creek, Schively is actually three different ranches. Schively itself, the Dry Head Ranch, and an irrigated farm located just east of Lovell, Wyoming. We provide beef cattle for the farm feedlots and tables of America. This magnificent facility is owned by the Bassett Family Partnership, who are delighted to share their exciting ranch life with you and your family. Leave your yearning for swimming pools, TV and tennis courts behind; but bring your love of the outdoors...you'll return to a life that still exists just as it did a hundred years ago.

We have been very fortunate to have ABC, ESPN, National Geographic and several TV stations come to our ranch to do programs and articles on what we do. These experiences have been special and we have available a video to enable you to better see what we offer. We will be glad to mail this video to you upon receipt of $12.00 covering cost and postage. (Foreign people, this is VHS, and will cost $18.00.)

Cattle drives start our season in mid April and continue for several weeks, until all cows with baby calves have been trailed into Montana. The cost of a cattle drive is $950.00 per person per week. We trail between 250-300 mama cows and their calves from Wyoming to Montana (about 50-55 miles.) Our fall drives begin mid October and continue for several weeks until we have trailed all the cows back to Wyoming for the winter. If weather permits we camp along the trail both in spring and the fall.

Our Ranch weeks start at the end of the cattle drives in mid May and continue until mid October. They include branding, vaccinating, roping, mini-veterinary needs, wrangling the cavvy, trailing to adjoining pastures, holding herd, salting, fencing, rotating pastures, bull gathering and rounding up strays. The cost of a Ranch week is $750.00 per person per week. We limit our weeks to 15-16 guests per week. We have a 10% discount for children under 12 years of age.

Being a cowboy has always been a secret dream of lots of men and women. The horse was not a pet but a companion and partner with whom he could work and accomplish great tasks. That's what we offer, the opportunity to fulfill that secret desire of Cowboy and horse succeeding together.

When staying at the Schively, you'll sleep in a rustic, comfortable cabin. While we're riding at the Dry Head Ranch, you'll stay in a log bunkhouse or the ranch house. We do have a few modern touches. There's telephone service and electricity and at both ranches we have central baths and showers.

Last but not least: As with any outdoor experience, weather is a large and deciding factor. We will proceed with all plans needed to accomplish our goals, but adjustments will be made when necessary and we reserve the right to do so. Also, we cannot assume liability or responsibility for personal injury or property damage, loss or theft while at the ranch. Now, with all that behind us, let's pick a week

Address: 1062 Road 15, Balz, Lovell, WY 82431

Telephone: 307-548-6688 (all year), 406-259-8866 (May - October); **Fax:** 307-548-6688

Internet: www.guestranches.com/schively/index.

E-Mail: schively@tctwest.net

Location: 2 hours south of Billings, Montana

Guest Capacity: 16

Season: mid-April through mid-November

Policy on Children, Pets: Children are welcome but are the responsibility of their parents. No pets.

Rates: $$ American Plan. Children's rates.

Credit Cards: Discover, MasterCard and Visa

(SEE PAGES 271 AND 272 FOR COLOR PHOTOS!)

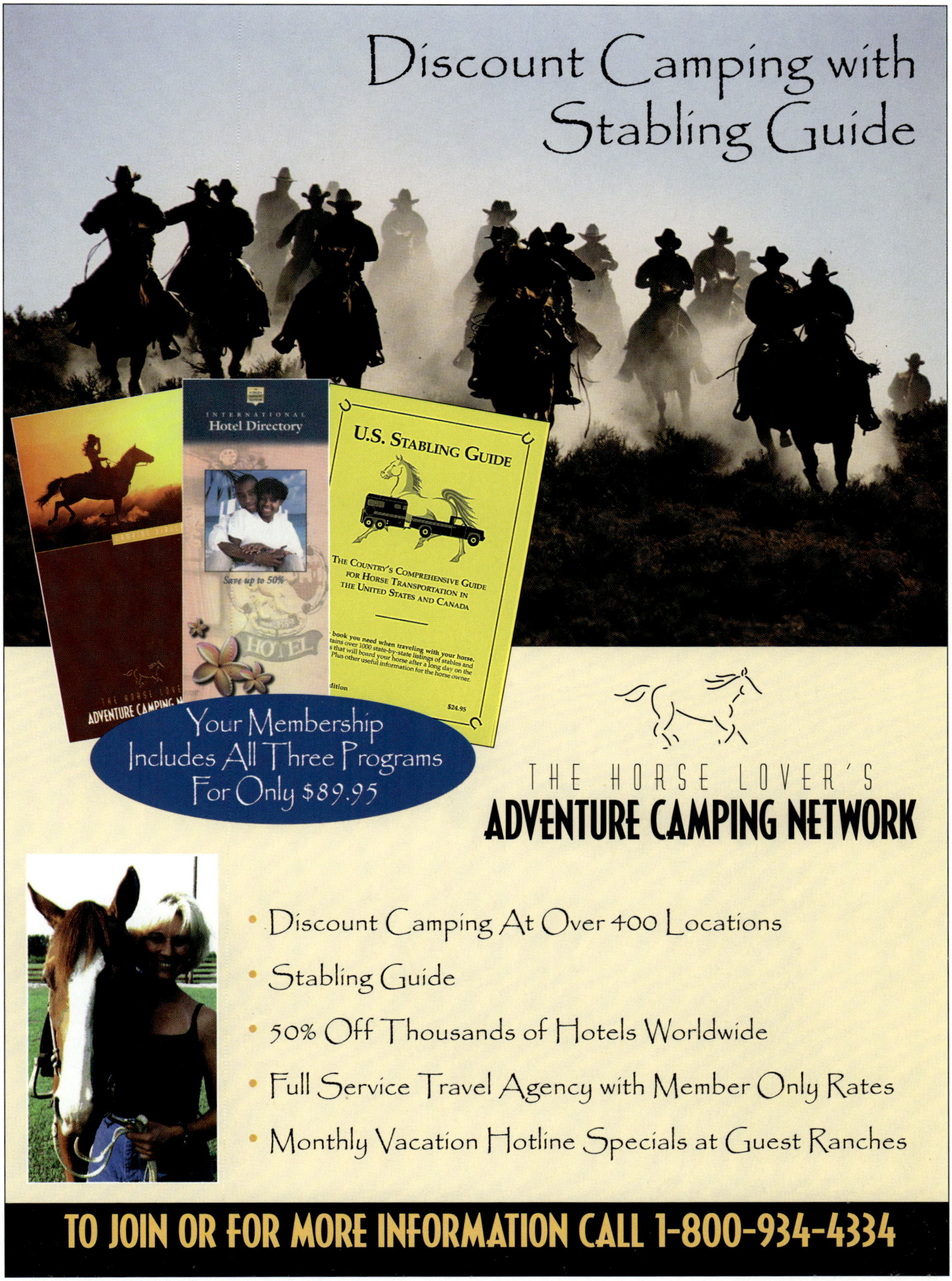
Discount Camping with
Stabling Guide
International
Hotel Directory
Save up to 50%
HOTEL
U.S. Stabling Guide
The Country's Comprehensive Guide
for Horse Transportation in
the United States and Canada
$24.95
The Horse Lover's
Adventure Camping Network
Your Membership
Includes All Three Programs
For Only $89.95
THE HORSE LOVER'S
ADVENTURE CAMPING NETWORK
• Discount Camping At Over 400 Locations
• Stabling Guide
• 50% Off Thousands of Hotels Worldwide
• Full Service Travel Agency with Member Only Rates
• Monthly Vacation Hotline Specials at Guest Ranches
TO JOIN OR FOR MORE INFORMATION CALL 1-800-934-4334

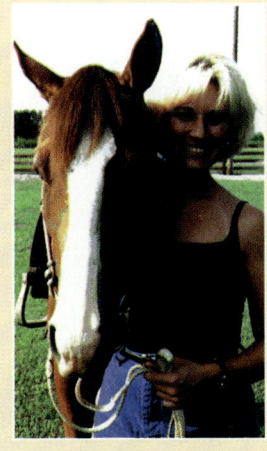

Arizona
Circle Z Ranch

RIDING THROUGH SCENIC CANYONS.

Arizona
Grapevine Canyon Ranch

A QUIET MOMENT.

RUSTIC LUXURY.

Arizona
Grapevine Canyon Ranch

A MORNING RIDE.

ARIZONA
IRONHORSE RANCH

A YOUNG COWBOY OUTSIDE THE CANDY STORE.

AN INDIAN TEEPEE AT THE IRONHORSE RANCH.

ARIZONA

IRONHORSE RANCH

VISIT THE OLD WEST SHOW IN NEARBY TOMBSTONE, ARIZONA.

THE WESTERN TOWN OF IRONHORSE RANCH.

Arizona
Lazy K Bar Ranch

COME ON, TURN THE PAGE.

SWIMMING IN THE ARIZONA SUNSHINE.

ARIZONA

WHITE STALLION RANCH

SOUTHWESTERN ADOBE.

NO WINTER BLUES AT WHITE STALLION RANCH.

ARIZONA
WHITE STALLION RANCH

RIDIN' AND ROPIN'.

GUESTS ON A MORNING TRAIL RIDE.

ARKANSAS

HORSESHOE CANYON RANCH

IT DOESN'T GET MUCH BETTER THAN THIS!

HORSES GRAZING.

Arkansas
Scott Valley Ranch

A BUCKET OF HORSE TIRES.

GIVING A YOUNGSTER SOME POINTERS.

CALIFORNIA
ALISAL RANCH

RIDERS CROSSING A STREAM.

California

Alisal Ranch

KIDS ON A FENCE.

GREAT FLYFISHING!

CALIFORNIA

Coffee Creek Ranch

YOU WON'T BELIEVE HOW GOOD THE FOOD IS HERE AT COFFEE CREEK RANCH.

CALIFORNIA
COFFEE CREEK RANCH

GREAT FISHING FROM OUR PEDAL BOATS.

AT NIGHT TIME, GUESTS GATHER AROUND THE CAMPFIRE TO ROAST MARSHMALLOWS. JOIN US!

CALIFORNIA
HUNEWILL RANCH

RIDING PAST SNOW-CAPPED MOUNTAINS.

MAKE SURE YOU BRING YOUR CAMERA.

CALIFORNIA

SHANNON RANCH

HEADING ON DOWN A DUSTY TRAIL.

MOVING CATTLE THROUGH THE TIMBER.

CALIFORNIA
SHANNON RANCH

"NEXT, I'M GOING TO BRAND YOU."

California
Wild Horse Sanctuary

A MAGNIFICENT WILD PAINT HORSE.

OUR HERD OF WILD HORSES.

COLORADO DUDE RANCH ASSOCIATION

Since 1934

OUR PLEDGE

For over 60 years the CDGRA (Colorado Dude and Guest Ranch Association) has maintained high standards and strict requirements for membership. Each ranch is regularly inspected for cleanliness, facilities, hospitality, quality of riding program, and honest representation in promotional messages. This assures that guest expectations are consistently fulfilled.

OUR SECRET

For pure value, CDGRA ranches offer America's best kept travel and vacation secret. Base prices typically include nearly every cost for an entire ranch visit – all lodging and meals, riding and fishing, and a smorgasbord of ranch activities – like Western entertainment, cookouts, hay rides, square dancing and more. Compare this to the typical pay-as-you-go ordinary vacation. At a cost you can confidently budget for and live within, your ranch hosts and new life-long friends provide you with virtually everything you will need for your entire stay. Plus, wonderful Western hospitality and lifetime memories.

CDGRA member ranches have proudly identified themselves in this book.

COLORADO DUDE & GUEST RANCH ASSOCIATION
P.O. Box 300
Tabernash, CO 80478
970-887-9248 • www.coloradoranch.com

Colorado
Aspen Canyon Ranch

FOREVER FRIENDS.

Colorado

Bar Lazy J Ranch

RIDING HIGH ABOVE A BLUE RIVER.

Colorado

Bar Lazy J Ranch

TAKING IT EASY.

RIDING DOWN THE STREAM.

Colorado

C Lazy U Ranch

AFTER THE WRANGLE.

DASHING THROUGH THE SNOW.

Colorado

Capitol Peak Outfitters

Capitol Peak Outfitters

HORSEBACK RIDING

Colorado

Cherokee Park Ranch

COWBOY DANCING.

WAITING FOR THE BIG BITE.

Colorado
Cherokee Park Ranch

CLOWNING AROUND.

Colorado
Coulter Lake Ranch

I DON'T WANT TO GO HOME!

Colorado
Deer Valley Ranch

GOD'S COUNTRY.

Colorado
Elk Mountain Ranch

LIFETIME MEMORIES FOR YOUR FAMILY.

Colorado

Focus Ranch

GIT ALONG, YOU LITTLE DOGGIE!

COME AND ENJOY THE WIDE OPEN SPACES.

Colorado

King Mountain Ranch

THE FINEST SKEET SHOOTING ANYWHERE.

THEY TOLD ME I WOULDN'T GET WET.

Colorado

King Mountain Ranch

RIDE WITH US THROUGH GOLDEN ASPENS.

Colorado

Mountain Meadows Ranch

RIDING THROUGH GOLDEN ASPENS.

Colorado
Peaceful Valley Lodge & Guest Ranch

FUTURE COWBOYS.

"WHEN'S DINNER?"

Colorado

Rawah Ranch

COME AND SET A SPELL.

WARM INTERIORS.

Colorado

Rawah Ranch

SPLASHING THROUGH THE STREAM.

UNSPOILED RIDING THROUGH THE RANCH WILDERNESS.

Colorado
San Juan Guest Ranch

COWBOY COOKING.

Colorado

San Juan Guest Ranch

A BEAUTIFUL SETTING.

COMING HOME.

Colorado
San Juan Guest Ranch

CREATE LIFETIME MEMORIES.

Colorado

Sylvan Dale Ranch

HANGING OUT WITH DAD.

COME ON OUT! THERE'S PLENTY MORE.

Colorado
T Lazy 7 Ranch

SUMMER FUN...

AND WINTER FUN.

Colorado
Weminuche Wilderness Adventures

RIDE INTO THE SPECTACULAR.

Colorado
Wilderness Trails Ranch

A SPECIAL TIME IN A SPECIAL PLACE.

Colorado

Wilderness Trails Ranch

MAKING A NEW FRIEND.

A PRISTINE MOUNTAIN POND.

Colorado

Wilderness Trails Ranch

FALL IN LOVE ALL OVER AGAIN.

MAKE YOUR DREAMS COME TRUE.

Colorado

Wind River Ranch

OUR HORSES LOVE CHILDREN.

COME VISIT US. IT'S PRETTIER IN PERSON!

Florida
Ace of Hearts Ranch

GIVE US A KISS!

A DREAM COME TRUE.

Idaho
Bar H Bar Ranch

I'VE ALWAYS WANTED TO DO THIS!

A FRIENDLY FACE.

Idaho
Diamond D Ranch

CAN I TAKE HIM HOME?

THIS ONE IS FOR THE PHOTO ALBUM.

IDAHO

IDAHO ROCKY MOUNTAIN RANCH

COME AND ENJOY OUR PEACE AND QUIET.

A GREAT PLACE TO WATCH THE SUNSET.

IDAHO

Hidden Creek Ranch

IDAHO

HIDDEN CREEK RANCH

You will find Hidden Creek Ranch in a private mountain valley undisturbed by the outside world at roads end. This is their home and from the moment you arrive you are treated like family. Their limited capacity of 40 guests allows them to provide special attention to all your needs and wishes. Hidden Creek Ranch combines the adventurous lifestyle of the Old West with the comfort and amenities of a world-class hideaway.

Idaho

Hidden Creek Ranch

IDAHO

HIDDEN CREEK RANCH

*T*rail rides deep into timbered forests or in grassy meadows are experiences you will long remember. Their rides are divided into small groups according to riding ability to ensure you a personal adventure every day. Their mountains are home to the Mountain Lion and Big Horned Owl. Along the trails you will have opportunities to spot Deer, Elk, Coyotes and maybe even a Black Bear. If you like to get off the beaten path and into the unspoiled splendor of the country, Hidden Creek Ranch is waiting for you.

Calm Your Soul

IDAHO

HIDDEN CREEK RANCH

HEADING OUT ON A SUMMER'S MORNING.

Idaho

Hidden Creek Ranch

YOUR FAMILY WILL LOVE OUR TEEPEE VILLAGE.

ENJOY THE BEAUTY OF OUR RIDING TRAILS.

Idaho

Hidden Creek Ranch

WHO RANG THE DINNER BELL?

GOURMET DINING AWAITS YOU AT HIDDEN CREEK RANCH.

IDAHO

HIDDEN CREEK RANCH

WINTERTIME BRINGS WINTER FUN TO
HIDDEN CREEK RANCH.

EXPLORE OUR BEAUTY ON SNOWMOBILES.

Idaho

Hidden Creek Ranch

A PERFECT ENDING TO A PERFECT DAY.

Idaho

Western Pleasure Guest Ranch

A PEEK INTO PARADISE.

OUR COZY CABIN NESTLED IN THE WOODS.

Kentucky

First Farm Inn

KENTUCKY HOSPITALITY.

SOUTHERN COMFORT.

Minnesota

Double JJ Ranch

TWO YOUNG COWBOYS MAKE PLANS.

WATER SPORTS AT THE DOUBLE JJ RANCH.

MINNESOTA

Double JJ Ranch

YOU DON'T HAVE TO GO ALL THE WAY TO ALASKA.

NIGHTTIME WARMTH.

Missouri

Meramec Farm B&B

YOU CAN LEAD A HORSE TO WATER...

Montana

63 Ranch

HEELS DOWN AND READY TO RIDE.

MONTANA
63 RANCH

"NELLIE, I'D LIKE YOU TO MEET CODY."

MONTANA
B Bar Ranch

BORDERING YELLOWSTONE NATIONAL PARK, B BAR RANCH OFFERS MANY HIKING OPPORTUNITIES ON ITS 20,000 ACRES.

MONTANA

BEAR TOOTH RANCH

FALL COLORS.

LEAVES BEGINNING TO TURN GOLD.

MONTANA

C-B Cattle and Guest Ranch

PRETTY AS A PICTURE.

Montana
Covered Wagon Ranch

ENJOYING THE SUN, SKY AND MOUNTAINS!

MONTANA
HARGRAVE CATTLE AND GUEST RANCH

YOUR HOSTS, LEO AND ELLEN HARGRAVE.

COME AS A GUEST, YOU'LL LEAVE AS A FRIEND.

MONTANA
TRIPLE JJJ WILDERNESS RANCH

RIDE TO YOUR HEART'S CONTENT.

COME VISIT BIG SKY COUNTRY.

Montana
Monture Face Outfitters

"THE CHINESE WALL" IN MONTANA'S BOB MARSHALL WILDERNESS.

A TOTALLY STRESS FREE TRAIL RIDE.

Montana
Parade Rest Ranch

A PERFECT PLACE FOR A PICNIC.

A WAGON RIDE IS A GREAT WAY TO SEE THE COUNTRY.

MONTANA
POWDER RIVER WAGON TRAINS

COME ON A REAL WAGON TRAIN RIDE.

KEEPING AN EYE ON THE HERD.

MONTANA
Powder River Wagon Trains

A GUEST LIVING HIS DREAM.

MONTANA
Roundup Cattle Drive

WHICH ONE IS MY TENT?

IT MUST BE LUNCHTIME!

MONTANA
SCHIVELY RANCH

WHO RANG THE DINNER BELL?

SO MANY GOOD CHOICES.

MONTANA
SCHIVELY RANCH

MOVIN' THE CATTLE OUT.

ROPIN'.

MONTANA
SWEET GRASS RANCH

BREATHTAKING COUNTRY.

Montana
Sweet Grass Ranch

READY TO RIDE.

SNOW-CAPPED PEAKS OVERLOOKING OUR RANCH.

Montana
Triple Creek Ranch

OUR LODGE OFFERS RUSTIC LUXURY.

Montana
Triple Creek Ranch

A PLACE TO RELAX AND UNWIND.

MONTANA

TRIPLE CREEK RANCH

A WINTER'S RIDE.

YOUR LUXURIOUS ACCOMMODATIONS.

MONTANA

TRIPLE CREEK RANCH

A TRANQUIL MOMENT.

ONE OF OUR CABINS NESTLED IN THE WOODS.

MONTANA
TWO MEDICINE TEEPEE ADVENTURES

YOUNG INDIAN DANCER AT INDIAN DAYS EVERY JULY.

Montana
Two Medicine Teepee Adventures

TEEPEE VILLAGE.

EVERYDAY SCENERY.

NEW MEXICO
GALISTEO INN

SOUTHWESTERN ADOBE.

New Mexico

N Bar Ranch

COME TO WHERE THE COWBOYS STILL LIVE.

RIDE THE WIDE OPEN SPACES.

NEW YORK
DIAMOND HORSESHOE RANCH

HEADING OUT FOR A RIDE IN THE ROLLING GREEN HILLS.

IT DOESN'T GET ANY BETTER THAN THIS!

NEW YORK

DIAMOND HORSESHOE RANCH

IS THIS MONTANA OR NEW YORK?

YOU WON'T WANT TO GO HOME.

New York
PINEGROVE DUDE RANCH

RIDING ON A COLORFUL FALL DAY.

OUR LAKE OFFERS GREAT FISHING FOR YOUNG AND OLD ALIKE.

New York
Pinegrove Dude Ranch

A PINEGROVE HAYRIDE.

New York

PINEGROVE DUDE RANCH

GUESTS READY TO RIDE IN OUR OLD WEST TOWN.

YOUR CHILDREN WON'T WANT TO LEAVE.

New York

Roseland Ranch Resort

BRINGING IN THE HERD.

DIG IN!

NEW YORK

ROSELAND RANCH RESORT

A PRETTY LADY ON A PRETTY HORSE.

North Carolina
Clear Creek Ranch

OUR BACKYARD.

NORTH CAROLINA
CLEAR CREEK RANCH

A PERFECT VACATION SPOT.

COMING BACK AFTER A GREAT RIDE.

North Dakota

Dahkotah Lodge

YOU'LL LOVE RIDING IN OUR UNIQUE COUNTRY.

SUMMERTIME IN NORTH DAKOTA.

Oregon
Rock Springs Guest Ranch

MAJESTIC SNOW CAPPED MOUNTAINS.

Pennsylvania
Mountain Trail Horse Center

COME ON IN. THE WATER'S GREAT!

SOUTH DAKOTA
TRIPLE R RANCH

BEST FRIENDS.

I SHOULDN'T HAVE EATEN SO MUCH.

SOUTH DAKOTA
WESTERN DAKOTA RANCH VACATIONS

I DON'T NEED THESE OTHER COWBOYS!

Texas

Bald Eagle Ranch

THIS SURE FEELS GREAT.

VISIT LIFE ON A TEXAS RANCH.

Texas

Leon Harrel's Old West Adventures

COME AND LEARN TO TEAM PEN WITH THE PROS.

YOU'LL BE PART OF THE FAMILY AT LEON HARREL'S RANCH.

Texas
Y.O. RANCH

A YOUNG COWBOY GIVES A KISS TO HIS FAITHFUL HORSE.

Utah

Best Western Ruby's Inn

BEAUTIFUL BRYCE CANYON NATIONAL PARK IS LOCATED TWO MILES FROM RUBY'S INN.

EXPLORE THE RED ROCK COUNTRY WHERE BUTCH CASSIDY USED TO HIDE OUT.

VERMONT
JACKSON HOUSE INN

A VERMONT SLEIGH BED.

COUNTRY ELEGANCE.

VERMONT

JACKSON HOUSE INN

MAKE OUR INN YOUR HOME AWAY FROM HOME.

VERMONT

KEDRON VALLEY STABLES

RIDERS ENJOYING A FALL DAY.

THE BEAUTY OF VERMONT.

Vermont

Mountain Top Inn

POETRY IN MOTION.

CROSSING A STONE BRIDGE.

Vermont

Vermont Icelandic Horse Farm

WINTER FUN.

VERMONT

VERMONT ICELANDIC HORSE FARM

RIDING IN A WINTER WONDERLAND.

COME SEE VERMONT ON HORSEBACK.

WASHINGTON
NORTH CASCADE OUTFITTERS

ON TOP OF THE WORLD.

RIDING UP TO SPANISH CAMP.

Wyoming

Blackwater Creek Ranch

WHITEWATER RAFTING THROUGH RED ROCK CANYONS.

WYOMING

BLACKWATER CREEK RANCH

BUFFALO GRAZING.

MEN TALK.

Wyoming

Brush Creek Ranch

OUR ORIGINAL BARN.

CREEK-SIDE COOKOUT.

Wyoming

Brush Creek Ranch

TAKING A BREAK DURING WINTER FUN.

LIVING THE DREAM.

Wyoming

Brush Creek Ranch

HERDING CATTLE.

Wyoming

Brush Creek Ranch

CHORES IN THE MAIN BARN.

WYOMING

DAVID RANCH

THE WIDE OPEN SPACES OF THE WEST.

SUNRISE IN WYOMING.

Wyoming
Flying A Ranch

THE BEAUTY OF WYOMING.

YOU'LL COME AS GUESTS BUT LEAVE AS FRIENDS.

WYOMING
GROS VENTRE RIVER RANCH

COME RIDE THE WEST.

BUILDING TRUST.

Wyoming

The Hideout at Flitner Ranch

A WESTERN CATTLE DRIVE.

RUSTIC ELEGANCE.

WYOMING

HIGH ISLAND GUEST RANCH AND CATTLE COMPANY

MOVIN' THE HERD.

RIDING ON MOUNTAINS.

Wyoming
Lazy L&B Ranch

YOUR HOSTS, LEE, BOB AND PIPER NAYLAN.

Wyoming
Lozier's Box "R" Ranch

THUMBS UP!

APPROACHING THE RANCH.

Wyoming

Rimrock Ranch

WESTERN LODGINGS.

Wyoming

Triangle C Ranch

GITTIN' THE HANG OF IT.

Wyoming
Triangle C Ranch

NOW, THAT'S HOW YOU ROPE A DOGGIE!

PRETTY LADY — PRETTY HORSES.

Wyoming
Triangle X Ranch

WAITIN' ON A HORSE.

WYOMING

TRIANGLE X RANCH

RIDING UP TO AN ELK HERD.

HORSES GRAZING.

ALBERTA, CANADA
WARNER GUIDING AND OUTFITTING LTD.

MOUNTAIN LAKE.

CANADIAN BEAUTY.

B.C., Canada
SPRINGHOUSE TRAILS RANCH

SERENITY.

THE MEETING PLACE.

IRELAND

Black Valley Equestrian

GETTING A DRINK.

IRELAND

BLACK VALLEY EQUESTRIAN

RIDING ALONG THE IRISH COAST.

THE BEAUTY OF THE COUNTRYSIDE.

IRELAND

BLACK VALLEY EQUESTRIAN

CROSSING AN ANCIENT STONE BRIDGE.

A PEMBROKE COWBOY.

POSSE WEEK AT THE
N-BAR RANCH

The HOTTEST Horseback Riding Vacation Adventure In The United States

JIM BALZOTTI

I never killed a man before. But this one was surely dead. I shot him as he broke cover and ran for it. I brought up my 45 Colt in one smooth motion and squeezed the trigger. The loud sound startled me as the gun jumped in my hand. He stopped running and looked down at the red stain spreading across his chest where his heart used to beat. He had a look of disgust and even sadness as he fell to his knees. He stayed like that for a moment or two, then his face crashed to the dirt. I must say, I felt no remorse. In fact, it felt pretty damn good.

By right, I should have never been in that gun battle. Hell, I would never have been with those outlaws if it weren't for my oldest boy, Jeff, now known as the infamous "Powder River Kid". His mama always said he was a good boy and I suppose he was, til he fell in with that no good, bank robbin', horse stealin' outlaw Preston Bates. Jeff wrote to us once explainin' he was just ridin' along with Preston when they came upon a stagecoach at the Powder River. Jeff said he never did know that Preston meant to rob that stagecoach all along, but I'm sure that's what they all say. The old timer ridin' up front fired off a blast from his twelve gauge shotgun hopin' to scare off these would-be robbers. But this was Preston Bates we were talkin' about.

Preston was the most cold blooded and feared outlaw in the whole Southwest, someone who believed that dead men do no talkin'. Preston shot that old timer right between the eyes and without a moment's hesitation. By the time the shootin' stopped, three good God-fearin' men were dead and our boy's face was plastered on "Wanted" posters all over the southwestern county.

I guess it was pretty foolish of me, but my wife, Margie, talked me into ridin' out and trying to find the outlaw band and try talkin' Jeffrey into turning himself in to the Sheriff. I managed to catch up to them by the third day of hard ridin' that dern near left my horse lame. Jeffrey told me where to meet him, makin' real sure that I was alone. Then, with Preston's blessin', he brought me into their camp, appropriately named "Incognito". I sat around the campfire and had coffee with some of the boys. Their names were well known in most of these parts for their string of bank robberies and killings. Some of them didn't seem so bad, almost even polite to me. But not Preston Bates. He had cold, piercing blue eyes, the kind that can make a man's blood run cold on a hot New Mexico day. They said he once shot one of his own men for no other reason than the man's snorin' kept him awake. Now, sitting across

the campfire, I was beginnin' to wonder if he wasn't going to shoot me too. "Dead men tell no tales".

Suddenly, one of the bandits perched up on a cliff gave a yell. "Posse's comin' and ridin' hard!" Preston shot me a look and said, "They must have followed you in." If all hell didn't break loose, he'd probably killed me on the spot. We ran for cover as bullets rained down on us. The posse had flanked us and gotten into the high rocks of the box canyon overlooking the outlaws' camp. I ran for cover in a group of juniper trees, bullets flying all around. I heard one of Preston's men scream, blood running down his belly - gut shot from a 30-30 Winchester. Not a pretty way to die.

It was then that I saw my son get killed. Shot in the head, he looked right at me, before he dropped to the ground. What was I going to tell his mama? Forty feet away, hiding in an outcrop of rocks, I saw the deputy sheriff who shot him. He broke from his cover and then I shot the man who killed my son. At that moment, I became an outlaw myself. We waged a fierce gun battle, running from rock to rock, tree to tree, but with Preston's guns blazin', one in each hand, we drove the posse back.

We had to break camp and do it quick. No tellin' when the posse would be back, especially now that they had picked up our trail. They also had Sheriff Diamond Jim with them. Jim was a feared lawman, second only to Wyatt Earp. They say Diamond Jim was once a bandit himself a long time ago, down Mexico way. That's where he got his nickname, robbin' banks' safety deposit boxes of their diamonds. He wore a large diamond on the pinkie finger of his right hand. In time, he settled down and became a lawman. Bandits either settled down or were killed.

There'd be no settlin' down for Preston Bates. He just wasn't the type. Bank robbin' is what he did though rumor has it he has a pretty young thing named Maggie down somewhere in Texas. Story is she was in a bank he was robbin' and he was so taken by her beauty that he didn't shoot the teller dead. Preston was having no feelings of love today however. Two of his outlaw band were killed outright and Preston shot the one who was gut shot himself. "Would have died anyhow." Anyway, he would have just slowed us down." The remaining six of us rode along the box canyon and through the creek to cover our tracks. We ran 'til the horses couldn't run any farther and then we tied them up in a bluff of towering Ponderosa pines.

"You damn fool, you led them right to us!" Preston, red-faced, hollered in my face. We had spread out our saddle blankets and Preston had two of his men stand watch. With all the excitement, shootin' and stuff, I hadn't had time to think of my own predicament. I had killed that deputy and now was a wanted man. If they caught me, they'd hang me for sure. What in blue blazes was I thinking? I figured I had no choice but to stay with Preston's gang. At least I'd be safe for the time being. As the sun disappeared behind the mountains, we made a campfire and had supper. I looked up from my plate and into the gun barrel of Preston's pearl handled pistol. "I guess this is the end of the line for you, cowboy." In the second before Preston pulled the trigger, I wondered how did I ever get here?

Well the answer to that is simple enough. Earlier this summer, my fifteen year old son, Jeff, our graphic designer, Cherry, and I signed up for "Posse Week" at the N Bar Ranch Outlaw Land & Cattle Company which has a fantasy adventure week that is the hottest horseback riding vacation in North America.

Imagine this, if you will: going back into time, to the 1880's Wild West to be exact. Preston Bates, the owner of the N Bar Ranch, has re-created how the authentic Old West was with "Posse Week". This is how it works.

The guests are divided into two groups, the Outlaws and the Posse. Both teams are given rolled-up leather maps that show the location of stashed loot bags from previous bank robberies. The Outlaws are given a two hour head start before the Posse is turned loose to hunt them down. Both teams are equipped with high tech paintball guns and only a saddlebag to carry your gear for the week. The idea is to collect all or as many of the loot bags as possible and to kill as many of the other team without getting killed yourself. You're riding on over 125 square miles of the most hauntingly beautiful land that was actually ridden by Billy the Kid, Butch Cassidy and the Sundance Kid, Black Jack Ketchum and was also the refuge for the Apaches. Geronimo was actually born not far from the N Bar Ranch. What you won't see is any street lights, McDonalds, telephone poles, or any other trace of civilization, for that matter. In fact, Preston searched for over three years to find the perfect location. Out on the trail, you might as well be back in the 1880's.

We flew into Albuquerque, New Mexico on Saturday and spent the day seeing all the sights of this delightful southwestern city. We strolled through the shops of Old Town where Cherry didn't waste the opportunity to do a little shopping. Authentic Indian jewelry, southwestern blankets in warm colors and other merchandise were available in the quaint shops.

Later that night we ate at a barbecue restaurant, ordering platters of ribs, chicken and pork that melted in your mouth. We were filled with nervous anticipation of beginning Posse week. I wouldn't be lying to say I don't think we slept much that night! The next day Preston came by the hotel to pick us up and the other guests who would be participating.

I knew from previous trips to New Mexico how pretty the countryside was. Most people wrongly think of New Mexico as only desert, high heat, and of course rattlers, but the N Bar is located high in the mountains of the Gila National Forest and is part of one of the world's oldest designated wilderness areas. Towering Ponderosa pines, firs, and lush grasslands are all in abundance. The ranch itself is located at 8200 feet, much too high for rattlers which are rarely found above 5,000 feet.

We arrived at the headquarters and met the other guests. There were two lawyers from Florida, a husband and wife from Germany, two high school girls from Michigan who were about to enter college, and a father with his three sons from Connecticut. We unloaded our gear (most of us packed way too much) and were shown to our cabins. After we had unloaded, we all gathered around the large cook tent for dinner. Heaping platters of steaks and chicken made me wonder if I really wanted to

ride off from this place. Over pie, cake and coffee we formed teams and Preston went over the very few rules there were:

- ☞ you must stay out on the trail for six days
- ☞ no returning to camp
- ☞ any gear you want must fit in the saddle or on your horse

You kept score two ways: first, by tallying the number of kills by each party, and second, by the total amount of loot recovered from the loot bags hidden on the leather map. Whichever team has the highest total, wins. Each bag represented a different bank or stagecoach robbery, thus different amounts. The team that had the highest total score won. Period. However, if one team recovers all of the hidden loot bags and brings them back to base camp on the final day, they are the automatic winners. Make no mistake, everyone there was playing to win. As for hardships, Preston had one piece of advice: "Toughen up, cupcake."

On the Outlaw team there were the two lawyers from Florida, Jeff, Cherry, the husband and wife from Germany, and myself, with Preston as our guide and his wife, Maggie. Ranch wranglers named Jim and Chris led the Posse, along with the father and his three sons from Connecticut and the two girls from Michigan . What makes this adventure so real is that the wranglers are very competitive and want to win as badly as their guests and they actually compete against each other!

We went down to the barn and picked out holsters and handguns. We were using paint ball guns fueled by a CO2 cartridge that delivered a round, soft, plastic ball filled with red paint. We then practiced shooting off round after round to get a feel for them. We were all impressed by how accurate they were. We were only given 40 rounds of ammunition as that was the average load carried by cowboys back in the 1880's. They would be refillable after any fight. Or should you "kill" an opponent, you were allowed to strip the ammo from him or her.

We then went on to the corral where we were matched with horses. I was immediately impressed with the conditioning of Preston's horses. These guys were in shape. I was given a magnificent four year old Paint gelding by the name of Gambler. Jeff was given a beautiful chestnut quarterhorse named Dulan. Cherry had a pretty bay by the name of Rocker. I mounted Gambler and looked around. The Outlaws, all decked out in cowboy hats, boots, bandanas with guns hanging from their hips, looked the part. We loped out of camp and headed into the wilderness, knowing that in just two short hours the Posse would be hot on our trail. I think probably the only thing that separated us from an outlaw bunch in the 1880's were the smiles on our faces.

We rode all day stopping only to study the leather map which held clues to where the hidden money might be. The first day we searched endlessly, but couldn't find a single bag. They weren't going to make this easy on us. At one point later in the day, we caught sight of where the Posse had been and we circled around trying to set up an ambush, but they were just too slick for us and got away. We ended up with darkness falling, making our way into our camp aptly called "Incognito". It was hidden in a box canyon, surrounded by high rock cliff walls with a small stream running through it offering water for our horses and with which to make coffee. We dismounted and took care of our horses - un-

saddled them, watered and fed them and put them into a makeshift pen for the night. Only then would we allow ourselves the comfort of the campfire and a hot meal.

We unrolled our sleeping bags and sacked out for the night. As corny as it sounds, the evening stars seemed to be in the reach of our fingertips. The climate was perfect, warm days and cool New Mexico evenings. No wonder they call this state the Land of Enchantment. We set up watches throughout the night in case the Posse tracked us into our own camp.

The rules of engagement were simply this: if shot, you were "dead" for 12 hours or until midnight, which ever came first. Thus, one of the strategies was to keep whittling down the numbers of the opposing team. If you were lucky enough to kill the opposing member who was holding the loot bag, you, of course, could take it as your own.

I was sound asleep in my sleeping bag, pleasantly tired from the day's ride when I was awakened to take my watch. I mumbled and groaned at having to leave my warm sleeping bag, and grumbled all the way up the hill to stand watch over my fellow Outlaws. The air was still and with a full moon I was able to see for miles. I wondered how many people were lucky enough to experience the serenity that unfolded before me. It was amazing to think that not all that long ago the Apaches camped on the very land that we were on.

By the time I wandered into camp, smells of fresh brewed campfire coffee and breakfast greeted me. It's amazing just how hungry fresh air and activity makes

you! We ate our breakfast and made plans for the day. The two lawyers were going to ride off together and try to find a loot bag in a far-off canyon. Jeff and I were going to saddle-up and see if we could see and then fight the Posse. Cherry, Preston and the others were staying in camp to shoe a horse that had dropped one during the night.

Jeff and I rode for miles without encountering anything or anybody. We rode across an expansive grassland that held some old bones and ended up at a small creek bed where we took our lunch out of our saddlebags, spread out our saddleblankets and rested up against our saddles, just like they do in the movies. It was great!

Upon returning to camp, we found that the lawyers had had no luck in finding the loot bags but they were discovered by the Posse who then gave hot pursuit. Shots were fired but they were able to escape alive. What excitement! They rode into camp covered with trail dust and happy as clams. By now, it was getting dark and we began the pleasant task of settling in for the night. We cooked canned ham and beans and it tasted as good as any four star restaurant's fare!

In the morning, we had barely lumbered out of our sleeping bags when a paintball gun went off, red paint splattering in the trees behind us. The Posse had slipped in during the early morning hours and had set up an ambush! Preston pulled on his holster and bolted to the rocks with the German couple right behind him. I crossed into a small thicket of trees, throwing myself as low as I could into the dirt. Jeff immediately engaged in a wild gunfight with two of the Posse deputies who promptly shot

We loped our horses out of camp, splashing through creeks and climbed high onto a ridge that afforded us greater visibility. We took a longer route riding right into the Posse's camp. The fire was still warm, but they were gone. We ate their cheese and crackers. That'll teach 'em!

The next day will go down as one of the most exciting days of my life, and believe me, I've had a lot of exciting days! We rose early, forwent morning chow, saddled the horses and rode back out to the Posse camp we had found the night before. We were feeling bloody and going to repay yesterday's ambush with one of our own. As we turned out of our box canyon we ran smack into the Posse heading into our camp! One of the few rules of Posse week was that you could not shoot anyone while on horseback. However, all participants in Posse week must wear an orange ribbon on their chest. If an opposing person yanks off the ribbon, it's the same as if you are shot with a paintball - you're dead. For a split second, the Posse and the Outlaws just looked at each other then someone gave a blood-curdling yell and everyone charged at each other! It was a scene out of the Wild West, Cowboys and Indians, going round and round in mortal combat. Horses and riders in a crazed frenzy, dirt flying, with Outlaws and Posse members reaching in trying to pull off ribbons. We had the Posse outnumbered as they had split their forces trying to box us in our camp. I was chased by Diamond Jim and one of his wranglers. I tried to evade them by galloping up into the rocks, but they quickly flanked me, one on each side, and pulled off my ribbon.

After they "killed" me, Diamond Jim set his sights and horse after Jeff. Jeff took off in a blazing gallop to es-

him. I swung around and shot the deputy who shot Jeff. Meanwhile shouts were coming from all around me. Preston and the Germans had succeeded in driving the Posse from the rocks, but not before Elsa was shot in the chest. One of the wranglers from the Posse ran across in front of me and I gunned her down. As quickly as it happened it was over. The Posse had hidden their horses on a bluff and were galloping away before we could even think of mounting up to give chase. We had two dead and had killed three Posse deputies. We decided to break camp, bundling everything we could on the horses and leaving what we couldn't carry. We rode for hours, going in a big loop hoping to get in ahead of the Posse. But their head wrangler, Diamond Jim, was no fool and headed out a different way. We now had to wait until 12 hours passed before Jeff and Elsa came "alive" again. Jeff was so upset at having been killed he would have been bouncing off the walls if there had been any there! He couldn't wait for the next chance to fight again. It would come. Diamond Jim had led the Posse to us once but we couldn't be sure he would do it again.

Just a few days into this action-adventure and we were really *into* it! We scowled, we cursed and swore revenge. We were dirty and dusty and we didn't care. I saw my fifteen year old son turn into an ornery young man, riding his horse better than any cowboy from a Hollywood movie, spoilin' for his next fight. The two lawyers were gone and in their place were two of the meanest bandits west of the Mississippi. Even Cherry, our feminine graphic designer couldn't wait to put a member of the Posse in her gunsights. We ate from metal plates, and washed down the chow with strong black coffee. We were bad!

cape his pursuer. As Diamond Jim pulled alongside, Jeff whipped his horse, Dulan, around in the opposite direction, leaving Diamond Jim wondering where the heck Jeff disappeared. By the time the battle on horseback was over, all five members of the Posse were killed and four loot bags captured. It had been the most one sided battle of the week and victory belonged to us - for the moment.

Having been victorious in battle, we headed back to camp to whip up a hot, hearty breakfast over the open campfire. Some of us jumped in the creek to enjoy an early morning New Mexican bath, while others wanted to take the opportunity to get some more riding in. This was living. It's hard to explain in words how alive and excited we felt. We spent the rest of the day taking leisurely horseback rides, hunting for loot bags, and just enjoying the warm sunshine and beautiful country. The horses were just great. Friendly and responsive, Preston did a great job matching them up with the riders. There wasn't a person in either camp that didn't just love his horse. We eventually headed back to "Incognito" with anticipation and growling bellies because tonight we were having our big steak cookout.

Jeff and I rode back into camp, watered and fed our horses, pulled off our cowboy boots and stretched out in front of the campfire. Preston and his wife, Maggie, were cooking monstrous size sizzling ribeye steaks over the open fire with all the trimmings. The late afternoon sun was setting and the sky was filling up with rich red hues. There wasn't a sound or sight of civilization for a million miles. We polished off every steak that was cooked and decided that after that morning's rout of the Posse, we didn't need to post watch for the night. We made plans to hit them again in the morning, and headed to the bedrolls for the night. With our bellies full, and our muscles pleasantly tired, it wasn't long before everyone was snoring away.

We never heard them coming. A little before midnight, the Posse rode up into the lower end of the box canyon, tied up their horses to the trees, and crept into our camp on foot. The next thing I

remember, I was rudely awakened by a gun pointed at my nose, letting me know I was now a dead outlaw. The Posse quickly surrounded our camp and captured most of us without a shot. Our two German Outlaws slipped into the darkness and one of the Florida lawyers woke up and engaged the Posse in a gunfight before getting flanked and killed himself. The Posse hooted and hollered, celebrating their night raid. We cursed them and ourselves for getting too complacent and not posting watch.

This is exactly what makes Preston Bates's Posse Week so great. The Outlaws and the Posse make all their own decisions. No one tells you what to do. We didn't post a watch, didn't change camp, and we paid for it. Lucky for us, they did not recover any of our loot bags as the Germans escaped camp with them. What was funny though, is that the Posse stole a pair of the lawyers underwear in the dark which was hanging from a tree thinking it was a loot bag. I can imagine the look on their face when they discovered what they really had!

After the Posse left, we broke camp at mid-morning and slept the rest of the night under the stars in a soft grove of towering pines. The next morning, we decided to attack the Posse in their new camp. We rode silently and in single file below the ridge of an outcrop of rocks to hide our presence. We left the horses in a draw and crept up the remaining 500 yards to the Posse camp. They had a fire going and appeared to be enjoying their morning breakfast. Well, we'd spoil that! We sent the two lawyers around the right side, Preston and the Germans flanked the right side, and Cherry, Jeff and I came straight down the middle. Guns blazing, we attacked the camp. Two of the Posse were killed outright, with the rest running for cover in the trees. The two lawyer outlaws were gunned down by a Posse member who had slipped around behind them. Another Posse deputy was hit by three paintballs at once. Jeff was engaged in a gunfight with one of the girls from Michigan (a coincidence? I don't think so! Jeff and the girls always seemed to be fighting each other) but

she got the better of him, hitting him square in the butt. Cherry unwittingly walked right in front of five members of the Posse who couldn't believe their good luck and promptly unloaded on her. I was in a shoot-out with one of the sons from Connecticut when the father slipped around the back of me and got the drop on me. Preston and two of the Posse were trading shots in the trees. When one of them attempted to slip in behind Preston, he shot her dead. Shortly afterwards, Preston drew a bead and shot the remaining deputy in the chest. The fight was over, with the Outlaws winning, but only by a few.

We all headed for the campfire and joined in the retelling of the battle and coffee, bacon and eggs. I think all of us were having such a great time, we were ready just to stay the month and do this over and over again. We ended up having one more big battle before it was time to mount up and ride back to the headquarters at the N Bar Ranch. We crossed through what appeared to be alpine meadows covered with every imaginable wildflower. This was not the New Mexico most people envision. We hated to head back. We stopped along the way to have lunch then rode on untill we reached the N Bar. We reluctantly dismounted and took long, luxuriously hot showers. It wasn't long before we all gathered around the cook tent and ate roasted turkey and beef with bowls full of fresh vegetables, baked breads, hot coffee, lemonade and everything else LeAnn, the cook, prepared for us. After dinner, Preston built a roaring campfire and we talked and laughed long into the night. The next morning we all exchanged phone numbers and addresses with our new friends and promised to do this again next year. Next year, I ride with the Posse!

Fot information on Posse Week and other exciting horsebacks riding vacations, call Jim Balzotti at 1-800-829-0715 or visit our web page at www.jimbalzotti.com.

The Photography Of
Jeffrey A. Goodman

THE PHOTOGRAPHY OF
JEFFREY A. GOODMAN

The Photography Of
Jeffrey A. Goodman

The Photography of
Jeffrey A. Goodman

I had the pleasure of meeting Jeffrey for the first time this past year when I flew to the White Stallion Ranch in Tucson, Arizona to shoot some of the covers for this edition. Though I had not met Jeffrey before, his reputation as a top notch photographer in capturing the essence of guest ranch and dude ranch landscapes, people, horses and life in general was well known to me. If I wanted someone to shoot the perfect photo, everyone told me that Jeffrey Goodman was my guy. I wasn't disappointed.

Jeffrey has been in the professional photography business for over 13 years, starting work at the early age of 14. At the age of 18, Jeffrey entered his first National Photography Competition against other well known professional photographers, some 30 to 40 years his senior, and won his first Blue Ribbon Competition Print. He was one of the youngest photographers to ever receive a blue ribbon in that competition.

Jeffrey graduated from college at age 21 and moved to Chicago to start his career in fashion and advertising photography where he worked under some of the top photographers in the field. At age 23, he started his own business and at age 24 he expanded his business to Arizona, where he currently lives.

His photography jobs have taken him all over the map from Hawaii to Alaska, from the Carolinas to the Virgin Islands and just about every other state in the U.S.

Jeffrey has been published in over 25 major magazines including American Cowboy, Western Horseman, Looking Fit, Vantage Island Sun and many more. He has also created cover images for some industry giants such as Wrangler as well as done a mulitude of fashion photography for clothing designers.

Jeffrey was good enough to share some of these photographs with us, giving us just a glimpse of his talent. He would like to offer both individuals and businesses a "custom full service" advertising and public relations service to produce brochures, print ads and PR packets to project a professional image to your clientele. For anyone interested in consulting with Jeffrey about his work, he can be reached by telephone at 602-788-7654 or by fax at 602-788-9876. Jeffrey may also be e-mailed at jagphotoinc@att.net.

MONTANA

SWEET GRASS RANCH
BIG TIMBER, MT

The ranch was built by a pioneer family and is listed on the National Register of Historic Places as the Brannin Ranch (Sweet Grass Ranch). We are at the end of six miles of private road and secluded in a beautiful mountain valley. There are deer, elk, antelope, bear, mountain lions, coyotes, an occasional wolf, eagles, hawks, and many songbirds.

We welcome riders of all levels and abilities. There are lessons available for those who want them and a variety of rides to accommodate all rider capabilities, as well as horses for respective levels. There are a variety of rides available daily: morning and afternoon half-day rides, a short all-day ride, and a long all-day ride. We also offer guests the opportunity to have hands-on experience with horses of all ages, from baby colts to grown horses, working with Rocco, our son who specializes in breeding, raising and training horses. He is gentle, patient and thorough with both horse and human.

Those who would like to bring their own horses for schooling should call to discuss particulars. We also offer bareback riding for those who like to ride bareback or those who would like to learn. We do not have an age limit on children for riding; that is up to their parents.

Sweet Grass is a working cattle ranch in connection with our winter ranch, Otter Creek, where our youngest son, Tony and his wife, Ellie, are raising Angus cattle year-round. The cattle summer in the mountains and winter in the lower country. This is the ranching operation that has been in the family over 100 years and now into its 6th generation. Our guests are welcome to take an active part in cattle work if they like: checking cattle, salting and/or doctoring as they may need it, and moving them when grass and water dictate.

There is challenging fly fishing in the creek for browns, brookies and rainbow trout; the lakes are good fishing too and the beaver ponds are great places for kids to learn or for those who like to use spinning outfits. All fishing is catch-and-release except in the beaver ponds. There is plenty of hiking country, magnificent photo-opportunities and great bird watching. Flowers are in their glory in June. We can arrange trips to Yellowstone Park, Indian ceremonials, historic sites and other points of interest.

The ranch is for those who want to live ranch life as we do. The cabins are log or frame to match the pioneer main ranch house. There is no hot tub, tennis court, or social organization of any kind. Our life offers people a chance to get away from the rush of modern life and retreat to a quieter time - time for quiet reflection, an appreciation of the beauty of God's creation, quality time with one's family, just plain old peace and quiet, close to land and livestock, without many people.

Address: HC 87 Box 2161, Big Timber, MT 59011

Telephone/Fax: 406-537-4477
Web: www.sweetgrassranch.com
E-Mail: sweetgrass@mcn.net

Location: 120 miles from both Billings and Bozeman airports

Guest Capacity: 20

Season: Second week in June through Labor Day

Policy on Children, Pets: Children welcome. No pets.

Rates: $$

Credit Cards: None

MONTANA
SWEET GRASS RANCH

ROCCO WORKING A COLT.

HORSES GRAZING.

MONTANA

TRIPLE CREEK RANCH
DARBY, MT

Triple Creek Ranch, an enchanting Montana guest ranch, is a place where riders can enjoy a breathtaking ride through the Bitterroot Mountains by day and then toast the day's activities in their own private hot tub at night. This is the ultimate retreat for both novice and expert riders. Choose from a lunch ride to a majestic view or an all-day trail ride among pristine forests and cascading streams.

Triple Creek is a small, adults-only luxury resort that caters to the world's most discerning vacationers. All riders will enjoy this romantic hideaway. Top off a fun-filled day of riding with a cocktail and a relaxing stroll under the dazzling Big Sky, or enjoy the weekly campfire and cowboy singalong. Luxury is not just a vision — at Triple Creek, luxury is a reality.

Well-trained western horses and private or group riding lessons are available for any caliber rider. When you are not in the saddle, enjoy fly fishing, hiking, tennis, and bird or wildlife viewing or photography. Relax in the outside pool or hot tub. Go for an exhilarating ATV or snowmobile ride. Polish your putting on the green, or catch your dinner from the stocked trout pond. Nearby, guests can go whitewater rafting, heli-hiking, heli-fishing, and flight-seeing.

Winter excursions support a number of inspiring activities. After a fresh early morning snowfall, go on a brisk horseback ride or take an invigorating snowmobile expedition through the glistening winter wonderland. Try downhill skiing in deep Montana powder at Lost Trail Powder Mountain just 28 miles away. Explore the ranch with snowshoes or cross-country skis. Take a trip to Chief Joseph Cross Country Ski Area, which features groomed and ungroomed trails for all ski levels. At the end of the day, snuggle up with that special someone and sip champagne in front of your private crackling fireplace.

Guests stay in one of the 18 custom cabins skirting a rushing creek or under whispering Ponderosa pines. Each cabin is outfitted with a fireplace, telephone with a data port, TV, VCR, and wet bar. The luxury cabins have separate living rooms, massive log beds, his and her baths, steam showers, and a private deck with hot tub.

At the center of Triple Creek sits a handsome log lodge. Within the lodge is a comfortable library/conference room ideally set for an executive retreat and a dining room with soaring windows framing majestic Baker Point. Dinner is a special occasion at Triple Creek. Guests choose the time they dine and their choice of entree from the menu. Dine by candlelight and sip a glass of fine wine while listening to a talented solo guitarist.

Breakfast features fresh fruits, juices, our homemade nutfilled granola, hot, scrumptious breads, specialty omelets, and fresh hot ranch meats. Lunches are tailored to guests' activities. Enjoy a tasty lunch in your own picnic basket, day pack, or saddle bag while on the waterfall hike, lake walk, or mountain top horseback ride. Or, enjoy a hearty gourmet lunch either in the dining room or on the deck.

Triple Creek is the only Montana member of Relais & Chateaux. Travel expert Andrew Harper rated Triple Creek one of the "10 Best Little Gems in the U.S." and one of the "Best Executive Meeting Sites in North America." *Country Inns Bed and Breakfast* describes Triple Creek as "far more than a humble mountain retreat . . . small luxuries heighten the wilderness experience . . . cuisine that is at once sophisticated and cosmopolitan . . . (the) perfect mountain hideaway." The price for this memorable Rocky Mountain luxury vacation includes all meals, snacks, refreshments, open bar and activities such as on-ranch horseback riding, tennis, swimming, hiking, and fishing.

Visit and see for yourself why Triple Creek has been described as "luxury in the midst of wilderness . . . roughing it in style."

Address: 5551 West Fork Stage Route, Dept. BP, Darby, MT 59829

Telephone: 406-821-4600;
Fax: 406-821-4666

Location: 42 miles south of Missoula

Guest Capacity: 42

Season: Year round

Policy on Children, Pets: No children under 16 in ranch unless entire facility is reserved by a group. No pets.

Rates: $$$ - $$$$ American plan.

Credit Cards: Amex, Discover, MC, Visa

(SEE PAGE 275-278 FOR COLOR PHOTOS!)

MONTANA
TRIPLE CREEK RANCH

THE DRIVE INTO THE RANCH.

WINTER FUN.

MONTANA

TWO MEDICINE TEEPEE ADVENTURES
BROWNING, MT

Two Medicine Teepee Adventure is an Indian owned enterprise. The president and owner is Terry Wellman. Terry was born and raised in Browning, Montana, and grew up on a ranch, attended Montana State University, and has been involved with cattle ranching, registered Quarter horses, steer wrestling competitions, and building construction. His father, Robert Wellman Sr., was in the dude ranching business in Glacier National Park for over 30 years and was a charter member of the Northwest Montana Outfitter and Guide Association. He brings valuable experience to the outfit by training our stagecoach drivers and wranglers as well as helping with the day to day operations. Terry's son, Terry Wellman Jr., is a very experienced ranch hand and competitive cowboy. He and Robert Wellman Sr., along with many other well trained ranch hands and camp staff, will take very good care of you.

Indian Resources International, owners of the WarBonnet Lodge and Two Medicine Teepee Adventures, welcome you to the beautiful 1.5 million acre Blackfeet Indian Reservation, located near Glacier National Park. Visit plains, lakes, rivers, streams and rolling foothills, set against the beautiful snowcapped Rocky Mountains. Triple Divide Mountain is the place where streams originate that flow into three oceans and the headwaters of the Missouri River begin here. This is the last best place Mother Nature has to offer. Montana is rich in history and we wish to share an exciting way of life with you.

Experience the peacefulness in miles of grass rippling in the wind. Ride through hills and grassy plains studded with blue ponds, ride along river banks, rock canyons, explore ancient Indian camps, buffalo jumps and old teepee rings. This beautiful land has remained in its natural state for thousands of years.

Smell the fresh air and pine trees along the many trails our expert guides will take you. See many varieties of wildflowers and wildlife such as deer, elk, eagles, coyotes, buffalo, bear and prairie dogs. You may see bands of horses with proud stallions moving with their mares and young ones to their favorite grass or water hole.

Back at camp, enjoy the sauna or Jacuzzi before a delicious dinner. You may participate in western sing-along music and listen to Indian storytellers around big campfires. Imagine yourself relaxing in your own cozy teepee fully equipped with beds, pellet stove, and all the furnishing necessary for your comfort. This adventure is like no other offered anywhere else.

Ride with Blackfeet Indian cowboys (they are some of the world's best horsemen) moving cattle, riding fence or looking for stray cattle. You will ride your horse every day and the stagecoach will move your possessions to your next camp and accommodate anyone who prefers the stagecoach to a horse. We accommodate 20 guests per teepee village so that each guest is able to participate fully and personally in all camp activities including, Native American dances, archery, crafts, storytelling, learn Blackfeet native language, cowboy rope tricks and much more!

Two Medicine Teepee Adventures is currently in the process of expanding a new and exciting venture to increase much needed tourism and economic development on the Blackfeet Indian Reservation. At this time there are no services of this kind being offered on the Blackfeet Indian Reservation or in the nation. We have a very unique situation here in the area with abundant natural beauty provided by Mother Nature. Several Blue Ribbon rated lakes and rivers for fishing are nearby. This project will create a situation where natural resources of the Blackfeet Indian Reservation can be utilized in an environmentally sound manner.

Address: 101 Pata Street, P.O. Box 1510, Browning, MT 59417

Telephone: 406-338-7574; **Fax:** 406-338-2672
Web: www.jimbalzotti.com
E-Mail: twomedtp@3rivers.net

Location: 120 miles from Great Falls airport, 90 miles from Kalispel

Guest Capacity: 120

Season: Teepees: June 1 through October 1, motel open year round

Policy on Children, Pets: Children ages 6 and older can ride. No pets

Rates: $$$

Credit Cards: MasterCard and Visa

Montana

MONTANA

X BAR A RANCH
McLeod, MT

The X Bar A Ranch is the oldest ranch in the Boulder Valley, since before the turn of the century. When the Clark family acquired it in 1925, they raised cattle, sheep, and hay and began taking a in few guests in the 1930's.There have been many changes over the years, but the ranch and the country are still beautiful and remote, unique and unspoiled. Guests love the way the X Bar A is so easy going and lends itself to the natural beauty that surrounds it. As there are no scheduled activities, guests come to horseback ride, fly-fish, hike on the numerous trails, observe the wildflowers and wildlife that make this their home, or just put their feet up and read a good book, letting the daily stress in their lives just melt away.

Accommodations are hand-hewn log cabins, the logs hauled down by horses from the mountain overlooking the valley. All the cabins and furniture were built by the family. The cabins have 2 large bedrooms with baths, and 3 cabins have living rooms with fireplaces. Three meals a day are served in the large dining room with the ranch having a reputation for preparing and serving gourmet food. After finishing one of our specialy prepared desserts, you may wish to retire to the hot tub to soak and look up at the evening sky to stargaze. Montana is known as "Big Sky Country" and you won't believe just how beautiful our nights really are until you come and see for yourself.

The Boulder River runs through the ranch, a blue-ribbon trout stream, with excellent fishing. Our guests say this is some of the best fly-fishing they have found. There is something special about being on the river in early morning, alone with your thoughts, when you hook into a trout. Other guests come to hike into the backcountry that surrounds our ranch and explore our country the way it was a hundred years ago. Our children's programs are all informal and mosts guests relish the ability to spend quality time with their children, something harder and harder to do in the fast paced times we live in. You'll give your children lasting memories that they will always cherish. There is a swimming pool, and a rustic vacation game room.

Horseback riding is available on numerous trails in the valley, there is an annual cattle drive and the opportunity to help with the ranch work.

There are many wonderful and exciting trails for hiking. Nearby is a 106-foot waterfall, Indian caves with ancient paintings, and a gold-mining town from the 1890's on the edge of the Absaroka-Beartooth Wilderness. Close by is Yellowstone Park and the Custer Battlefield. You can enjoy a rodeo or a float trip on the river.

Wildlife abounds with deer, elk, antelope, bears often seen and wildflowers are in glorious abundance.

Activities are unstructured, allowing you to do as much or as little as you like, but we can point you in many directions to help you discover the natural beauty or our ranch and region.

The movie, "The Horse Whisperer", was filmed on the ranch next to us and we have the same wonderful scenery. Everyday is an adventure. We feel that this is life as it was meant to be!

Address: Box 5, McLeod, MT 59052

Telephone: 406-932-6108

Location: 22 miles south of Big Timber.

Guest Capacity: 20-30

Season: June-September

Policy on Children, Pets: Children welcome. Call regarding pets.

Rates: $$

Credit Cards: None

MONTANA
X BAR A RANCH

COMING HOME.

ON THE TRAIL.

NEW HAMPSHIRE

HORSE HAVEN BED AND BREAKFAST
SALISBURY, NH

In a quiet little town northwest of Concord, New Hampshire, a unique Bed & Breakfast has entered its 2nd year of operation. HORSE HAVEN of Salisbury, NH offers the opportunity for people to bring their horses and dogs to a comfortable farmhouse on an unpaved country road where the driving is just fantastic and the riding trails varied and superb. What makes HORSE HAVEN even more unique is that now folks with horses can combine the riding and driving opportunities with being able to stay overnight in a comfortable bed and breakfast with their horses housed in a safe environment on the property. The costs are moderate and since HORSE HAVEN B&B is only about a two hour drive from Boston, a quick weekend or weekday trip is always possible and practical.

Located on a gravel road on a peaceful ridge with views of Ragged & Tucker Mountains, with the owner's thoroughbreds romping in the 35 acres of hillside pastures surrounding the almost 200 year old farmhouse, Horse Haven provides a retreat for those seeking solitude and quiet country roads and trails, yet Salisbury is within easy reach of the Lakes, the NH Speedway or the many attractions of the Concord, NH area. The area abounds with other recreational possibilities including fantastic fishing in nearby rivers and cross-country skiing and groomed trails for snowmobiles. Bicycles, canoes and a rowboat are available for guests. A connection to statewide snowmobile trails runs through the farm's property.

Horse Haven B&B is a three season Bed & Breakfast, fully open from May through the early winter months, but may be opened at other times for groups taking three or more rooms.

The comfortable 1805 farmhouse features an immense deck overlooking the pastures. Bountiful continental breakfasts are provided and gas grills and other cooking facilities are made available to guests. Rooms are large and comfortable, some with fireplaces or a wood-burning stove. Rather than being elegant and formal, guests agree that Horse Haven B&B lives up to its brochure's claim of being "country-comfortable" and the guest book shows the repeated phrases of "comfortable" "friendly" and "home-like".

Velma Emery, the owner of HORSE HAVEN came up with the idea of a horsemen's bed and breakfast because of her strong interest in driving and because of the need of summer pasture for the thoroughbred foals she breeds at her Wilmington, MA (Butters Farm) location. And the idea to allow people to bring their dogs just naturally developed when Velma thought about the many times she would have liked to stay at a bed & breakfast, but couldn't because she had her constant companion friends with her.

At Horse Haven, there are sturdy pipe corrals that are available at no charge for overnight stabling or guests can stable their horses in well-bedded stalls in the renovated barn for $15.00 per night. Midweek, the third night in a stall is free. Also, guests staying two nights or more, midweek, are extended a discount, as are club, social or family groups that stay at least two nights.

Address: 462 Raccoon Hill Road, Salisbury, NH 03268

Telephone: 603-648-2101

Internet: www.bbonline.com/nh/horsehaven

Location: 35 miles north of Manchester, 18 miles north of Concord

Guest Capacity: 10 or 12

Season: May through January

Policy on Children, Pets: We cater mainly to adults. Pets welcome.

Rates: $

Credit Cards: MasterCard and Visa

NEW MEXICO

CEDAR CREST COUNTRY COTTAGE AND STABLE
CEDAR CREST, NM

Cedar Crest Country Cottage and Stables offers charm and hospitality. A home away from home nestled in the evergreen forests of the beautiful Sandai Mountains. The breath-taking Alpine views make this a storybook getaway - ideal for a romantic rendezvous. The cottage is also ideal to be your headquarters for a weekend of relaxation or outdoor activities. It is perfect for the vacationer or the business traveler. This mountain country home has three bedrooms and a self-sufficient kitchen. There are many wonderful restaurants nearby and a grocery store for stocking the kitchen if you would like to prepare your own meals.

Though the setting seems remote, it is only 4 miles north of I-40 on Hwy. 14, a 4-lane highway. It is 15 minutes from Albuquerque and 45 minutes to Santa Fe. Located on the historic Turquoise Trail in the Sandia Mountains, the ranch is adjacent to Cibola National Forest.

Cedar Crest can be your headquarters for the many activities and attractions offered in nearby Santa Fe and Albuquerque plus the historic sites, galleries, and shops along the Turquoise Trail. Nearby Sandia Peak Tramway, the world's longest, offers a beautiful view of Albuquerque and the Rio Grande Valley. The Sandia Ski Area is only 15 minutes from the cottage via the National Scenic Byway. This offers great central New Mexico skiing from December to April. The ski area offers five chairlifts and can accommodate skiers from the beginner to the pro. Ski school and equipment rentals are available. You may also go a half-hour north to the Santa Fe Ski Area. The ski lift is open in the summer also to see the colorful evergreens, oak trees, wildflowers and animals. You can also ride the lift up and come down the many trails on mountain bike. In the evening, you'll enjoy the peace and tranquility of the New Mexican sky, with its' million of stars that appear close enough to touch. What a wonderful place to come on a second honeymoon!

In October, over 800 hot air balloons participate in the Albuquerque International Balloon Fiesta and assend every morning. Balloons glow at night with fireworks.

Wonderful skiing, hiking and horseback riding are nearby. Remember we also have a stable so bring your own horse and ride the nearby trails. Wildlife West is another attraction which is 122 acres of non releasable wildlife. Indian Pueblos with gambling casinos are 1/2 to 3/4 hour drive. Cedar Crest Country Cottage and Stables is the perfect location for all kinds of activities, from trail riding on nearby trails in pretty country, to hiking, skiing, hang gliding, hot air ballooning, or just putting your feet up and relaxing. The shopping in Albuquerque or Santa Fe will be unique and offer you the best in southwestern merchandise and antiques. We recommend advance reservations as this is a popular vacation destination. We are located in the Sandia Mountains and are adjacent to the Cibola National Forest. The Sandia Ski Base is only a convienient 15 minute drive away. Shopping in Albuquerque is a 15 minutes drive, and Santa Fe is 45 minutes away by car.

 Address: 47 Snowline Road, P.O. Box 621, Cedar Crest, NM 87008
Telephone: 505-281-1915
Web: www.southwesterninns.com/cedar.htm
E-Mail: Dracccc@aol.com

Location: 7 miles from Albuquerque

Guest Capacity: 8

Season: Year-round

Policy on Children, Pets: Children welcome. No pets.

Rates: $ -$$

Credit Cards: Visa and Mastercard

NEW MEXICO

DRIPPING SPRINGS GUEST RANCH AND SPA
MOUNTAINAIR, NM

Welcome to Dripping Springs Ranch Guest Ranch & Spa. We are a small, exclusive family-run operation, offering a rich variety of experiences and located one hour South of the Albuquerque International Airport at the edge of the Cibola National Forest. Our guest ranch and spa is nestled between the Los Pino and Manzano mountains at 5,800 ft., on 7,000 acres of rolling hills and mesas, dotted with Pinon, Cedar, and Juniper trees. The vistas, sunsets, and night canopy of stars are all breathtaking.

Recently, we have built four new buildings and converted the old ranch house into The Lodge with a two-story Great Room, dining room, game room (pool & Ping-Pong), and seven private guest rooms. Outside areas include decks, relaxing patios, a children's play area, horseshoe pits, tee-pees and storytelling fire ring. Guest rooms have king or queen beds and a space for a roll-away. Our extensive southwest art collection contributes to the "Old West" atmosphere.

By advance appointment, we offer an excellent spa experience. Enjoy the quiet of our private garden or a few minutes in the sauna or hot tub. Schedule a massage with one of our licensed therapists. Our spa experience is available for both "day" and overnight guests.

We offer guided horseback adventures on our "smooth as glass" Missouri Fox Trotters. Daily rides of varying lengths are available. As a working cattle ranch, we also offer periodic cattle drives and round-ups. Our Trotters are exceptional, truly the "cadillacs" of the horse world, providing a smooth bounce-free ride even for the inexperienced rider. The riding is incredible and the easy-going temperament of our horses and wranglers makes the entire experience memorable for everyone.

Many miles of rolling ranch terrain offer exceptional guided hiking and riding for our guests. The arroyos and mesas are filled with a variety of scenery, including old settlers' ruins, old mines, dripping springs and watering holes. Ranch wildlife ranges from black bear and mountain lion to red-tailed hawks and endangered Rocky Mountain Big Horn Sheep. From wildflowers to fossils and petroglyphs, trekking through the territory is an enjoyable experience.

Exciting opportunities for photography exist on all rides and hikes. Photograph a Santa Fe train as it passes through the canyon, deer or antelope grazing in the valley, or capture friends and family enjoying a "wild west" moment at the corrals.

Gourmet meals ranging from authentic Thai to Mexican cuisine are prepared from the freshest of ingredients. Our chef is primed to satisfy your appetite and delight your palette. Meals are served to our guests in our Lodge dining area. Saddle-bag lunches, picnic hikes, and outdoor BBQ are optional treats.

Our staff is ready and waiting to provide you with excellent service, delicious food, an incredible spa experience and the southwest ranch adventure of a lifetime! We look forward to hearing from you to schedule your visit to come and share the spirit of western adventure in the land of enchantment. HAPPY TRAILS!!

 Address: H.C. 66, Box 607, Mountainair, NM 87036

Telephone: 505-423-3214; **Fax:** 505-423-3360

Internet: http://www.rt66.com/vacation

E-Mail: vacation@rt66.com

Location: one hour south of Albuquerque International airport

Guest Capacity: 44 to 50

Season: Year round

Policy on Children, Pets: Children welcome 6 years or older. No pets.

Rates: $ to $$$

Credit Cards: None

NEW MEXICO
DRIPPING SPRINGS GUEST RANCH AND SPA

WANT TO GO FOR A RIDE?

A COW KISS.

NEW MEXICO

THE GALISTEO INN
GALISTEO, NM

The Galisteo Inn is situated on 8 serene acres in the historic village of Galisteo, just 23 miles southeast of Santa Fe. Close to the diverse seasonal splendors and multicultural events of northern New Mexico, yet off the beaten path, the Inn provides the ideal environment for a memorable horseback riding vacation and getaway.

The 240-year-old hacienda and grounds belonged to one of the region's original settlers. In less than an hour, you can stroll the village of Galisteo, founded as a Spanish colonial outpost in 1614. Now home to writers, artists, entrepreneurs, and healers, among others. The community still retains a strong sense of its Native American heritage and Spanish traditions.

What makes The Galisteo Inn so remarkable as a horseback riding vacation is its unique offering of scenic horseback riding and luxury accommodations for the discriminating rider who wants it all. Outside, the setting is magical — ancient cottonwoods frame the gentle lawn, portales edge the hacienda, while horses graze nearby. A quiet pond sets off the far edge of the property. The light and air are enchanting, the vistas remarkable. The Inn's 8 acres are bordered by stone walls, inside of which sheep and horses graze in the pasture and ducks swim in the spring-fed pond. The Inn is a place for sheer comfort and enjoyment. It is one of the finer examples of New Mexico territorial architecture built in the eighteenth century, and many original features have been preserved, including hand-sawn vigas and lattillas. Corner fireplaces, plank and Mexican tile floors, and thoughtfully decorated rooms add charm to its adobe heritage. The ample living room is ideal to talk about that day's horseback ride or just enjoy solitary activities, and the dining room is perfectly scaled for intimate, friendly meals. Guests may relax in the library or by the pool, and always the delicious aromas of roasting peppers, herbs, and freshly baked pastries being prepared for dinner invade the air. In winter, the scent of pinyon pine and cedar burning in The Inn's eight fireplaces adds to the sparkling high desert air.

The guest rooms at The Galisteo Inn are varied, and each is decorated with a Southwestern feel. Some rooms have fireplaces and sitting areas; some have views of the pasture and lawn, others have views of the back courtyard and pool. All are designed to give you the feel of being in a spacious, nurturing home, a place to enjoy a quiet horseback ride, and then retreat to a peaceful environment.

Guided horseback rides are provided for the beginning rider to the more advanced. You can also enjoy other activities such as mountain biking and hiking. After a morning horseback ride, you may want to spend time in the indoor sauna or outdoor hot tub, or perhaps just lie out in the sun by the 50-ft. outdoor heated lap pool. A therapeutic massage is available for an additional fee. The local mountains are known for their petroglyphs — ancient stone carvings that tell of the many hundreds of Indians that once lived in the Galisteo Basin.

Given the stresses and demands on so many of our lives, travellers and horseback vacationers alike have high expectations, and they deserve the best. At The Galisteo Inn, they are committed to providing an unusual haven, for people who want to be able to get away and enjoy a horseback riding vacation, but in a relaxed, refreshing environment that bridges the nostalgic integrity of the past with the personalized conveniences of the present.

The spectacular setting, restful ambiance, congenial staff, quality cuisine, and excellent horseback riding, all combine to make your horseback riding vacation truly unforgettable.

Address: 9 LaVega, Galisteo, NM 87540; Mailing Address: HC 75, Box 4, Dept. BP, Galisteo, NM 87540

Telephone: 505-466-4000; **Fax:** 505-466-4008

Location: 23 miles southeast of Santa Fe on Highway 41

Guest Capacity: 24

Season: February through December

Policy in Children, Pets: Children must be 12 years old. No pets.

Rates: $ American plan.

Credit Cards: Discover, MasterCard, Visa

NEW MEXICO

NEW MEXICO

LOS PINOS
TERERRO, NM

Los Pinos Ranch...Where the road ends and the trails begin...

Nestled in the heart of the Sangre de Crist Range of the southern Rockies, Los Pinos Ranch recaptures the serenity and ambiance of an earlier, simpler time. The main lodge, built as a summer residence, shares its origins with New Mexico statehood. A guest ranch since the early 1920's, Los Pinos continues to be a relaxing haven for its visitors and has been operated by the McSweeney family since 1964.

Situated at an elevation of 8,500 feet, the ranch is perched above the wild and scenic Pecos River. A profusion of wildflowers and diverse species of birds and animals abound in the sub-alpine setting. Summer temperatures range from 75 degrees to 85 degrees at midday then dipping into the 40's at night.

At Los Pinos, riding, fishing, hiking and birding are the order of the day. Horseback riding is the main activity at the ranch and guided rides accommodate both the beginners and experienced rider. Instruction is offered each morning. Trails wind through coniferous forests, aspen groves and across the spectacular, high mountain meadows of the Pecos Wilderness in the Santa Fe National Forest.

The sparkling waters of the upper Pecos and its tributaries provide excellent fly fishing. Fishing gear is not provided by the ranch and the McSweeneys will direct you to a local store where the fishing permit can be obtained. So bring your own equipment and be ready for great trout fishing.

Hikers will enjoy trails that begin at the ranch or from nearby trailheads and provide the opportunity to enjoy fresh air, brisk exercise and views of the local wildlife and scenic New Mexico countryside. Be sure to pack your camera and field guides for bird and plant identification.

For those interested in touring, the historic city of Santa Fe is 45 miles from the ranch (about an hour's drive). Pueblos, mountain villages, and other attractions of northern New Mexico offer a variety of sightseeing options.

Ranch guests are housed in one of four aspen log cabins. Each cabin has a private bath, front porch, and wood burning stove. There are no telephones or televisions in the cabins - a refreshing change for those seeking a bit of solitude.

Limited to 12 guests, the atmosphere at Los Pinos is that of being a guest at a private country house. The front porch of the lodge is a gathering place for guests and a good spot for reading that book you haven't been able to get to yet or for having afternoon tea. If you didn't bring a book, the McSweeney's library is stocked with numerous books on nature and local history. The living room provides a place for fireside conversation on cool evenings.

Each week's menu offers a variety of flavorful well-presented dishes, including freshly baked breads. Dinner and breakfast are served in the lodge dining room. Lunches are packed for trail and streamside picnics. At Los Pinos, the staff takes great care and pleasure in the preparation and service of our meals. Due to county laws, we have a BYOB policy on alcoholic beverages.

Address: Route 3, Box 8, Tererro, NM 87573

Telephone: 505-757-6213, **Winter:** 505-757-6679

Location: 1 hour drive from Santa Fe

Guest Capacity: 12

Season: June 1 through Labor Day

Policy on Children, Pets: Children welcome over 6 years old. No pets.

Rates: $

Credit Cards: None

New Mexico
Los Pinos

BOY'S BEST FRIEND.

SOUTHWESTERN COUNTRYSIDE.

NEW MEXICO

N BAR RANCH
RESERVE, NM

The N Bar Ranch is a working cattle and horse ranch located high in the mountains of western New Mexico. Surrounded by the Gila National Forest, it encompasses 78,000 acres and adjoins the world's oldest Wilderness Area. Still virtually untouched, the area offers unsurpassed beauty and diversity of both ecosystems and wildlife.

The N Bar isn't about kidney-shaped swimming pools, shuffleboard courts, or gift shops. What it has to offer is real Western adventure that'll take you back to a time when luxury meant a good horse and the freedom to do as you pleased.

Rich in western history, this high and lonesome country was a refuge for the Apache Indians led by such chiefs as Victorio, Mangas, Loco, and Geronimo, whose birthplace is a short dozen miles away. Billy the Kid rode these trails, as did Butch and Sundance. In keeping with this colorful past, each of the ranch's horse programs takes you back in time to capture the real excitement and adventure of the Old West.

Consider "Posse Week." This high-action, low-comfort adventure game involves two teams: the outlaws and the posse. The outlaws want the gold, and the posse wants the outlaws. Both sides are armed with laser tag guns, hard-riding horses, and not much else. Food is scarce, and sleep comes only after your midnight watch is up. Necessary skills include orienteering, tracking, and a crafty mind. Each team travels light and fast, carrying everything needed for the trail, sleeping wherever night finds them. Billy the Kid could do it. Can you? The winner gets a real treasure — a free trip!

Or maybe you'd like to try a little cattle work. In the spring and fall, there are roundups and drives; in the summer, there's branding and moving the cattle between pastures. Experience the real cowboy life for a week, riding the range, rounding up strays, fixing fence, checking water. Out at cow camp, a chuckwagon is your kitchen, serving up range feasts to keep you fueled for the next day's work. Nights are spent on a bedroll under the stars or in range teepees, and a good night's sleep is just about guaranteed.

Or maybe you'd like to experience the Gila Wilderness. A pack trip down the Middle Fork of the Gila offers one spectacular sight after another; cliff dwellings, impossible rock formations, secluded hot springs, and icy cool swimming holes.

For the more laid-back individual, General Ranch Week is your chance to relax around the ranch, exploring the area on horseback or on foot, staying in the wall tents or rustic cabins, and joining the crew for three hearty meals a day. You're welcome to sleep in, kick up some dust with the cowhands, or just ride the porch with a good book and plenty of sunscreen. The N Bar has some of the finest horses south of the North Pole, whether you're a beginner wanting to take it slow and easy or an experienced rider looking for a real cowhorse. There are no nose-to-tail trail rides. With the riding as limitless as the horizon, you find your own path.

Guests stay in rustic cabins or wood-paneled wall tents. Each has two beds, table with a kerosene lamp, wool rugs for that first step in the morning, reading material, and a small woodstove. The tents are watertight and quite cozy, with plenty of headroom, screen doors, and a window. Cabins are preferable for winter use, of course. Food is abundant and delicious at the ranch, served buffet-style three times a day. Between meals, guests are welcome to raid the fridge, and the cookie jar is always chock full.

So, whether you choose to take part in an authentic 1890s round up and cattle drive or become an outlaw trying to elude a persistent posse, the past is anything but forgotten at the N Bar. Whatever your fancy, the N Bar Ranch will offer you the richest, most unforgettable vacation of your life.

Address: P.O. Box 409, Dept. BP, Reserve, NM 87830

Telephone: 800-616-0434, 505-533-6253
Web: www.jimbalzotti.com

Location: 250 miles from Albuquerque

Guest Capacity: 16

Season: May through October

Policy on Children, Pets: Children must be 12 years old. No pets.

Rates: $$ American plan.

Credit Cards: MasterCard, Visa

(SEE PAGE 282 FOR COLOR PHOTOS AND SPECIAL FEATURE STORY!)

NEW MEXICO
N BAR RANCH

GUESTS RELAXING AFTER A HEARTY DINNER.

HORSES AT PLAY.

NEW YORK

THE BARK EATER INN
KEENE, NY

The Bark Eater is a gracious 18th-century country inn located on a spacious farm. Set in a grassy meadow in the famous Adirondack Mountains, formerly a stagecoach stopover, the inn is minutes from the Olympic Village of Lake Placid. Emphasis at the inn is on horseback riding in summer and cross-country skiing in winter.

The inn has three buildings for lodging: the inn itself with shared bath; the carriage house - a step away - with private baths; and a unique hand-hewn log cottage minutes from the inn located in the woods, literally beside the riding and ski trails, with two suites.

A scrumptious breakfast sets you up for the day. A delightful trail lunch will help you keep going mid-day, if needed, and a country-style gourmet dinner will fine tune the evening.

The relaxed atmosphere surrounding the riding stable is a compliment to the staff who will attend your needs, and is reflected in the manner and training of our horses. Even if you have never ridden, we can make you comfortable in the saddle. For the advanced rider, there are miles of trails and back roads and there's even polo when you're ready.

Surprisingly, there are unlimited opportunities for horseback riding in the Adirondacks. For some reason, people seldom think of "wide open spaces" for riding when the eastern part of the U.S. The fact is, there is a rich history and tradition of horseback riding in the Adirondacks. Thousands of small farms in the 1700's and the 1800's were worked with horses and one of the biggest industries, logging, was forced to rely on horses.

For cross-country skiing, the Lake Placed area is world famous. Ski right at the inn on a fun trail system developed by the owner, a former Olympic skier and coach, or minutes away enjoy several other cross-country ski areas, the Olympic trails at Mt. Van Hoevenburg, the famous Jackrabbit trail, or miles and miles of wilderness backcountry.

Lake Placid is an exciting place to visit. In addition to the traditional swimming, golf and tennis, one can mountain bike, kayak and rock climb (instructions available from local guides) or ascend some of the challenging and remote Adirondack Mountains. The area boasts one of the top ten fishing streams in the United States.

In winter try some ice skating on the Olympic ice rinks, or if that isn't exciting enough, try some downhill skiing on the 1980 Olympic downhill trails (beginner and intermediate downhill skiing is plentiful also) or, bobsledding and luge anyone? Yes, it is possible to do both. It is one of the few places in the world where this is possible.

Naturally, there is plenty of relaxing and some plain good old "rockin on the front porch" for those who wish quiet time. Winter or summer, one of the unique aspects of the Lake Placid Olympic Region is the combination of its history, its art and museums, its dining, its spectacular scenery and its international sports events that are breathtaking to watch.

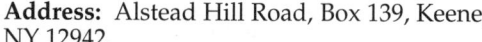

Address: Alstead Hill Road, Box 139, Keene, NY 12942

Telephone: 518-576-2221; **Fax:** 518-576-2071

Web: www.jimbalzotti.com

E-Mail: barkeater@tvenet.com

Location: Minutes from Lake Placid Olympic Village

Guest Capacity: 36

Season: Year round

Policy on Children, Pets: Children welcome. No pets.

Rates: $ to $$

Credit Cards: American Express, Discover, MasterCard and Visa

New York
The Bark Eater Inn

POLO!

A WINTER SLEIGH RIDE.

NEW YORK

DIAMOND HORSESHOE RANCH
ELKA PARK, NY

Nestled in the great northern Catskill mountains of New York is a real ranch run by real cowboys. The ranch is located on 150 acres adjoining 12,000 acres of state land making spring, summer and fall a horseback wonderland loaded with wildlife, streams, fields and fabulous views. Winter is a skier's paradise. Three marvelous ski areas encircle the ranch: Hunter mountain, one of the snow making capitols of the world; Ski Windham, a great upscale ski center; and Cortina Valley, including a children's program and night skiing. Horseback riding continues all winter long. As an added attraction, we offer horse drawn sleigh rides to thrill young and old alike.

Diamond Horseshoe Ranch prides itself on great food featuring at least 10 choices served nightly, sirloin steak, snow crab, catch of the day, center cut medallions of pork, Canadian baby back ribs, boneless breast of chicken and four specially prepared pasta dishes every night. Breakfast is always a country feast: eggs & omelets any style, thick cut Texas toast (French toast), flap jacks (plain, blueberry, strawberry, apple or banana), bacon, sausage, ham, homefries, fresh fruit, hot & cold cereals, bagels, super ranch muffins (with egg, cheese, ham, bacon or sausage), and helps our guests ready in the morning for a great day - any time of the year.

Accommodations range from our basic ranch room for two to six people, ranch room deluxe from two to six people with Jacuzzi, or our ranch "super suite" with two sleeping areas, patio sitting room, fireplace and Jacuzzi. All rooms are fully carpeted with private bath, individual heat control/A/C units and color cable TV.

There is full camp programming for the younger wranglers called Little Spurs day camp. Activities abound all year long with all the standard resort features as well as heated indoor pool, Jacuzzi and steam rooms, outdoor pool, tennis, basketball, volleyball, softball, archery, skeet shooting range and two game rooms, fireplace lounge and full nightclub facilities where you can enjoy the best in country western dancing, shows, and Vegas nite for fun-filled evenings of events. Diamond Horseshoe Ranch is proud to offer the finest in vacation time for singles, couples and families all year long in the most scenic area of the northeast.

Summer rodeo, horseshows, winter festivals and carnivals are all here to enhance the great accommodations, superb food and a fabulous vacation atmosphere. Accommodating up to 200 guests gives us the ability to personalize your vacation. Individual attention is given on all levels of horsemanship.

You are only a stranger once at the Diamond Horseshoe Ranch.

Address: Dale Lane, Elka Park, NY 12427

Telephone: 518-589-5197; **Toll Free:** 800-926-2771; **Fax:** 518-589-5278

Web: www.diamond horseshoe ranch.com

Location: 125 miles north of New York City

Guest Capacity: 200

Season: Year round

Policy on Children, Pets: Children welcome. Pets welcome with security deposit.

Rates: $$$

Credit Cards: All major credit cards accepted.

(SEE PAGES 283-284 FOR COLOR PHOTOS!)

NEW YORK
DIAMOND HORSESHOE RANCH

OUR WINTER RIDING CAN'T BE BEAT!

OK. LISTEN UP. THE FIRST RULE IS FUN, FUN, FUN!

New York

Diamond Horseshoe Ranch

BEAUTIFUL RIDING IN THE CATSKILLS

NEW YORK

NEW YORK

PINEGROVE DUDE RANCH
KERHONKSON, NY

Pinegrove Dude Ranch is nestled in the peaceful, gentle, rolling hills of upstate New York's Catskill Mountains with miles of scenic mountain trails for horseback riding. Dick and Debbie Tarantino started the Pinegrove Dude Ranch in 1971. Today, it is one of the best family-oriented resort ranches in the Northeast. In 1992, the ranch won the *Family Circle* "Ranch of the Year" award.

Pinegrove is family-owned and -operated, and down-home friendliness is one of their first concerns. The ranch is nestled in the Shawangunk Mountains on 600 beautiful acres rambling over fields, brooks, and woodlands. You will want to bring your camera to photograph the abundant resident wildlife — wild turkey, coyote, fox, raccoon, quail, pheasant, Canadian geese, turkey vulture, plus a multitude of songbirds.

At Pinegrove, all in one day, you can saddle up for a two-hour cattle drive, swim, hike, play tennis, and take a snooze in the afternoon before the evening entertainment begins with line dancing. Pinegrove remains the only ranch with eleven indoor sports facilities, as well as three all-you-can-eat meals daily, a daily country-style cocktail party in the enormous lobby, singalongs, and much more! Relax in jeans and slacks, day or night, in the informal, rustic atmosphere. No fancy duds needed, ever! The ranch specializes in children's activities, and the excellent full-time children's program so parents can enjoy themselves as well.

In the summertime, guests will be offered free beginner and "junior wrangler" lessons and will be able to go right from lessons to riding the trails. For the more advanced rider, specialized instruction is offered to improve riding skills. You'll be able to ride on miles of trails over acres of picturesque rolling hills. You can ride deep into the forest for miles on secluded trails, crossing gentle mountain streams and enjoying distant views. Most rides can accommodate 15-20 people. Horseback rides go out on the hour, all day long, so there is ample opportunity to ride at your leisure.

You can be part of a "City Slicker" cattle drive and feel the excitement of being on a good horse, cattle ranging ahead of you, out in wide open spaces. Without ever leaving the ranch, you can fish for bass in the stocked lakes, enjoy boating, and hiking on beautiful New York paths. Swim in the heated pool or play tennis. Miniature golf, bocce, archery, basketball, volleyball, paddle and handball, table tennis, and shuffleboard are some of the other activities available at any time for your enjoyment. In the winter, there's downhill skiing on two slopes and ice skating. They've thought of everything — all you need to do is show up and enjoy yourself!

The Lil' Maverick Day Camp and the Belle Starr Nursery are where our younger guests can expect the utmost in TLC while parents go off on their own. At Day Camp, the older "colts" and "fillies" enjoy arts & crafts, storytime, cartoons, and the indoor pool. Or they'll spend time in the playground featuring natural-wood, junior-size choo-choo train, bi-plane, pirate ship, and fire engine. And, of course, pony rides!

With all these going-ons, you're going to work up one mighty appetite! Spectacular views are yours to enjoy while eating in the Phoenix and Scottsdale dining rooms. Three all-you-can-eat country-cooked meals are served daily by the friendly staff. And for those real hungry wranglers in your family, the free Chuck Wagon Snack Bar is open all day to late at night serving grilled cheeseburgers, hot dogs, french fries, iced tea, and coffee.

At Pinegrove Dude Ranch, your happiness is of prime importance. Pick your season and have yourself a country time of your life.

Address: Box 209, Dept. BP, Kerhonkson, NY 12446

Telephone: 914-626-7345, 800-346-4626; **Fax:** 914-626-7365; **Web:** www.pinegrove.com or www.jimbalzotti.com

Location: 100 miles northwest of New York City, 1 mile west of Kerhonkson

Guest Capacity: 350

Season: Year round, including holidays

Policy on Children, Pets: Family-oriented resort, children an integral part. No pets.

Rates: $$ American plan. Children 4-6 half price; under 4, free. Family specials and single parent, group, and senior discounts.

Credit Cards: Amex, MasterCard, Visa

(SEE PAGE 285-287 FOR COLOR PHOTOS!)

New York
Pinegrove Dude Ranch

ROWING ON A LAZY SUNNY AFTERNOON.

CHILDREN LOVE PINEGROVE.

New York
Pinegrove Dude Ranch

MAKING NEW FRIENDS.

A NEW COWBOY.

NEW YORK

RIDIN – HY RANCH
WARRENSBURG, NY

Ridin-Hy is quite a spread! Located in the Adirondack State Park (largest park in the United States at 6 million acres), Ridin-Hy covers 800 acres of mountains and fresh water lakes. Water sports abound. Sherman Lake offers water skiing, rowboats, paddle-boats and speed boating. Enjoy leisurely scenic cruises on our pontoon boat. Fish for northern pike, small and large mouth bass, brown and rainbow trout in Sherman Lake, the Schroon River and Burnt Pond.

Horseback riding consists of daily mountain trail rides and are paced according to the riders' abilities. Lessons are offered by our experienced wranglers. This is a great place to get on a horse for the first time under the watchful eyes of our wranglers, or come and improve your riding skills for those who haven't ridden for awhile. We offer different paced rides to accomodate all levels of experience. Fast, intermediate and slower paced rides are available on our mountain trails. Pony rides are available for the little ones. The trail rides in the fall are especially enjoyable during the colorful foliage season. Rides are offered daily, and our guests love the weekly guest "rodeo". We also have daily wagon rides for the young and young at heart.

Ridin-Hy offers a comfortable, informal, friendly atmosphere. The beautiful main log lodge with its huge fireplaces offers a spacious living room, TV, card tables, cocktail lounge, game room, indoor pool and lakeside dining room.

Mealtime means homemade breads and desserts, specially planned lunch and dinner menus with a choice of menu. Enjoy our famous weekly Smorgasbord plus lakeside steak barbecue. All you can eat, piping hot and farm fresh. After dinner, you may want to take a stroll or just put your feet up and relax while you star gaze.

You'll be amazed at the countless variety of other things to do besides fishing and horseback riding. Our social director organizes a full schedule of activities each day, year-round for adults and children alike. Sprawl on our sandy beach (the little ones love it). Swim in crystal clear Sherman Lake or our heated indoor pool. Play tennis, volleyball, softball, shuffle board, horseshoes and archery. Area golf includes four eighteen-hole courses and two nine-hole courses. Evenings are filled with music and entertainment, masquerade shows, hayrides, rodeo and much more. There is a children's activities director and babysitting is available. Ridin-Hy Ranch Resort offers the perfect place to bring your family to create lifetime memories that your children will never forget.

Relax or play, Ridin-Hy is casual, easy going with old time western hospitality. Accommodations include chalet cabins, lakeside main lodge rooms and one story motel style units. All are easy distance to the main lodge and lake with full private baths.

Winter at the ranch means snow skiing (over beginner-intermediate slope with snowmaking) makes skiing at your door a reality. Snowmobiling, ice skating, horse drawn sleigh rides, sledding, and horseback riding.

Truly a year-round fully inclusive vacation, with the kind of care and service that can only come from a family-owned and operated ranch resort. Begun in 1940 by Ed and Orabel Carstens and operated today by their children and grandchildren.

Address: Sherman Lake, Warrensburg, NY

Telephone: 518-494-2742

Location: 1 Hour North of Albany

Guest Capacity: 220

Season: Year round

Policy on Children, Pets: Children welcome.

Rates: $

Credit Cards: American Express, Discover, Visa and Mastercard

(SEE PAGES 283-284 FOR COLOR PHOTOS!)

NEW YORK

ROCKING HORSE RANCH
HIGHLAND, NY

Welcome to the perfect family ranch-resort, voted "Best Family Resort-Ranch in U.S." by readers of *Family Circle* magazine and similarly hailed by *Better Homes & Gardens* magazine. In 1996, New York's First Lady Libby Pataki recognized the ranch with a "Why I ♥ NY" award and named it 1 of 12 of the best values in family getaways.

The Rocking Horse Ranch, a AAA 3-Diamond resort, is a place where the whole family will find lots to do, together or apart. The complete children's program gives parents that option. Kids love the independence of taking themselves to the barn for a visit with the friendly wranglers, stopping at the archery range, playing with the llama in the petting zoo, or going for a swim in the three heated pools (one just for kids). Parents love knowing there is safe supervision for all children's activities. And for "late niters," the ranch's security patrol will periodically check the kids while parents enjoy a show or dance band.

But it is horses and riding them that so intrigued "Bucky" and "Toolie" Turk that they decided in 1958 to build Rocking Horse Ranch. The ranch is located in the scenic, historic Hudson Valley in the foothills of the Catskill Mountains, just 90 minutes north of New York City. On its 600 acres, the brothers have created one of the largest and best-equipped dude ranches east of the West.

The ranch is a paradise for horse lovers and urban cowboys. The stable boasts more than a hundred well-groomed horses. They're available for riders at all levels of ability — from pony ride or cart for the very young, to gentle and well-mannered for the novice, to frisky for the experienced rider. Several one-hour horseback rides are scheduled each day, and the rides are categorized as to experience. The rides are over simply gorgeous upstate trails and are a year-round activity at the ranch.

To maintain the traditional Western flavor of the ranch, there are also hayrides, bonfires complete with cowpoke "tall tales," group singing, and square dancing. In addition, as a year-round resort, there are the three swimming pools, fishing and paddle boating in the private 10-acre lake, tennis, and waterskiing in warmer weather and sleigh riding, ice skating, tobogganing, and skiing in the winter. Mini-golf, volleyball, archery, and riflery are a few of the literally dozens of sport activities. Hiking is always on the menu. There are scheduled activities, crafts, and organized sports if you care to join. Or you can choose to spend the day relaxing with a good book in a secluded spot in a comfortable lounge. There is also nightclub entertainment and dancing. Guests can truly design their own vacation.

The surrounding region has an abundance of scenic views and historic sites. It is apple-growing country, the second largest in the country; there are numerous vineyards and wineries or you might choose to tour West Point U.S. Military Academy, located on a nearby bluff. The Franklin Roosevelt and Vanderbilt mansions are only minutes away.

The ranch is ideal for groups and can accommodate any size for reunions, senior citizens, social and religious groups, school outings, day trips, and business conferences.

The Turk family is dedicated to providing a safe, clean, and wholesome experience. For decades now, families have traveled from all parts of the world to spend their vacations in Western style at the Rocking Horse Ranch. The ranch may have hundreds of acres, but the hospitality is very personalized and down home. Taking good care of visitors has been the trademark of the ranch since its inception. And it is that that brings those visitors back year after year.

Address: 600 Route 44-55, Dept. BP, Highland, NY 12528

Telephone: 800-647-2624, 914-691-2927; **Fax:** 914-691-6434

Location: 90 minutes north of New York City

Guest Capacity: 350-400

Season: Year round

Policy on Children, Pets: Family-oriented resort; terrific children's program. No pets.

Rates: $$ Modified American plan (breakfast, dinner). Children's rates; no charge for infants under 4.

Credit Cards: All major cards accepted.

NEW YORK
ROCKING HORSE RANCH

HORSEBACK RIDING IN THE SNOW.

ROSELAND RANCH RESORT
STANFORDVILLE, NY

Just 90 minutes north of New York City, Roseland Ranch Resort is family-owned and operated. Roseland is the oldest, continuously operated Western style dude-ranch resort in New York state's mid-Hudson Valley. The area is well known for its scenic beauty and the wide variety of activities it offers. Each season offers its own unique activities for you and your family to enjoy.

Open year-round, the ranch is casual, relaxing and family oriented. From the moment you arrive, our friendly and courteous staff welcomes you as one of the family. You'll meet new friends and create many memories. We are proud to say that most of our guests feel like part of our family when they leave.

On its 800 acres, the Fichera Family has captured the Old West and brought it to the East. Whether it is a pony ride for the very young, a leisurely walk through the woods, or a gallop into the sunset, our well-mannered horses and experienced wranglers can be accommodating to your level of ability. For the early-risers in the family, Roseland has an early morning breakfast ride that is an exhilarating experience for both the riders and non-riders. A horse or hayride can be taken through the woods where you can take in the crisp country air and experience nature's wildlife. You will arrive in an open field where a nice warm hearty breakfast and the smell of fresh brewed coffee can be enjoyed.

A getaway at the Roseland Ranch is an all-inclusive stay with hearty family-style country meals served each day plus late night buffets on the weekends and freshly made cakes, muffins and coffee around the clock.

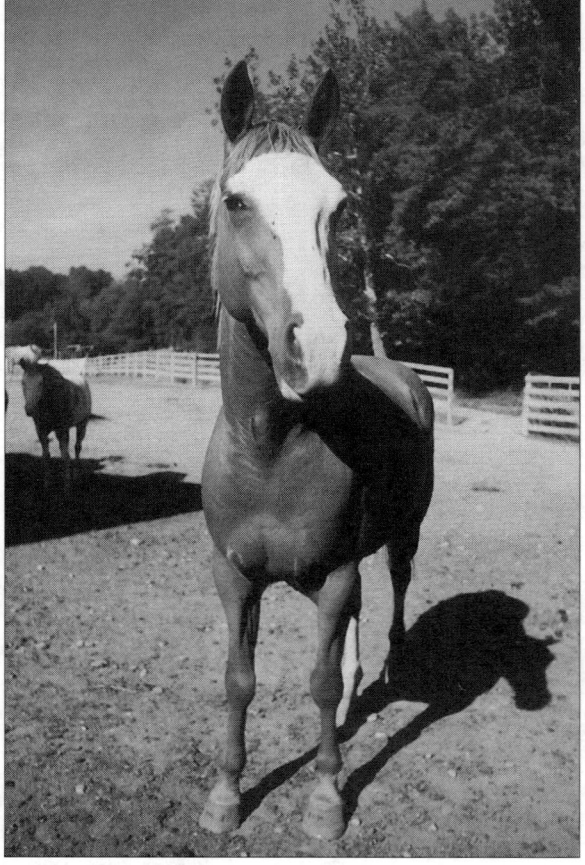

In addition to the horseback riding, the ranch maintains the western attitude with hayrides and marshmallow roasts. As a year-round resort, there is child care, arts & crafts, a petting zoo, indoor and outdoor heated pools, fishing, paddle boat, tennis, volleyball, shuffle board, bocci and arcade room. At Roseland Ranch you can do as much or as little as you like.

Winter brings skiing, snowmobiling, snow-tubing and our snow-making equipment which insures snow on our immaculately groomed trails with our own lifts and lighted for night skiing. The modern ski equipment - boots, poles and skis are available free of charge at our ski shop. Wintertime in New York is a special time not to be missed. It is a perfect time to escape from the hustle and bustle of city life and come and put your feet up with a good book and let us take care of you.

After a day on the trail, enjoy a soothing sauna or relax by the fireplace and then come join in the nightlife. Dancing, excellent bands, magic shows, musical chairs and pizza party...every night is a new and exciting experience.

Roseland is ideal for groups, school outings, day trips and business conferences. The Fichera Family is dedicated to giving you enjoyable memories that bring you back year after year.

Attractions that are nearby include, scenic views, historic sites, Franklin Roosevelt's home, Vanderbilt Mansion, antique shops and the Bardavon Music House.

Address: Hunns Lake Road, Stanfordville, NY 12581

Telephone: 800-431-8292; **Fax:** 914-868-7203

Location: 90 minutes north of New York City

Guest Capacity: 280

Season: Year round

Policy on Children, Pets: Children welcome.

Rates: $$

Credit Cards: Visa, Mastercard, Discover and American Express

NEW YORK

ROSELAND RANCH RESORT

I LOVE THIS GUY!

NEW YORK

ROSELAND RANCH RESORT

THERE'S ALWAYS ROOM FOR MARSHMALLOWS!

(SEE PAGES 288-289 FOR COLOR PHOTOS!)

New York

NORTH CAROLINA

CLEAR CREEK RANCH
BURNSVILLE, NC

Clear Creek Ranch is nestled in the lush green mountains of western North Carolina, just off the famous Blue Ridge Parkway. The scenery is unsurpassed and the weather is usually cool and comfortable, even in mid-summer. Built in 1995 by Rex and Aileen Frederick, the ranch realizes their dream of creating a bit of the "Old West" east of the Mississippi.

Buildings at the ranch are a perfect blend of new comfort and old charm, both spacious and cozy. The buildings are constructed of log siding, with ponderosa pine interiors, and offer a comfortable and relaxed feel of "rustic luxury". The main lodge features a huge stone fireplace which often is still roaring in mid-June, and a spacious dining room where mouthwatering homecooked meals are served. The rooms are decorated with country quilts and lodge pole pine beds and all have spacious porches with highbacked rockers and a spectacular view of the mountains. The main lodge, with its large stone fireplace, comfortable furnishings and relaxed atmosphere, always seems to draw a crowd, where new friends are made and old acquaintances are renewed.

It's often difficult to get guests off the porch to participate in the many activities offered each day. An excellent string of sure-footed mountain horses will provide the perfect mount for each rider and chosen to suit the individual's riding ability. There are trail rides offered each morning and afternoon. Aside from horseback riding, there is an outstanding mountain golf course just a mile up the road. Guests can also opt to take a ranch van trip to one of our local artisans....wood carvers, glass blowers, weavers, potters, quilters....or they might want to visit Linville Falls (where The Last of the Mohicans was filmed), or Linville Caverns or the Gem Mines.

Weekly whitewater rafting trips are offered. "Tubing" on the Toe River is a highlight with the wonderful sound of children laughing with delight as they sit in the big rubber tubes and float down the river!

Meals are sumptuous feasts of homemade breads, pastries and "downhome" cooking, and are served family style. A weekly steak cookout on the banks of the Toe River and Saturday nights are "rib and chicken cookout" on the ranch, followed by a "Hootenanny" presented by our talented staff of college students. A favorite with the children is the Monday night hayride and marshmallow roast down at the Clear Creek Ranch "Teepee" and everyone loves the Friday night line dancing!

Between activities, there is always time for a dip in the heated pool or hot tub. If you are so inclined, casting a line in the stocked trout pond will almost always produce a nice fat trout that the chef will prepare especially for you! Our private pond and nearby South Toe River offer fishing at its best. The Rainbow and German trout always seem ready to do battle. Whether you are an experienced "old-timer" or a beginning fisherman, come prepared to spend some marvelous hours testing our waters.

The most unique feature of Clear Creek Ranch is not all the funfilled activities, although they are great...it's the warm family atmosphere that is projected by the hosts and their staff. If the rate of returning guests is any indication, the folks at CCR are really doing it right!

Credit Cards: Visa and Mastercard

Address: 100 Clear Creek Drive, Burnsville, NC 28714

Telephone: 828-675-4510; **Fax:** 828-675-5452
Web: www.clearcreekranch.com
E-Mail: CCRDUDE@prodigy.net

Location: We offer pick-up and return from Asheville airport.

Guest Capacity: 50

Season: April 1 through December 1

Policy on Children, Pets: Children welcome. No pets.

Rates: $$

(SEE PAGES 290-291 FOR COLOR PHOTOS!)

North Carolina
Clear Creek Ranch

REFLECTIONS ON A SUMMER DAY.

COME AND SIT A SPELL ON OUR PORCH.

NORTH DAKOTA

DAHKOTAH LODGE
MEDORA, ND

Loren and Sherri Ross operate the Dahkotah Lodge Guest Ranch on their family ranch in the savagely beautiful Badlands of Western North Dakota. Five generations have raised cattle and led the true cowboy life on over 13,000 acres of fantastic buttes, ridges, bluffs, and canyons located along the Little Missouri River.

Once you experience this scenic land horseback, you'll know why guests from around the world have joined the family for a once in a lifetime adventure vacation. Guests ride well-trained horses to help work cattle or just enjoy the scenery. Photographers, be prepared to use a lot of film! Wildlife-viewing rides in the early morning and late evening will bring sightings of our bighorn sheep, eagles, hawks, mule deer, whitetail deer, wild turkeys, pronghorn antelope, and many species of birds.

The ranch offers horseback riding for any level of experience, from beginners to advanced riders. You will ride daily, with more experienced riders allowed to move along a little faster in safe areas. Also available are roping lessons, hiking (unguided) and wrangling horses in the morning and evening.

The second week of each month (June - August) will be reserved for folks staying a full week, Sunday - Sunday. Guests will participate in rounding up and branding calves, riding herd during the move to summer pastures and an overnight pack-trip while following the cattle. Guests flying in to the ranch will be picked up at the airport (Bismarck), invited to tour Fort Lincoln (Custer's last post), then transported to the ranch in time for a steak supper. After supper, you'll meet your horse and go for a ride to enjoy your first North Dakota sunset. A campfire follows with a bit of cowboy poetry and singing old cowboy songs. Each morning, either before or after breakfast, depending on the weather, you'll ride for a couple of hours to watch for wildlife and see the Badlands come to life. Some days will involve cattle work, others will be strictly for enjoying the view. On some afternoons van rides will be available to tour the nearby Theodore Roosevelt National Park and the restored western town of Medora, the Fort Union Historic site, or taking in a local rodeo. An evening at the famous Medora Musical (not included in price) can also be arranged. We limit this package to ten people. For guests not wishing to stay a full week, three-to-six-day vacations are available. They feature two to four hours of horseback riding daily, all meals, and private cabin.

The cabins are rustic, featuring one room with carpeting and electricity. Each cabin is assigned a private bathroom in a separate building. Bedding, including hand stitched quilts, and towels are provided. Meals are mouthwatering, good old-fashioned home cooking. Don't plan to lose weight while you're here!

The Rosses cordially invite you and your family to turn your dreams of being a cowboy into reality, or just enjoy true peace and quiet under the big western sky of North Dakota!

Address: P.O. Box 465, Medora, ND 58645-0465

Telephone: 701-623-4897;
Toll Free: 800-508-4897

Location: 160 miles west of Bismarck

Guest Capacity: 15, special arrangements available for larger groups

Season: May 1 through October, hunting in November

Policy on Children, Pets: Children welcome. No pets.

Rates: $ to $$ American Plan

Credit Cards: Amex, MasterCard and Visa

(SEE PAGE 292 FOR COLOR PHOTOS!)

North Dakota

OHIO

McNUTT FARM II/OUTDOORSMAN LODGE
BLUE ROCK, OH

Blue Rock State Forest is one of southeastern Ohio's most scenic preserves — and the location of the McNutt Outdoorsman Lodge. It is surrounded by thousands of conifers and hardwoods, and you'll enjoy horseback riding in an area with rolling hillsides and deep valleys. Patty and Don McNutt own and operate the lodge on their sprawling farm, which lies just at the edge of the forest. The farm trails join with the forest paths, and you can ride or hike for hours without crossing the same spot twice. The modern, rustic, lodge-type farmhouse and guest cabins lend themselves to the landscape as well.

Patty McNutt is a horsemanship instructor, a teacher of outdoor skills, a publisher and outdoor writer, a full-time farmer, and an animal behavior consultant. Patty believes her love of the outdoors and her skills developed naturally from her Cherokee heritage and from observing her father and grandfather train horses for the military. She studied horsemanship and hunting dog training from the Canadian military and adopted their philosophies on hunting, horses, dogs, and even shooting. Since the 1940s, Patty has intensely pursued research in these fields of outdoor endeavors and she enjoys passing her knowledge on to her horseback riding guests.

An unusual feature of McNutt/Outdoorsman Lodge is that Patty and Don allow you to bring your own horse, and they'll help you map out trails to ride in southeastern Ohio Hill Country. You can also go off on your own and explore the Blue Rock State Forest. Riders should be at an advanced level.

The facilities and activities at the McNutt Outdoorsman Lodge are personalized for each guest and include instruction on archery and rifle shooting safety and proficiency, an abundance of horseback riding on farm and forest trails, an outdoor sleeping deck for guests who really want to "rough" it, and comfortable lodging. There are guided hikes through the Blue Rock area where you'll enjoy the sounds and silence of nature, the reflections of trees in silent ponds, the wing beats of wild turkeys becoming airborne, a silent deer sniffing the air trying to identify you. Naturally, they have beef cattle, a milking cow, calves, dogs and kittens, and working horses and guests are welcome to join in their care and feeding. You may be inclined to do little or nothing but have a relaxing vacation — and that's just fine too.

There are completely equipped kitchens in the guest cabins and outdoor grills where you can prepare your own meals as you like. In the evenings, there are campfires, good conversation, and tall tales.

A popular day trip away from the lodge may take you to visit the Amish, just an hour's drive north. They still do all their farm work with horses, do not fertilize with chemicals, have no electricity, and utilize buggies and horse-drawn carts as their means of transportation. It is a trip worth taking, and one you'll always remember. You can find an Amish auction going on somewhere almost every day of the week except Sunday.

Farms in this region are unique in the sense that many of them are located on ridges and employ bench-farming techniques, plowing and planting hilltops and hillsides ridges to support beef cattle, sheep, and small farm operations along with orchards, vineyards, and walnut groves. This is no ordinary horseback riding vacation, but one that will interest and educate you and your family in a different lifestyle — all while you are enjoying some of the prettiest horseback riding east of the Mississippi.

McNutt Farm II/Outdoorsman Lodge is a friendly place for you and your horse to visit where the services are geared to your personal needs.

Address: 6120 Cutler Lake Road, Dept. BP, Blue Rock, OH 43720

Telephone: 614-674-4555

Location: 1.5 hours from Columbus

Guest Capacity: 10

Season: Year round

Policy on Children, Pets: Children welcome with special permission. No pets.

Rates: $ No meals provided.

Credit Cards: None

McNutt Farm II/Outdoorsman Lodge

COME ENJOY LIFE AS IT USED TO BE.

TRANQUILITY IN RURAL OHIO.

Ohio

Smoke Rise Ranch Resort
Murray City, OH

Nestled in the rolling hills of southeastern Ohio is Smoke Rise Ranch Resort. The resort is a 2,000-acre working cattle ranch that offers guests an opportunity to experience the adventures of the West, just not so far from home. The ranch is located just 7 miles from Burr Oak State Park. The ranch's 2,000 acres are bordered by a wildlife management area and the Wayne National Forest, resulting in over 30,000 continuous acres available for ranch operations.

The geography of the property varies, from rock bluffs to lush green bottomland. There are ponds where you can water your horse or just fish the day away. The national forest and wildlife management area mean that there are plenty of opportunities to view the abundant wildlife in the area. Birds, beaver, deer, turkey, raccoon, grouse, and squirrel, just to name a few, share the land with the ranch animals.

The ranch is owned and operated by the Semingson family. Walt and Helen and their four children have been in the ranching business all of their lives, from the Dakotas to Oregon, to Arizona, and now Ohio. Breaking horses is what they do best. Walt and the "boys" participate in roping, cutting, and team-penning competitions here in the East and in the West.

Smoke Rise invites you to spend a week at the ranch and experience the exciting and challenging life of a real ranch hand. All levels of experience are welcome. Learn to work your horse or ours on cattle! As a ranch hand, you can expect to spend time in the saddle rounding up strays, doctoring sick calves, checking fences and water tanks, and driving the herd from pasture to pasture.

Private and group training sessions are available for both horses and riders of any experience. Natural Horsemanship clinics are offered at certain times with certified Pat Parelli instruction. Clinics in penning and cutting are also offered. The large indoor and outdoor arenas are used for the training sessions.

Whether you prefer to ride at your leisure or be part of the action — with cattle drives, team penning, cow cutting, and guided trail rides — Smoke Rise is sure to give you a taste of the Old West and the famed cowboy lifestyle.

After a long day in the saddle, you can refresh yourself in the heated swimming pool or relax in the hot tub while the children enjoy themselves in the playground. Special events are planned throughout the year and include cattle roundups, team penning competitions, Fun Day Ranch Rodeos, 20-mile trail rides, pool parties, barbecues, hayrides, music and dancing.

Accommodations at Smoke Rise Ranch Resort include six cabins (each of which sleeps 6), which are equipped with kitchenettes. If you prefer not to cook, Smoke Rise also offers meal packages where you can enjoy the hearty, home-cooked meals (breakfast and dinner). Guests are welcome to bring their RVs along and stay in the campgrounds (primitive or full hookups offered). In the evenings, enjoy the warmth of an open campfire, gazing at the stars, and singin' them campfire songs.

So, come to Smoke Rise Ranch Resort and enjoy a full-fledged Western vacation in southeastern Buckeye country!

Address: P.O. Box 253, Dept. BP, Murray City, OH 43144-0253

Telephone: 800-292-1732, 614-767-2624

Location: 65 miles southeast of Columbus, 20 miles north of Athens

Guest Capacity: 36 in cabins, 150 in campgrounds

Season: Year round

Policy on Children, Pets: Children and pets are welcome.

Rates: $ Modified American plan. Horse rental or boarding extra.

Credit Cards: MasterCard, Visa

Ohio

Smoke Rise Ranch Resort

MARK, BRENT & LYNN (LEFT TO RIGHT).

Oklahoma

Lotspeich Cattle Co.
Rosston, OK

Lotspeich Cattle Co. is a historic cattle ranch that has been in the Lotspeich family for nearly one hundred years. The ranch was once the headquarters of the Old Chain C Ranch, which in the late 1800s had herds of cattle grazing this entire open range area. Lotspeich Cattle Co. is located halfway between the original Fort Supply Army Post and the renowned Fort Dodge Army Post, which later became known as Dodge City.

The ranch is bordered on the north by the Oklahoma-Kansas state line. To the west lies what was once known as No Man's Land, infamous for its outlaws and its gunfights.

Through the middle of the ranch runs the historic Cimarron River, which, during the days of the "Wild West," marked the beginning of No Man's Land. About 10 miles north of the ranch is an old ghost town where, folklore has it, the "baddest hombres in the West" used to hang out. This small town afforded the outlaws a perfect hideout from lawmen such as Wyatt Earp. If they got wind of the law coming, all they had to do was ride south about 10 miles to the safety of No Man's Land. The only law enforced there was survival of the fittest.

Lotspeich Cattle Co. is a working ranch covering about 10,000 acres. There are approximately 1,000 head of cattle on the ranch. Those who wish to experience the life of a real cowboy are invited to spend three days, or more, on the ranch. There are no set schedules here, but if you wish you can ride, rope, and work the cattle alongside the ranch's cowboys and cowgirls. You can also just ride along as they drive the cattle down the historic Cimarron River and retrace the footsteps of the cowboys, outlaws, and Indians of yesteryear.

The ranch has great horses for both skilled and unskilled riders. The horses have temperaments that range from very gentle to spirited, and the Lotspeichs take great care to match each rider with a horse that reflects that rider's level of ability. If you wish, you may even bring your own horse. A number of "little Lotspeichs" ride with the cowpokes daily during the summer, so children of all ages are welcome on the ranch.

Accommodations for guests are in two bunkhouses. A ranch hand's day consists of hard work and long hours. For this, he is rewarded with great food. Our guests are treated no differently. You'll enjoy hot and hearty, home-cooked meals featuring ranch-style meats, vegetables, salads, breads, and delicious desserts. The cook believes that a full ranch hand is a happy ranch hand — and the same goes for the guests!

For guests who get all the trail dust they can handle or those who'd like a diversion from the riding and cattle work, the ranch has a swimming pool, tennis court, basketball court, and a driving range.

If you'd like the opportunity to see what it was like in the days of the Old, Wild West, visit Lotspeich Cattle Co.!

Address: R.R. 1, Box 117, Dept. BP, Rosston, OK 73855

Telephone: 405-533-4718, 800-769-8495; **Fax:** 405-533-4719

Location: 185 miles northwest of Oklahoma City, 60 miles south of Dodge City

Guest Capacity: 12

Season: May through October

Policy on Children, Pets: Both children and pets are welcome.

Rates: $$ American plan.

Credit Cards: None. Personal checks accepted.

OREGON

PONDEROSA RANCH
SENECA, OR

If you are looking for something new, something challenging, an experience to remember for a lifetime, search no more! The guests at the Ponderosa Ranch ride with the cowboys. We run up to 4000 herd of cattle on 120,000 acres which means that there is something new to do each day of your vacation at the ranch.

You would probably like to know just exactly what you will do for a week in a place with no phone in the room, TV only in the main lodge and no pool or tennis. So here is a brief idea of what happens from May to October on a cattle ranch:

May: We calve a little later than a lot of folks to avoid the severe winter temperatures, so in early May we are still having lots of calves born. If the cow needs help, the Buckaroos are ready to do whatever is necessary - even a Cesarean section if it needs to be done. We also have to get the cattle ready to go to the mountain pastures. This means it's branding time. We are old-fashioned and "head 'um and heal 'um" and use a hot iron to brand. They get vaccinations and earmarked at the same time. You can take part in as much of the branding as you want to.

June: There is still a little branding to be done but mostly we are moving onto the forest. You will be moving the cattle into the mountains through breathtaking scenery. There will be good long days in the saddle. Boy, will that hot tub look good at the end of the day!

July & August: The cattle are in the mountains and you will be helping move them between pastures or "scattering" them around the current pasture. Cattle like to congregate at the same spot (just like we folks do) and we have to move them around so they get the best use of the pasture. This also helps us to maintain the forest pastures in great condition.

September is the month we start moving them home to the Valley pastures. Those cows will hide from us which means lots of days in the saddle looking for them. The days are crisp in the mornings and warm in the afternoons. I think it's the best time of year to ride. In **September & October** we start the fall corral work, vaccinating the cows and weaning the calves. You might get to learn a lot about cow cutting this time of year. That's a sport that is gaining popularity now and has its roots in the everyday life of the ranch hand.

You might get to lend a hand sorting and shipping a load or two of cattle any time that you are here because we have yearlings, cows and calves on the Ponderosa. If you've been yearning to swing a rope, the fellas will be glad to give a little instruction in the evenings after a days ride.

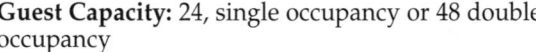

Address: P.O. Box 190, Seneca, OR 97873
Telephone: 541-542-2403, **Fax:** 541-542-2713
Internet: http://www.PonderosaGuestRanch.com
E-Mail: seeyou@PonderosaGuestRanch.com
Location: 38 miles north of Burns, 31 miles south of John Day

Guest Capacity: 24, single occupancy or 48 double occupancy
Season: Year round
Policy on Children, Pets: Guest 18 years or older only. No pets.
Rates: $$
Credit Cards: American Express, MasterCard and Visa

Oregon

Rock Springs Guest Ranch
Bend, OR

Rock Springs is a beautiful guest ranch located in the foothills of the Cascade Mountains in Central Oregon. The Summer American Plan is an all-inclusive program that runs from the middle of June through August. Guests enjoy week-long stays from Saturday to Saturday. The program includes three delicious meals each day, comfortable lodging in tastefully appointed cabins, a fantastic youth program for children ages 3 to 5 and 6 to 12, a second- to-none horseback riding program, and great outdoor activities such as tennis, swimming, fishing, volleyball, hayrides, and much more! The ranch accommodates a maximum of 50 guests so the atmosphere is always warm and the service is always personal. In fact, the staff to guest ratio is nearly one to one and that is what keeps people coming back year after year.

Accommodations are provided in warm, well-appointed duplex and triplex cabins. All cabins are finished in knotty-pine and all have private entrances, baths and sun decks. Most also have fireplaces and wet bars and serviced daily by our housekeeping staff.

All meals are served buffet style, either in the ranch dining room or out on the deck. Children enjoy meals with their new friends in a separate dining area adjacent to the main dining room. There is a special gourmet dinner for adults the night of the children's camp-out. Hors d'oeuvres are served each evening prior to dinner. Dinner can offer such entrees as fresh Northwest seafood, mouth-watering prime rib, to fresh vegetables and health-conscious choices. For snacks, we always have fresh baked cookies and fresh fruit available.

Rock Springs maintains a herd of 65 horses on a year-round basis. Because all of the horses are owned by the ranch, the Rock Springs' barn staff knows their dispositions and can match horses to riders based on personality and skill level. Unless there is a problem, guests ride the same horse all week, allowing both horse and rider to become acquainted and comfortable with one another. Rides lead to scenic areas with panoramic views of the Cascade Mountains are areas where wildlife are often found. Cantering and off-trail riding is possible over varied terrain on the edge of the high desert of the mountain foothills.

Children 6 and over can participate in the riding program and those under 6 are led on a lead-line around the barn area. An all-day luncheon ride is offered and individual instruction is also available as staffing permits. For those guests who show sufficient riding ability, the terrain is also well suited to cantering.

At Rock Springs, the youth program is a major part of the vacation experience. Rock Springs believes that the quality of the vacation experience for the young guest is as important as that of the adults. Children participating in the youth program will ride a hay wagon out to the backcountry for an "old west cook out" and sleepover. Other activities include riding, outdoor games, nature walks, swimming, arts and crafts, folklore and storytelling.

From September through June, Rock Springs operates as a conference center offering exclusive use to groups of 20 or more for corporate training and retreats.

Address: 64201 Tyler Road, Bend, OR 97701

Telephone: 541-382-1957; **Toll Free:** 800-225-3833

Web: www.rocksprings.com

Location: 16 miles from Bend/Redmond airport

Guest Capacity: 50

Season: Year round

Policy on Children, Pets: Children welcome.

Rates: $$ to $$$

Credit Cards: American Express, Diner's Club, Discover MasterCard and Visa are accepted for final payment

(See Page 293 for Color Photos!)

OREGON

ROCK SPRINGS GUEST RANCH

AFTER A HEARTY BREAKFAST GUESTS LEAVE
FOR A MORNING TRAIL RIDE.

OREGON SPLENDOR.

PENNSYLVANIA

BUCK VALLEY RANCH
WARFORDSBURG, PA

Buck Valley Ranch is located in the Appalachian Mountains of south-central Pennsylvania and surrounded by 2,000 acres of Pennsylvania state game lands. Despite its secluded setting, the ranch is only 120 miles west of the Baltimore-Washington-Northern Virginia area and 120 miles east of the Pittsburgh area.

The state game and forest lands that surround the ranch provide thousands of acres for some terrific trail riding. The terrain ranges from open fields to mountain tops. Owners Leon and Nadine Fox even offer a lunch ride where you'll stop for a picnic atop one mountain. The trail rides are all guided and at a walking pace due to the mountain terrain, but even the most avid rider will enjoy the spectacular views and wooded trails. Each season in the Appalachians has a unique "feel," and the views of the flora and fauna will be different each time you come. The ranch's well-trained horses will make even first-time riders feel comfortable and provide them with a wonderful experience on horseback.

Buck Valley Ranch is a great place to enjoy the outdoors in a private setting. The guest house is a large 4-bedroom farmhouse filled with Western memorabilia and antiques. For the exclusive use of guests, it has a modern kitchen, 1-1/2 bathrooms, air conditioning (a luxury seldom needed here thanks to the cool mountain air!), and a large porch where guests can enjoy a peaceful view of the farm or listen to the whippoorwills on a calm summer night. There is also a large in-ground swimming pool for those sultry summer days and a relaxing sauna for the winter nights. Guests can also play volleyball, horseshoes, and shuffleboard.

The Foxes offer beans and franks only at your request! The specialties of the house are healthy meals prepared from their home-grown vegetables, homemade desserts and bread, and locally raised meats. Special dietary requirements can be accommodated with advance notice.

Buck Valley Ranch is perfect for singles, couples, and groups, and there are special family packages during the week. The ranch's small size and secluded setting make it possible to tailor outings to suit individual needs and desires. It's a great place for friends and families to get together for vacations, and special occasions. Small businesses can use the ranch for company picnics and conferences.

Leon and Nadine work at making your stay a special experience. The casual atmosphere makes guests feel right at home immediately. The ranch is a member of the Pennsylvania Farm Vacation Association, and the owners never tire of answering questions about ranch life, the surrounding mountains, and the animals they have on the farm. Children and adults can help gather eggs, feed the peacocks, and play with the Australian Heeler dogs and the Manx cats. Here, you'll forget about the hustle and bustle of city life and take time to "smell the roses." There is no television in the house (they do not get any reception!), but there are plenty of board games and books to read. It's a perfect place for couples and families to spend time together and get back in touch with each other.

The Foxes' policy is to offer riding that has as little impact on the environment as possible. They practice recycling and are environmentally friendly to their surroundings.

For those who have the need to go exploring, nearby attractions include fishing (with a state license), antique and craft shows, the C&O Canal Towpath, tennis, golf, skiing, canoeing, bicycle riding, and outlet shopping. There is also the opportunity to take train rides in the mountains, receive a massage at a nearby spa, visit Civil War battlefields, amusement parks, and Bedford Village.

For a vacation that offers wonderful opportunities for scenic and challenging riding, visit the Buck Valley Ranch.

Address: Rte. 2, Box 1170, Dept. BP, Warfordsburg, PA 17267-9667

Telephone: 800-294-3759, 717-294-3759; **Fax:** 717-294-3759, **Web:** www.jimbalzotti.com

Location: 120 miles east of Pittsburgh, 120 miles west of Baltimore-Washington area

Guest Capacity: 8-10

Season: Year round

Policy on Children, Pets: Children must be age 7 to ride on the trails. No pets.

Rates: $ American plan. Children half-price during the week.

Credit Cards: Discover, MasterCard, Visa

PENNSYLVANIA

BUCK VALLEY RANCH

"I THINK YOU'RE SUPPOSED TO GET ON!"

Pennsylvania

Mountain Trail Horse Center
Wellsboro, PA

The home base of Mountain Trail Horse Center is nestled within the Tioga State Forest, which has over 160,000 acres of some of the most pristine tracts of forest lands in the mid-Atlantic region. Wellsboro is a landmark in its own right, famous for its gaslit boulevards and genuine small-town charm. It also plays host to Pennsylvania's Grand Canyon — the Pine Creek Gorge, a National Natural Landmark, with depths in some places exceeding 1,000 feet.

The terrain in north-central Pennsylvania is mountainous and rugged. With the Tioga State Forest adjacent to two other state forests, Mountain Trail Horse Center has access to some genuine back country and can provide true wilderness adventure trips. Each season in Pennsylvania provides its own unique beauty. The spring time envelops the forests with lush greens, new buds, and young wildlife sightings. Late June is when the mountain laurel blooms and fills the woods with large expanses of pink and white. The mountains and foliage provide a comfortable canopy even in the summer, and the hues in the fall are unparalleled, making autumn one of the busiest seasons.

Mountain Trail Horse Center, custom outfitters since 1982, offers half-day and day-long trail rides plus overnight expeditions. Although somewhat rigorous, the half-day rides are at a walking pace and are nice for the less adventurous or for those folks who want to warm up with a shorter ride. Actual riding time is about three hours. Full-day rides cover twice the miles and offer a variety of pace and terrain. and an opportunity to canter and gallop.

The overnight trips are very popular and can vary in length from two days to five. You'll go where motor vehicles can't, through country you didn't know still existed. You'll journey through vast forests, explore old logging roads, climb mountain trails, descend into valleys, and ford rushing streams. Guests have the option of bringing their own camp gear or of having Mountain Trail provide it. Mountain Trail offers a choice of economy, basic, and deluxe trips, which vary in the gear provided, food served, and the services provided. Regardless of the level chosen, Mountain Trail provides guide service, horses, and tack. With a deluxe package, you arrive at an already set-up camp equipped with large four-person canvas tents, a varied meal you don't have to cook, hors d'oeuvres, and beverages. Basic packages include food and gear, but the fare is simpler and there are less comforts at camp. On economy trips, guests provide their own camping gear and food. Mountain Trail will take care of the logistics and horses.

Trips can be scheduled through most of the year. There are also rides that combine camping with a stay at a quaint, country inn and rides specifically designed for families.

The horses are fine mountain horses, and Mountain Trail maintains a variety of spirit in its string to accommodate different tastes and riding experiences. Riding experience is helpful. On the overnight trips, you'll travel through different areas each day, going through a variety of terrain, with constantly changing ecosystems and landscapes.

Mountain Trail Horse Center also features custom-built, covered wagon rides that travel through the Pine Creek Gorge. Rides are commonly two and three hours, but day rides and an overnight wagon ride can be scheduled. Mountain Trail Cottage, furnished and equipped with an outdoor spa, is available for rental at any time.

Mountain Trail Horse Center provides a western flair in the East. Owners Patrick and Jill Maier are serious about riders' safety, serious about the quality of their horses, and passionate about the beautiful country they ride in!

  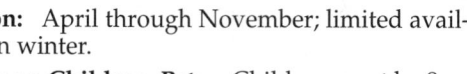

Address: R.R. 2, Box 53, Dept. BP, Wellsboro, PA 16901

Telephone: 877-376-5561; **Fax:** 877-376-2453

Web: www2.epix.net/~mthel or www.jimbalzotti.com

Location: 1 hour from Elmira, New York; 1 hour from Williamsport, Pennsylvania

Guest Capacity: 24

Season: April through November; limited available in winter.

Policy on Children, Pets: Children must be 8 years old for family rides, 12 years old for others. No pets.

Rates: $ - $$. Weekday rates. Meals provided depend on package chosen.

Credit Cards: MasterCard, Visa for deposits only. Personal check or cash preferred.

PENNSYLVANIA
MOUNTAIN TRAIL HORSE CENTER

SETTING UP CAMP.

TAKING THE GUESTS FOR A RIDE.

SOUTH DAKOTA

TRIANGLE RANCH B & B
PHILIP, SD

The Triangle Ranch B & B is owned and operated by Kenny & Lyndy Ireland. It lies in the midst of rolling prairie lands, grazing cattle and horses, lush river/hay bottoms dotted with ancient cottonwood trees, and rough river breaks to the East. The scene is much the same as H. H. Williams saw when he came west of the Missouri River in 1903 to purchase a relinquishment on the 160-acre homestead of Smokey Stearns, owner of the Triangle horse ranch.

H. H. Williams was Lyndy's great-grandfather. In the spring of 1904, he brought his wife Grace and his family out here to live. As was the lot of most Dakota prairie homesteaders, they faced hardships. H. H. was a land "locator," and he and his family ranched and farmed here through drought, epidemics, and good times. At first, they lived in a creekbank "dugout," then in a log home. It was in 1923 that they built the "Sears & Roebuck Honorbilt Home," loved by six generations and now housing the Ireland's B & B.

The "Alhambra" and matching garage were recently placed on the National Register of Historic Places because of their Sears & Roebuck origin, unique Spanish style, and significance to the history of the area. The facade is hand-thrown stucco with mission-style parapetting trimming the dormers, attached stairway, and front porch. It features numerous French windows and doors and has a grand total of 409 window panes! The house arrived in two railroad cars with the rafters and other pieces precut and numbered. The cabinetry was already assembled.

The ground floor is oak, mission or prairie styling, with crystal chandeliers, bay windows, and leaded glass bookcases. Comfortable, overstuffed furniture echoes the 1900-1920s decor throughout the house. Family heirlooms, such as the sleigh bed and matching dresser from Grandma Grace's log home, blend with acquired antiques that have been "adopted" by the family. Except for the Sony in the living/common room, you could feel you've gone back in time 70 or 80 years.

Upstairs, greeting the morning sun, are the Blue and Red rooms, with their connecting bathroom. Lace curtains blow in the breeze as wild turkeys "putt" their morning calls and whitetail deer graze just across the river bottom. Sleepyheads can snooze in the pristine White room, bathe in the clawfoot tub, or relax in the rustic Cowboy Room with two bunks.

In the early morning, smells reminiscent of Grandma's house drift up the stairwell tantalizing you to sit down at the round oak table and enjoy the generous ranch breakfast and warm conversation.

Overnight horse stabling consists of both indoor and outdoor tie stalls, indoor box stalls, and outdoor plank and pole corrals. As Lyndy's grandfather raised Registered Hereford cattle, Arabian horses, and held livestock auctions here, there are numerous stalls and pens that are no longer used for working cattle.

Lyndy's father, Vint Williams, is an accomplished guide and unsurpassed historian who is also active in running the ranch. Ride your horse, accompanied by a host, for a ranch experience to be remembered. Also offered is hiking, fishing, bird-watching, campfire entertainment, vehicle ranch tours, and a base camp for area game hunting. Artists and photographers are welcome.

Area attractions include Badlands National Park, 1880 Town, Pioneer Auto Museum, Petrified Gardens, Pine Ridge Indian Reservation, and Wounded Knee Memorial. The world-famous Wall Drugstore is just 30 minutes away.

Kenny and Lyndy were both born and raised on ranches in this area. They love the lifestyle and are interested in livestock, riding, music, reading, poetry, art, woodworking, sewing, fishing, and family (especially grandchildren). And they're quick to say, "Come visit us, the coffee's always on!"

Address: HCR 1, Box 62, Dept. BP, Philip, SD 57567

Telephone: 605-859-2122

Location: 82 miles east of Rapid City

Guest Capacity: 10

Season: Year round

Policy on Children, Pets: Children are welcome. No pets.

Rates: $ Full ranch breakfast provided. Children's rates. Horse boarding extra.

Credit Cards: MasterCard, Visa

South Dakota

SOUTH DAKOTA

TRIPLE R RANCH
KEYSTONE, SD

The Triple "R" Ranch is located in the heart of the beautiful Black Hills of South Dakota, rich in history and natural beauty. The ranch is also near one of the great wonders of the world, Mount Rushmore. The Black Hills, much more than mere hills, boast of Harney Peak, the highest point between the Atlantic Ocean and the Rocky Mountains. This peak, which rises upt 7243 feet, is located only a few miles from the Triple "R" Ranch nestled in the hills at a mile high elevation. General George Custer's discovery of gold in the Black Hills attracted Wild Bill Hickok, Calamity Jane and a host of others to the area.

The friendliness of this area is extended to guests by the owner/hosts, Jack and Cherrylee Bradt, who choose to keep the ranch small, thus assuring the personal attention that they feel is so important in creating their "ranch family".

Horseback riding is the main activity at the ranch. Each rider must be 7 years or older, and will be given their own horse to keep for the entire stay. Morning and afternoon rides take you into the surrounding Black Hills National Forest where wildlife is frequently seen. The all-day ride, with a picnic on the trail, is highlighted by a visit to a ghost town.

Riding through mountain terrain challenges even the experienced; however, the combination of a good horse, conscientious wranglers, and an orientation and safety program assure the confidence of beginners. Mt. Rushmore is often viewed while riding the Triple "R" trails.

Accommodations are not luxurious. However each duplex cabin or room is quite comfortable, quaintly furnished, and has its own private bathroom. Our sit-down, pass-around, family style home cooked meals are sure to please and no on goes away hungry. There are also breakfasts and picnics on the trail and a chuckwagon supper. Ranch guests may dine on barbecued ribs and thick-cut steaks from our native state. You'll find our meals hearty and stick-to-your-ribs good!

The schedule allows guests to set their own pace and to be as active or inactive as they choose. The relaxing, quiet atmosphere is conducive to vanishing stress. A variety of activities include trips to Mount Rushmore and Crazy Horse Memorials in ranch vehicles, a breakfast ride, picnic lunch on the trail, chuckwagon supper, swimming, fishing, hiking and birdwatching -- something for everyone! The nearby Black Elk Wilderness Area and the Norbeck Wildlife Preserve offer guests a great opportunity to see both native wildlife and flora.

The Triple "R is open year around. During the main season, (Memorial through Labor Days), guests are hosted at the original ranch where a minimum of 4 days stay is required. During the off-season, (minimum of 3 days), guests are accommodated at the horse ranch.

The Triple "R" is not unique because it is located in scenic country, with fresh air and star filled skies; nor because it has friendly hosts, great horses, good food, adventure, history, wildlife, nor the fact that you don't want to leave when your stay is over. Why is it unique? Well, pardner, you'll just have to come to see for yourself.

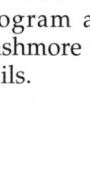 

Address: P.O. Box 124B, Keystone, SD 57751-0124

Telephone: 605-666-4605; **Toll Free:** 888-RRRANCH; **Fax:** 605-666-4896

Location: 28 miles from Rapid City, SD

Guest Capacity: 12

Season: Year-round

Policy on Children, Pets: Children welcome, must be 7 years old to ride. No pets.

Rates: $$

Credit Cards: Visa and MasterCard

(SEE PAGES 295 FOR COLOR PHOTOS!)

South Dakota
Triple R Ranch

LOVE AT FIRST SIGHT.

SOUTH DAKOTA

WESTERN DAKOTA RANCH VACATIONS
WALL, SD

Welcome to Western Dakota Ranch Vacations and the Shearer Ranches. These are working cattle and horse ranches located along the Cheyenne River north of Wall, South Dakota (and the world-famous Wall Drug). The Shearers are also outfitters for cattle drives, wagon trains, "posse-bandit" treks, and horseback rides in the Badlands National Park and Pine Ridge Indian Reservation.

Here, you'll see country like that depicted in *Dances with Wolves* — the Cheyenne River, open prairies, and cedar-tree covered canyons. The Shearers have, in fact, been outfitters for motion pictures, television movies, and commercials with a Western theme: *Wyatt Earp*, *Crazy Horse*, *Return to Lonesome Dove*, *Far and Away*, *Geronimo*, plus others... and of course *Dances with Wolves*. Out on the prairie are historical wagon-wheel ruts that date back to the days of the Deadwood to Fort Pierre stagecoach trail. See the crossings on the Cheyenne River and the old buildings of a stagecoach stop on the river.

During your stay, you can choose to ride with the ranch hands and experience the life of a cowboy. You might find yourself checking on the Texas Longhorn, Black Angus, and Crossbred Commercial cattle herds. Perhaps you'll help move cattle from one pasture to another, ride a fence line, or check water and mineral supplies out in the pasture. Or, just stand back and watch.

The Shearers have horses to match every ability or experience their riding guests have. Or, you can bring your own horse. There are corrals and box stalls for traveling horses.

Guests have several choices of lodging. The air-conditioned ranch home has five sleeping rooms and lots of cots for the large recreation room, which has a built-in 7-ft. x 10-ft. trampoline. The Bunk House sleeps four, has a kitchen, and a bathroom with a tub for soaking. The rustic Log Cabin is a duplex and sleeps twelve in bunk beds. It has lanterns, wood stoves for cooking and heat, an outdoor grill, and "outdoor" plumbing. For the adventurous, other, more unusual lodging includes Indian teepees, a sheepherder's wagon, and cowboy tents.

Guests can also camp at the primitive cowboy camp area along the Cheyenne River. There is a collection of old wagons and buggies on display — the "circle of wagons" — with a fire pit in the middle. The chuckwagon, complete with its fire pit, tripods, iron cauldrons, squaw-coolers (for shade), and log seating is at one end of the area. An Indian teepee is set up along the river. The area has cottonwood trees to park under, and here too there is "outdoor plumbing." Guests can swim, wade, and fish in the river.

The Shearers offer a unique 1880-style wagon ride and chuckwagon supper. Guests are taken on a covered wagon ride while the chuckwagon cook prepares a "cowboy meal" in cast-iron pots over an open fire. The menu is usually son-of-a-gun stew, cowboy beans, sourdough fry-bread, and whipped honey served on tin plates with coffee from an old tin can!

Many guests make this their "home base" for exploring. The Shearers will gladly help you plan a tour of the area. The Badlands National Park is only 20 minutes away, where you can experience the views, feel the soil and climate, and see the plant and animal life. It's just a 1-1/2 hour drive to Mount Rushmore, Crazy Horse Mountain and Museum, and Wind Cave. Beautiful Spearfish Canyon is 2 hours away, and Homestake Gold Mine is in the same area.

Some people believe that you can only get a "cowboy experience" out West. Western Dakota Ranch Vacations will help you plan a Western vacation that is everything you dreamed it would be!

Address: HCR 1, Box 9, Dept. BP, Wall, SD 57790

Telephone: 605-279-2198, 605-279-2300;
Fax: 605-279-2250

Location: 9 miles north of Wall, 60 miles east of Rapid City

Guest Capacity: 24 in lodging, up to 150 for camping

Season: Year round

Policy of Children, Pets: Children are welcome. No pets preferred.

Rates: $ American plan for ranch stays; breakfast provided for B&B stay.

Credit Cards: None

(SEE PAGES 296 FOR COLOR PHOTOS!)

SOUTH DAKOTA
WESTERN DAKOTA RANCH VACATIONS

HORSES AND CATTLE.

MOVIN' THE HERD ACROSS THE STREAM.

TENNESSEE

GILBERTSON'S LAZY HORSE RETREAT
TOWNSEND, TN

A warm welcome awaits you and your horse at Gilbertson's Lazy Horse Retreat, which is located in scenic Dry Valley within 1-14 miles of several of the major trail heads leading into the Great Smoky Mountain National Park. The Smokies have some of the finest horseback riding in the South. Trails extend for miles and are some of the most scenic to be found. The area is abundant with wildlife, so make sure you pack a camera in your saddlebag. You'll find the riding as easy or challenging as you'd like, and you can ride for miles without crossing your own tracks twice.

At Gilbertson's Lazy Horse Retreat, you can choose any one of four separate cabins for your accommodations. Each cabin is in its own secluded setting and is completely furnished— equipped for your "home away from home." The three main cabins at the retreat — "The Rancher," "The Treehouse," and "The Hideout" — each sleep six people comfortably with a queen-sized bed in each of the two bedrooms and a sofa sleeper in the living room. These cabins are decorated with a Western flair and have cozy fireplaces and two-person Jacuzzis, which makes for a totally relaxing, not to mention romantic, atmosphere. The cabins have fully equipped kitchens, washers and dryers, central heat and air conditioning, telephones, and color TVs with VCRs. The fourth cabin is the one-room "Cowboy Cabin," which sleeps two to four people with a queen-sized bed and a queen-sized futon. It too has a fully equipped kitchen, telephone, color TV, heat and air conditioning. After a "hard" day on the trail, come back and relax in the rocking chairs or porch swings on the front porch of your cabin, enjoy the beautiful mountain views and the peacefulness of the woods. The charcoal grills at each of the cabins are just waiting for a good ol' family cook-out! If you don't feel like cooking, there are a number of restaurants in Townsend where you can get anything from pizza to burgers to steaks to seafood.

Horses visiting Gilbertson's Lazy Horse Retreat will also feel pampered in the stable and paddocks. The stable, which is easily accessible to accommodate large horse trailers, provides 11 indoor box stalls, two 80-ft. x 100-ft. paddocks with wooden fences, and two round pens. The stalls are cleaned daily, and fresh wood shavings are provided daily for your horse's comfort. You should, however, bring your own feed or hay.

When you're not out on the trail, there are plenty of other activities to keep you very busy. Nearby you can fish for trout, play golf, or swim in or go tubing down the Little River. Tuckaleechee Caverns, a popular tourist destination, is just one mile from the Retreat. If all this isn't enough to keep you entertained, action-packed Pigeon Forge and Gatlinburg are approximately 20 miles away. There you can enjoy the Dollywood amusement park or stroll through the many shops and outlet stores. But most of the area's best craft shops and restaurants are located right in Townsend.

If you are out sightseeing, a definite must-see is Cades Cove, in the Great Smoky Mountain National Park. Cades Cove is an outdoor, historic museum depicting life in the mountains in the mid-1800s. You can drive or bicycle around this picturesque valley on the Loop Road to see deer, bear, turkey, and other mountain wildlife roaming the fields and grazing peacefully in the pastures. It's also the perfect place for a picnic!

Come to the lovely Smoky Mountains and enjoy the Southern hospitality. And don't forget to bring your horse!

Address: Dept. BP, 938 Schoolhouse Gap Road, Townsend, TN 37882

Telephone: 423-448-6810

Location: 20 miles from Gatlinburg and Pigeon Forge

Guest Capacity: 22

Season: Year round

Policy on Children, Pets: Children are welcome. No pets (except horses).

Rates: $ No meals. Winter rates available.

Credit Cards: American Express, Discover, MasterCard, Visa

402

Tennessee
Gilbertson's Lazy Horse Retreat

BACK AT THE STABLE AFTER ANOTHER GREAT RIDE.

STOPPING FOR A DRINK AT A BEAUTIFUL
MOUNTAIN STREAM IN THE SMOKIES.

TENNESSEE

NAMASTE ACRES BARN, BED & BREAKFAST
FRANKLIN, TN

Namaste is an Indian word meaning "we welcome you and the peace that resides in all men." And that's just what owners Lisa and Bill Winters do at Namaste Acres Barn, Bed & Breakfast in central Tennessee.

Namaste Acres is located in Leiper's Fork in a beautiful valley 20 miles southwest of Nashville in an area that lies adjacent to the famous Natchez Trace. The "Trace" was built by the U.S. Army in 1801, through the 500 miles that separated Nashville from the new Mississippi Territory. The trail became the Natchez Trace Parkway in 1938, and in 1991 a scenic 26-mile horse trail was blazed to parallel the parkway. Parts of that trail are actually the original Natchez Trace.

The Winters family moved to the area in 1993. When Lisa saw the Trace, she said "We've got to do something. It's so beautiful out here. We've got to share this with the world." So they turned their home into a bed and breakfast — and barn — and got permission from the Department of the Interior to blaze an access trail from Namaste to the trail head.

Situated on 12 acres, Namaste Acres is a warm, inviting Dutch colonial home with four guest suites, each with private entrance, private bath, TV/VCR, fridge, coffee maker, and telephone. Each suite is decorated in a particular theme. The "Tennessee Bunkhouse" is a favorite. Its Old West flavor includes rough-sawed lumber walls, handmade furniture, a claw-footed bathtub, and horse-related decorations. The "Carriage House," "Frontier Cabin," and "Lodge" suites round out the accommodations. Guests awake to the sound of a crowing rooster and a nineteenth-century dinner bell calling them to a full country breakfast.

Namaste Acres caters to people who want to leave the hubbub of cities and work and get away to the peace and quiet of the woods — and who want to bring their horses along. For those who don't have their own horses, arrangements can be made for guided trail rides. In addition to the Natchez Trace, other riding opportunities available to guests include a local nature park with 800 acres of trails, and Percy Warner equestrian park. Children 10 and older may ride with parents.

As a change from riding, guests can hike in the valley or along the Trace. Guests who bring their own bikes will find good riding in the area. Namaste Acres also has an indoor exercise room that includes a cross-country ski machine, computerized stationary bike and free weights. The hot tub is available for soaking tired and sore muscles in all seasons.

The outdoor swimming pool area has deck chairs and sun loungers. Comfy rocking chairs, swings, and hammocks on the various decks around the guest suites round out the rest-and-relaxation amenities. There're also horseshoe pits, outdoor gas grills, and a fire pit for evening campfires, weiner roasts, or just jawin'. Guests, especially children, enjoy playing with the Winters' animals, which include dogs, cats, horses, and rabbits.

Central Tennessee is a historic area, and there are numerous things to do and see close by. Guests can go on their own day trips. Franklin is just 20 minutes away. Its entire 15-block downtown is on the National Register of Historic Places. Franklin, voted the best small town in the U.S.A., is filled with Civil War homes, sites, cemeteries, and battlefields. Its picturesque town square has a variety of great little shops and unique eating establishments., and the town has one of the largest groupings of antique dealers in the nation.

Nashville, "Music City, U.S.A.," is located 45 minutes to the northeast. There you'll find the world-renowned Grand Ole Opry, Opryland, The Hermitage of Andrew Jackson, Music Row, The Parthenon, the Hard Rock Cafe and Planet Hollywood, and any number of night spots. Numerous other attractions for a variety of interests are only minutes away.

Address: 5436 Leipers Creek Road, Dept. BP, Franklin, TN 37064

Telephone: 615-791-0333; **Fax:** 615-591-0665

Location: 12 miles west of Franklin, 20 miles southwest of Nashville

Guest Capacity: 12

Season: Year round

Policy on Children, Pets: Children must be age 10 or older. Some pets accepted with prior permission.

Rates: $ Country breakfast provided.

Credit Cards: American Express, Discover, MasterCard, Visa

TENNESSEE

SERAPHIM STABLES
HURRICANE, TN

Seraphim Stable is located one hour west of Nashville in the heart of middle Tennessee. With over 300 acres of lush fields, miles of quiet countrylanes, and both indoor and outdoor arenas, Seraphim provides *"Heaven on Earth for Horse & Rider"*.

Open year round, Seraphim provides an oasis for horseback adventurers, whether overnight guests, vacationers, or monthly boarders.

The atmosphere at Seraphim is far removed from the hustle and the bustle of modern life. An aura of peace and tranquility surrounds the property as our broodmares graze in the emerald fields. The Owens' family home for over one hundred years sits graciously beneath towering trees. No smokestacks or industrial sites mar the pristine blue of the Tennessee sky, and the air is rich with the songs of the birds who nest in the surrounding woodlands.

Miles of great trail riding is accessible directly from the barn and the variety and challenge of trails at two nearby commercial trail ride facilities are available seasonally to those who choose to trailer their horses a short distance. Both the 60' X 100' enclosed arena and the adjoining 100' X 200' outdoor arena feature sand footing and full overhead lighting. The 50' round pen facility is also well lit and has # 10 lime footing.

Equine amenities include 12' X 12' box stalls, each with an ample, attached run, rubber floor matting beneath high quality saw dust, and individual fans. Both indoor and outdoor wash stalls are safely designed, secure and easily accessible. The indoor wash stall also features on-tap hot water and a heat lamp to facilitate timely drying. Both barns feature security lights and radios (the horses love music!). A free-standing two stall facility is also available with the same features as found in the large barn.

Both the well-designed tack room and the barn office are fully air conditioned and heated. The tack room offers ample storage for tack and grooming tools, while the office provides toilet and kitchen facilities, a comfortable sofa and the Seraphim Stables' collection of equine related videos, books, and magazines.

A range of guest facilities and restaurants are available within just two miles of Seraphim Stables. Overnight accommodations include both national chain motels and campground accommodations. The Seraphim Stables staff can provide you with telephone numbers for these facilities.

Riding lessons aboard your own horse are available, based on the understanding that riders of all abilities are most comfortable on their own horse. Both independent and guided trail rides on the fertile Buffalo River Valley farmlands can often be arranged. Poles, barrels, and cavalettis are available for contest training.

Seraphim Stables provides the perfect location for a relaxing get-away with both your human and horse friends. Whether taking a relaxing ride down a shady country lane, training in the indoor arena, reviewing ground work in the round pen, or enjoying a moonlight ride in the outdoor arena, Seraphim enables both you and your horse to relax in safety, luxury and convenience at a first rate equine riding and boarding facility.

Address: 125 Owens Cove, Hurricane, TN 37078
Telephone: 931-296-1967
Location: 1 hour west of Nashville
Guest Capacity: None — Lots of Motels near by

Season: Year round
Rates: $
Credit Cards: None

TEXAS

THE BALD EAGLE RANCH
BANDERA, TX

The Bald Eagle Ranch - where quality is more important than quantity - is the only Texas ranch approved as an active member of the prestigious Dude Ranchers' Association. The land was purchased in June of 1994 and the construction of the buildings was completed by April of 1995. The goal was to create a first-class facility allowing our guests to enjoy "a unique Western experience in a setting which involves a livestock oriented way of life and a sincere concern for the visitor's comfort and pleasure." (The last quotation is the description of a dude ranch by Dude Ranchers' Association).

Horseback riding is our main activity. As a rule, we ride in the mornings and in the afternoons, but we always stay flexible according to our guests' ability and experience as well as weather conditions. Our terrain is rocky and steep and for the safety of our guests, we do not do fast riding. To quote Western Horseman magazine (September 1997) "Take lots of film to Bald Eagle, because the leisurely pace of the riding and the changing beauty of the rocky, rolling, brush-dotted hills will give you lots to shoot. The combination of the ranch and the adjoining Hill Country State National Area provide close to 6,000 acres of Texas to ramble over, served by almost 40 miles of trails. These provide a nice mix of meadows, occasional high points, trees, water, and some old ranch buildings in the park". In addition to this, our two head wranglers have only 60 years combined riding experience, and they are still learning! We work very hard at the Bald Eagle, but what is a "working ranch"? We breed the purest known Texas Longhorns in the world, but we do not consider ourselves a "working ranch". Our definition of a "working ranch" is that at least 50% of the revenues should come from raising cattle. We do not qualify for this.

Our culinary trained chef makes sure that eating time is one of the daily main events. While most meals are served buffet style, weekly dinner events such as an outdoor Texas barbecue, hamburger cookouts and formal plated dinners all contribute to the excitement of the meals. Breakfast always includes a "cold bar" accompanied by an additional hot dish with a casual brunch on Sundays. Lunch and dinner feature New West Cuisine, a blend of Texas, Mexican, Native American and Southern American influences. With advance notice the ranch can accommodate vegetarians or those with special dietary concerns.

Rates include accommodations in comfortable modern Western style rock and cedar fourplex cabins. All guest units are suites consisting of a bedroom with either king size or queen size beds, a second room with a Futon which may be used as a sitting area or a second bedroom, a bath and private porch. All suites have two ceiling fans and are individually air-conditioned. Daily maid service is provided. Also included are 3 great meals each day, horseback riding and lessons, entertainment and other ranch activities.

Address: P.O. Box 1177, Bandera, TX 78003

Telephone: 830-460-3012; **Fax:** 830-460-3013
Web: www.baldeagleranch.com
E-Mail: info@baldeagleranch.com

Location: Less than 1-1/2 hours from San Antonio

Guest Capacity: 16

Season: Year round

Policy on Children, Pets: Children under 16 are accepted only when the ranch is reserved by a family/group. No pets.

Rates: $$

Credit Cards: Visa, Mastercard and Amex

(SEE PAGE 297 FOR COLOR PHOTOS!)

Texas

The Bald Eagle Ranch

TEXAS HOSPITALITY.

ENJOY A COOL DIP AFTER A DAYS' RIDE.

Texas

Leon Harrel's Old West Adventures
Kerrville, TX

Located in the picturesque hill country of Kerrville, Texas, is the ranch of World Champion Cowboy Leon Harrel — and the most unique horseback riding vacation in the world. On his ranch, Leon runs his famous "Old West Adventures," with the help of his son Lance (a champion cowboy in his own right) and his professional staff. Make no mistake about it, this is no ordinary dude ranch! Written up twice in *Western Horseman* magazine and featured on CBS News, Leon sets the pace in a Western ranch vacation.

Leon Harrel is a seven-time National Cutting Horse Champion and was inducted into the National Cutting Horse Hall of Fame in 1989. He is considered an institution at the National Futurities, where he has competed for twenty-three years and made the finals an unbelievable twenty times. He is America's most well-known and respected living cowboy.

Whether you have never ridden a horse before or are an experienced rider, Leon Harrel's Old West Adventures offers the most unique opportunity to learn horsemanship and cowboy skills from a World Champion. It's like spending a week at Larry Bird's home to learn basketball. You will be expertly trained in all the necessary skills of the cowboy way, gaining knowledge and expertise while having the time of your life! Best of all, you will have accomplished things you never would have thought possible and make friends for life. Leon will turn you into a real cowboy! You will receive instruction on catching your horse plus saddling, mounting, bathing, and grooming. You will learn basic riding skills for pleasure or competition. You will be doing a lot of real riding on real horses. Get ready for a week of fun, trail riding through the beautiful hills of Texas, as well as rounding up cattle, sorting, and team penning.

For those of you who have never been on a cutting horse, hang on to your hat because you are in for the ride of your life! Leon and Lance will put you on real competition cutting horses and teach you the skills of cutting cattle. It will be unlike anything you have ever done before, yet all the while you'll build up your confidence level and riding ability.

Cowboying builds big appetites. So you'll be served a hearty breakfast, lunch, and dinner cooked at the campsite located on the range. A chuck wagon that was actually used in cattle drives in the 1800s, open-fire cooking, a hand-built cook shack, and authentic cowboy recipes make mealtime a favorite of guests. You will be treated to everything from steaks and fajitas to fresh fruit and delicious desserts. It might be the atmosphere or the company, but it sure seems to be the best dern food south of the Red River!

As the sun begins to set, you'll all gather around the campfire to listen and sing along with cowboy artists as they share with you genuine cowboy songs. Guitars, fiddles, and haunting melodies echoing in the canyons as night falls makes a perfect ending to your Old West Adventure.

After a long day on the ranch, nothing feels better than a hot bath and a soft bed. You will have accommodations at the world famous Y.O. Ranch Hotel in Kerrville. This luxury hotel is rich in the character and hospitality of the Old West.

If you're looking for something a little different — and absolutely wonderful — look to the Lone Star State and Leon Harrel's Old West Adventures.

Address: 1120 Spur 100, Dept. BP, Kerrville, TX 78028

Telephone: 210-896-8802, 800-362-7095; **Fax:** 210-792-4292, **Web:** www.jimbalzotti.com

Location: 50 miles from San Antonio

Guest Capacity: 50

Season: Year round

Policy on Children, Pets: Children must be at least 9 years old. Pets welcome.

Rates: $$$$ Corporate/group rates.

Credit Cards: MasterCard, Visa

(See Page 298 for Color Photos!)

Leon Harrel's Old West Adventures

LEON HARREL INSTRUCTING A YOUNG COWBOY.

DOIN' SOME CUTTIN.'

Texas

PRUDE RANCH
FORT DAVIS, TX

Prude Ranch is located in an area known as "The Last Unspoiled Frontier," a vast, open land known for its beauty, clean air and beautiful mountains. The ranch was established in 1897 and in 1921, the Prude family started to share the beauty of the land and the cool summer months with city dwellers. Now guests enjoy coming to the ranch anytime of the year. The Prude family and their staff treat guests right and that's why they keep coming back.

The Prude Ranch offers you the perfect place to host your family reunion, business workshops and retreats or tour groups.

Elevation at the ranch is 5,000 feet. In addition to the many activities on the ranch, there are three major attractions all within ten miles: McDonald Observatory which offers the largest telescope complex in the Northern Hemisphere; Fort Davis National Historic Site which is an 1850's Calvary fort; and the Chihuahuan Desert Research Institute specializing in hiking and plant life.

Horseback riding is at the top of fun things to do at the ranch. Other ranch activities include swimming in the enclosed and heated pool, hiking on well-marked trails, enjoying various court games of tennis, volleyball and basketball. Hayrides, chuckwagon cookouts, overnight trail rides and camp fires complete with country music and tall tales are all available by special arrangements. Boat trips on the Rio Grande are available and an 18-hole golf course is nearby. Because of our size, we are able to accommodate and tailor-suit your vacation for you.

Bird Watching is a popular activity in Texas, and birders cannot find a better place for this than the many birding trails on the ranch. There have been 60 species found on the ranch and an estimated 130 found in other parts of the Davis Mountains.

Accommodations: Guest Lodges are located away from the ranch headquarters on the mountainside. These accommodations have two double beds, dressing and shower bath. Family Bunk Rooms may have one double bed and twin size bunk beds - up to as many as six sets. Bathrooms may be "dormitory type" with multiple facilities in each room. The Ranch Bunkhouse can house from eight to twenty persons per room. All beds are bunk beds. Guests provide their own bed and bath linens.

As a working guest ranch, Prude Ranch has on-staff real live cowboys that entertain with music, camp fire cookouts, and trail rides. We have over 100 head of cattle that roam over 3,000 acres. We can choose from over 50 different head of horses to select the perfect saddle horse for you. Guests are welcome to join in and be a "cowboy" for the week. The ranch has an indoor heated swimming pool, tennis courts, corrals, RV Park, over forty motels and family cabins plus nine bunkhouses.

Meals are all served buffet style and no one goes away hungry. The ranch has the best cooks in the territory. All meals are reasonably priced and meal tickets can be purchased in the Ranch Office or may be charged to the room. Come to the Prude Ranch in West Texas and enjoy our homestyle hospitality. You'll be glad you did.

Address: P.O. Box 1431, Fort Davis, TX 79734

Telephone: 915-426-3202; **Fax:** 915-426-3502

Location: 220 miles from Midland/Odessa airport

Guest Capacity: 300

Season: Year Round

Policy on Children, Pets: Children welcome, must be 6 years or older to ride. Pets allowed if you are a RV camper only, and must be on a leash.

Rates: $

Credit Cards: American Express, Mastercard, Visa and Discover

Y.O. RANCH
MOUNTAIN HOME, TX

The Y.O. Ranch and Brand remain as part of 550,000 acres acquired in 1880 by Captain Charles Schreiner, a native of Alsace-Lorraine who immigrated to Texas with his family in 1852. He bought the Y.O. Ranch and its brand with profits from driving more than 300,000 head of Longhorn cattle along the Western Trail to Dodge City. Four generations later, the Schreiner family, including Charles Schreiner III and his four sons, Charlie IV, Walter, Gus and Louie still proudly own and operate the sprawling 50 square mile ranch located in the ruggedly beautiful Texas Hill Country just 100 miles Northwest of San Antonio.

The ranch is home to the nations largest collection of natural roaming exotic animals. Over 10,000 animals make their home on the Ranch and include 55 different species; approximately 25 of which can be hunted.

Children especially love coming to the Y.O. to participate in the Y.O. Outdoor Awareness Programs as well as the Safari Club International Apprentice Hunter programs both of which are designed to teach an appreciation for and dedication to the future of wildlife.

With forty thousand acres of Texas Longhorns, cowboys and exotic animals, the Y.O. Ranch is a throwback to our roots - a modern working ranch and the home of southwestern hospitality for over 110 years. Spread over the unspoiled beauty of the Texas Hill Country, the Y.O. Ranch is a favorite destination for the stars of Hollywood, Nashville and Wall Street. It's an adventure camp for youth and a true "out of your rut" getaway for adults.

The horseback riding program offers instruction to children and novices and children 6 years and older can ride by themselves. Daily rides are not "nose to tail" and the guided trail rides are for 1-20 people. You will horseback ride in the unspoiled beauty of the Texas Hill Country, known for being as exotic as it is beautiful. Cattle drives and "showdeos" also take place on the ranch.

A natural style swimming pool provides relaxation in a Hill Country setting as exotic as it is beautiful after a long-day's horseback ride, hike or off-road bicycling excursion.

Year-round hunting is offered in a friendly and uniquely unforgettable atmosphere enequaled by other look-alike ranches in Texas. Aggressive ranch management philosophies and the vast natural habitat of the Y.O. have resulted in an impressive collection of game animals with some of the finest hunting to be found anywhere.

Above all, the Y.O. Ranch is a real 40,000 acre working ranch. It's an experience that takes the visitor back in time - to a simpler lifestyle, a clear work ethic, and an involvement with the land and nature unique in today's busy cocrete world. The Y.O. is home to the largest herd of Texas Longhorns. It's a ranching experience for people of all ages. We also offer the "Old Cowboy Way" program to corporate groups for teaching teamwork. Groups and individual parties are lodged in Old West 1880-era cabins. And guests dine in the Ranch's Chuckwagon. Meals are included with all accommodations. We promise you a unique vacation experience that you and your family will always cherish.

Address: Hwy. 41 W, Mountain Home, TX 78058

Telephone: 830-640-3222; **Fax:** 830-640-3227

Web: www.jimbalzotti.com

Location: 100 miles NW of San Antonio

Guest Capacity: 35

Season: Year round

Policy on Children, Pets: Children welcome. Pets welcome, but must be on leash.

Rates: $ and $$

Credit Cards: Yes

(SEE PAGES 299 FOR COLOR PHOTOS!)

Texas
Y.O. Ranch

YOUR HOSTS WALTER AND TERI SCHREINER...
THE RANCH HAS BEEN IN THEIR FAMILY SINCE 1880.

UTAH

ALL 'ROUND RANCH
JENSEN, UT

Make no mistake. This is no dude ranch. "Dudes" get waited on. All 'Round Ranch is about becoming an all-round rider — taking care of yourself and your horse, doing hands-on cattle work, and pushing your personal limits along the way. All 'Round Ranch features four- and six-day cowboy adventures with the emphasis on instruction and participation. You'll develop your horsemanship through a variety of activities.

To be a cowboy, you must first become a rider. Whether an experienced rider or novice, there are plenty of skills to learn and improve. The ranch's pride and joy are its American Quarter Horses — quick, agile, strong, intelligent, and calm. The horses have personalities of their own, and much of the fun during the week is getting to know them. Al Brown, one of the owners, will match horses to riders, taking into account temperament and skill. The horse selected for you becomes your primary responsibility. Don't be surprised if you develop a bond that is part concern, part respect, and part real affection. Safety is emphasized and that emphasis results in a four-to-one instructor ratio, which translates into close supervision and individual attention. You can expect a healthy balance between teaching and learning by doing.

All 'Round Ranch offers genuine adventures; the challenges evolve naturally through the week-long journey and the chores necessary to care for horses, cattle, and each other. A prominent feature of the adventure is group living and learning. On the range, there is strength in numbers, in helping and looking after others, working together on the daily chores. There is always something to do — riding herd, culling strays, doctoring, branding calves, and moving a herd from one pasture to the next.

The riding adventures primarily take place spring through fall in eastern Utah and western Colorado on the Blue Mountain Plateau adjacent to Dinosaur National Monument. The terrain is rugged with a maze of red rock canyons, dense thickets of sage, and mountain tops covered with aspen and pines. The area abounds in wildlife. Butch Cassidy and the Sundance Kid hid in these canyons and rode these trails.

You'll camp at working cattle camps, sleeping in spacious canvas cowboy teepees. Rather than using pack horses, All 'Round Ranch utilizes the "chuck wagon" for trucking gear as well as food when camp is moved. Discover the delights of Dutch oven cookery and your untapped culinary prowess should you choose to assist with the cooking. The ranch provides all the necessary gear for the week riding this beautiful country: equipment, food, horses, tack and instruction for the adventure. And there is the relaxing evening campfire with plenty of cowboy poetry and stories. You may even be inspired to write your own!

All 'Round Ranch offers several special adventures. The Family Adventure brings parents and children (minimum age 12) together in a shared adventure to provide priceless memories and keep the lines of communication open. The Alumni Adventure is for returning guests and is structured around cowpunching. Riders will cover a lot of ground as they move the herd to holding pastures. Alumni and experienced riders can also participate in the Horse Camp adventure, where they'll get the opportunity to do some intense riding and delve more deeply into horsemanship skills and theory. Guests for the Rider-Rider-Rider adventure will help get the ranch's horses ready for the summer and fall work by riding for three, four, or five days. Adventures can also be customized for special groups.

With a staff of exceptionally qualified wranglers, owners Al and Wann Brown share a love of horses, a kinship with the land, and a desire to provide participants in these adventures with a positive experience.

Address: P.O. Box 153, Dept. BP, Jensen, UT 84035
Telephone: 800-603-8069, 801-789-7626;
Fax: 801-789-5902
Location: 3-1/2 hours from Salt Lake City
Guest Capacity: 12

Season: May - September
Policy on Children, Pets: Children must be 12 for Family adventure; other adventures age 16. No pets.
Rates: $$ American plan.
Credit Cards: American Express, MasterCard, Visa

UTAH
All 'Round Ranch

MOSEY'IN DOWN A TRAIL.

YEE — HA!

Utah

BEST WESTERN RUBY'S INN AND RUBY'S RV PARK AND CAMPGROUND
BRYCE, UT

From its early beginnings in a simple log structure named "Tourist Rest" to the current modern facilities, Ruby's Inn has been a haven in the wilds of southern Utah. Ruby's Inn is located next to Bryce Canyon National Park, famous for its intricate limestone spires. Brilliant in color and contrast, this is red rock country at its best. It is also mountain country. At 7,600 feet, the days are never hot and the nights can be brisk. The air is so clear, you can see more than a hundred miles. The skies are the truest blue, and at night millions of stars brightly hover overhead as if defying the poets to capture their description.

Ruby's Inn is known for its selection of Western activities. You can explore the mountains and meadows on horseback as you travel through giant ponderosa forests, view the beauties of Bryce Canyon from the rim trails, or relive the outlaw days on the old Butch Cassidy trails. Trail rides range from one- and two-hour rides to half-day and all-day rides. Overnight trips may also be arranged.

After a day on the trail, you can relax in the comfort of your room or in the cool, modern campground. Enjoy the swimming pools, spa, and hot tub. Join the fun at the Bryce Canyon Country Rodeo, presented nightly, June through September. Ride in a covered wagon to the Cowboy Cookout and Western Hoedown, or tour Bryce Canyon National Park by van or by helicopter or by biplane. Spend some time visiting Bryce Canyon and hike one of the trails, take some photos, and just enjoy the beautiful scenery that makes the park so unique.

Mountain bike rentals and shuttles are available at Ruby's Inn, and guided ATV rides along trails winding through the Dixie National Forest are offered. The animals at the free petting farm bring smiles to everyone's faces, and free cowboy entertainment amuses children and adults alike. And you can even try your hand at panning for gold and sapphires.

Ruby's Inn General Store is the gateway to a popular activity — shopping! You'll find a large selection of Western apparel including top name cowboy boots and hats, regional books, music, photo supplies and groceries. Visit the Western Arts Gallery, featuring one of the finest collections of Southwest Native American Arts and Crafts. The gallery also has collectibles of the American West and fine art from local and nationally known artists. You can stroll through the Old Bryce Town Shops after the sun has set for a perfect way to end the day. The shops feature a wide selection of regional gifts, Christmas items, and the Canyon Rock Shop.

If you want to take in some additional sightseeing on your visit to Ruby's Inn, Zion National Park, Capitol Reef, Lake Powell, and the Grand Canyon North Rim are a short distance away. Several state parks — Kodachrome, Anasazi Indian Village, and Escalante Petrified Forest — are located along Utah's famous Scenic Byway 12, which has been rated among the ten most scenic roads in America according to *Car and Driver* magazine. The surrounding Dixie National Forest offers plenty of recreational opportunities including fishing, hiking, mountain biking, and of course, horseback riding.

For more than 80 years the Syrett family has been hosting visitors to the Bryce Canyon area. Join them for a truly memorable vacation!

 Address: Box 1, Dept. BP, Bryce UT 84764

Telephone: 800-468-8660;
Internet: http://www.rubysinn.com

Location: 5 hours north of Las Vegas, 5 hours south of Salt Lake City

Guest Capacity: 369 guest rooms

 Season: Year round; RV park and campground and horse programs March - October.

Policy on Children, Pets: Children must be 5 to ride trails; smaller children can ride with adult. Pets allowed with a deposit.

Rates: $ - $$ No meals or activities included.

Credit Cards: All major credit cards accepted.

(SEE PAGES 300 FOR COLOR PHOTOS!)

UTAH

BEST WESTERN RUBY'S INN AND RUBY'S RV PARK AND CAMPGROUND

SPECTACULAR BRYCE CANYON.

ON A TRAIL RIDE.

Utah

Dalton Gang Adventures LLC
Monticello, UT

This is a family operation. Our ranch has been a place where we have raised our children to work hard, ride hard, do their share of work, use good judgment and have a lot of wholesome fun. You will become part of the family while you're here.

Saddle up and come ride with the 3rd and 4th generation cowboys in a unique area of Abajo (Blue) Mountain to the Canyonlands Country, bordering Canyonlands National Park and the Colorado River. Our operation runs from a 10,000 feet altitude on the winter range and the ranch covers about 200,000 acres of a beautiful wild and rugged mix of private, U.S. Forest Service and BLM land.

Our outfit is what they call "a real cowboy's outfit," one that cowboys dream about.

The ranch is connected by a natural driftway from summer to winter range and is run pretty much the way it was run 100 years ago. There are long hours in the saddle and a lot of range to cover. Each day brings new challenges and that's the way we like it.

The ranch cabins and camps have no electricity. Two of the camps have modern bathrooms and running water. The rest you have to run and get it. The cattle are range cattle and are not fed hay, but instead forage on the range from summer to winter. The wildlife are plentiful including beaver, bear, mountain lion, deer, elk, antelope, and desert bighorn sheep.

This is a Working Ranch. The work schedule may vary due to range conditions and weather, but regardless, there will be a lot of riding. You need to be an experienced rider and able to ride a full day, 10-20 miles or more a day.

We are inviting you to come and ride with us and share a working ranch experience that will truly be an adventure. This is not a dude ranch. You will be treated as much as possible like a working ranch hand. You will be assigned a personal horse or horses and equipment for your stay. We'll do our best to make your stay an enjoyable and educational experience. Ranch stay is from 3 to 6 days or more, and there is a limit of six guests at a time, including families and children 12 years and up. We seek guests who value good common sense and obedience to rules of the Ranch Boss. We will camp out in ranch cabins or under the stars, depending on the time of year. Pack trips are occasionally a part of our ranch work, winter and summer.

The meals will be western style: good old Dutch Oven cooking, homemade breads, meat, potatoes and a variety of delicious desserts. Menus will vary, but they will be hearty family meals.

Sunset will end your day's journey spinning tall yarns, and listening to stories and history of these lands and their people. But most of all you'll find enjoyable peace and quiet, few people, fresh air and nature, Cowboy Style.

Address: P.O. Box 8, Monticello, UT 84535

Telephone/Fax: 435-587-2416
Web: http://www.daltongangranches.com
E-Mail: dgangadv@sanjuan.net

Location: Southeastern Utah, bordering Canyonlands National Park

Guest Capacity: 6, for larger groups call

Season: March through November

Policy on Children, Pets: Children must be 12 years or older. No pets.

Rates: $$

Credit Cards: None

Utah
Dalton Gang Adventures LLC

GUESTS WRANGLIN'.

CATTLE ON A DUSTY TRAIL.

Utah

ROCKY MEADOWS ADVENTURES
BLUEBELL, UT

Duchesne County in northeastern Utah is located under the majesty of King's Peak, the state's tallest mountain (13,582 feet), and is perhaps the best-kept secret in the state. The county boasts the only major mountain range in America that runs east-west. The Uinta Mountain range contains one of our nation's first primitive-wilderness areas. Its snow-capped peaks feed rivers and streams that course their way through high mountain valleys and beyond to meet the Green and Colorado rivers. Stories of Spanish conquistadores and hordes of lost gold are woven with eerie Indian legends through the history of this area, Utah's "last frontier."

It is here, in these high mountain valleys that you will find Melvin White's White Ranch and Alan White's Rocky Meadows Ranch. The two working cattle ranches have separate and shared holdings of 12,000 acres of private land and grazing permits on 50,000 acres of state and national forest land, spread over a 60-mile radius.

Against these magnificent mountain backdrops, the ranching activities take place. Guests are invited to take part in those day-to-day tasks, which might include riding and mending fences, scattering salt, checking cattle for sickness or other problems, and trailing cattle. The first cattle drive at Rocky Meadows Adventures is in June, when the cattle are trailed or driven around the 10,000-foot summit of Tabby Mountain to the higher ranges of State Trust Land. The scenery along the way and in the uplands is spectacular — for as far as the eye can see. The fall roundup at the end of September takes place just as the aspens transform their bright green leaves into neon yellow and orange. The breathtaking colors and the crisp mountain air surround you as you help gather the cattle and move them down to winter pastures.

In addition to the working ranch vacations, Rocky Meadows Adventures offers trail rides that can last a few hours, a whole day, or several days and give you the opportunity to enjoy the solitude and splendor of your surroundings. The landscapes vary from irrigated meadow and cropland to desert terrain, to grassy aspen- and pine-covered mountains. Rocky, steep mountains define the rugged splendor of Rock Creek in the Wasatch National Forest. The high primitive area, accessible only on horseback or on foot, offers a unique experience away from the hubbub of civilization.

If you wish, you can combine trail rides with camping, fishing, photographing, or with hunting. Moose, beaver elk, deer, bear, and mountain lion, are abundant; the rivers and the well-stocked ponds provide fishing to your heart's desire.

Accommodations at Rocky Meadows Adventures range from a primitive, rustic cabin at Rock Creek to the ranch house at Sand Creek that has semi-rustic amenities. There's even a teepee for guests who wish to camp out. Home-cooked, hearty meals are provided with special dietary considerations on request. Guests make their own entertainment in the evenings, and storytelling, singalongs, and conversation with newly made friends are always on tap.

At Rocky Meadows Adventures, you choose the activities you want to take part in. Regardless of what you choose, this little-known area of Utah will leave you with an experience that will last for years.

Address: HC 65 Box 165, Dept. BP, Bluebell, UT 84007

Telephone: 801-454-3176

Location: 80 miles west of Vernal, about 100 miles east of Salt Lake City

Guest Capacity: 1-15

Season: May to November

Policy on Children, Pets: Children must be age 8 for horseback riding, 12 for cattle drives. No pets.

Rates: $$ American plan.

Credit Cards: None

Utah

Rocky Meadows Adventures

STRINGIN' ALONG A FEW PACK HORSES.

FALL COLOR.

Vermont

Firefly Ranch
Bristol, VT

Firefly Ranch in Vermont is, as its name states, tiny as a firefly. But its small size makes for a personalized riding experience. This long-established, low-profile horse farm sits high in the foothills of Mount Abraham. Marie-Louise "Issy" Link, the owner, keeps several trail horses for her guests to ride through the tranquil New England countryside.

In 1980, with lots of people experience in her previous business as a restauranteur in New York state, Issy decided to make a change in her life. She learned all about horses and created the Firefly Ranch. The ranch caters to a maximum of six persons and has a casual, contemporary, country atmosphere. Issy wants her guests to think of it as a "home away from home."

At Firefly Ranch, guests can ride to their heart's content. There is excellent riding for both the intermediate and the experienced horseback rider (one who is able to trot and canter) with English, Western, or Australian stock saddles. Saddle up and discover beautiful country roads and tree-lined trails with breathtaking views in the foothills of the famous Green Mountains. Each season provides its own surprises for riders — the tender greening of early spring, the lush emerald tones of full summer, the magnificent and vibrant colors of the famed New England fall foliage season.

This is an outdoor-activity-oriented ranch. If guests aren't riding, they can walk to the renowned New Haven River for some excellent fly fishing, hike up to the famous Long Trail, or ride a light mountain bike through the delightful rural area. After a day's outing, luxuriate in the indoor five-seat spa/hot tub or take a refreshing dip in the private spring-fed pond. There are also three golf courses within 20-30 minutes north or south of the ranch.

This is a historic area of New England, and there are museums and galleries, shops both quaint and crafty, and fantastic shopping in Middlebury and Burlington.

After a day of riding or other activity, return to the Firefly for a four-course, delectable gourmet dinner, complete with wine. The cuisine is continental but it is geared to today's health-conscious consumer — low fat, low salt, as well as vegetarian. Guests can enjoy their favorite beverages on the verandah and drink in the panoramic vistas as the sun sets on their day.

Firefly Ranch has three guest rooms with either private or shared bathrooms. The rooms are spacious and decorated in "contemporary country" style. If there is "no room at the inn," guests are welcome to set up a tent on the beautiful grounds. Guests who do so may take advantage of all meals and outdoor activities with no charge for camping. And a family of four or more may reserve the entire place for themselves.

Firefly Ranch is also open in the winter, and guests can go cross-country skiing from the back door. Downhill skiing and snowboarding are available at Mad River Glen and Sugarbush resorts, just a half-hour's ride away. Snowmobilers can take advantage of the many marked trails around Lincoln.

The atmosphere is relaxed, the hospitality warm and inviting in this peaceful Vermont mountain valley setting.

Address: P.O. Box 152, Dept. BP, Bristol, VT 05443

Telephone: 802-453-2223

Location: 27 miles south of Burlington, 16 miles north of Middlebury

Guest Capacity: 6

Season: May 1 to October 30; Winter Season: December 1 to March 31

Policy on Children, Pets: Children must be 10 years old and intermediate or experienced riders. Pets welcome with prior notice.

Rates: $$ American plan. Other packages available.

Credit Cards: None. Checks, money orders, and cash welcome.

VERMONT
FIREFLY RANCH

LOCATED IN BEAUTIFUL VERMONT.

TRANQUIL RIDING IN THE VERMONT COUNTRYSIDE.

VERMONT

THE JACKSON HOUSE INN
WOODSTOCK, VT

It is with pride and enthusiasm that we welcome you to the Jackson House Inn.

We have sought to create an ambiance of great comfort and rare experience, with a tempo of casual friendliness.

The Jackson house was built with devotion in 1890 by Wales Johnson, a local saw-mill owner. His aim was to erect the finest example of late Victorian architecture possible. Floors of cherry and maple complement fine furniture; walls and moldings are of the finest workmanship. The many exterior eaves, decorative woodwork, and twin chimneys are a statement of pure classic design and dignity of workmanship. The building is now listed in the National Register of Historic Places.

In 1940 the building was bought by the Jackson family, who opened it as The Seven Maples Tourist Home. In 1983, Mrs. Jackson sold the property to its third owners in 100 years, Jack Foster and Bruce McIlveen, who carefully, who carefully renovated the house, meeting Wales Johnson's high level of quality. Furnishings are only of the genuine stamp: fine European antiques, carved Oriental rugs, French-cut crystal, a library of the classics, and a parlor with a welcoming fire.

We acquired this elegant property in 1996, committed to maintaining the Jackson House tradition of excellence and personalized service. In 1997, we opened the Jackson House Restaurant, serving elegant new American cuisine prepared by Executive chef Brendan Nolan, previously of Aujourd" hui at the Four Seasons Hotel, Boson.

Our cathedral-ceiling dining room has a striking granite open hearth fireplace and magnificent views of the four-acre garden. Chairs are hand-crafted by Charles Shackleton, a local furniture maker.

Additionally, the Jackson House has grown to 15 rooms, including four new one-room luxury suites with fireplaces, thermal massage tubs, and splendid garden views. In keeping with the Jackson House tradition, the new rooms are furnished with period antiques, accented with fine Italian fabrics by Anichini.

The natural outdoor beauty of these four landscaped acres augments this outstanding Victorian structure. Huge ancient trees, masterpieces in themselves, cause the visitor to stop in wonder. Winding through the spacious gardens, you cross an arched bridge over a brook to our pond, home to the resident snow geese, ducks and trout.

Breakfast is a memorable occasion: warm fruit compote with champagne, almond-crusted, French toast, ricotta pancakes, frittatas with fresh asparagus and oven-cured tomatoes. Join us for evening libations, a complimentary glass of wine or champagne hors d'oeuvres.

Staying at the Jackson House Inn places you in the centre of all the wonderful things Vermont has to offer: antiquing, art galleries, glass blowing, biking, hiking, horseback riding, canoeing, swimming, museums, golf, tennis, theatre, skiing, sleigh rides, hot-air ballooning, stargazing and moonlight. We look forward to welcoming you to the Jackson House.

Address: 37 Old Route 4 West, Woodstock, VT 05091

Telephone: 802-457-2065; **Toll Free:** 800-448-1890; **Fax:** 802-457-9290

Web: www.jacksonhouse.com

E-Mail: innkeepers@jacksonhouse.com

Location: 153 miles from Boston, 280 miles from New York, 145 miles from Montreal

Guest Capacity: 30

Season: Year round

Policy on Children, Pets: Children 14 years or older welcome. No pets.

Rates: $$ to $$$

(SEE PAGE 301-302 FOR COLOR PHOTOS!)

VERMONT

THE JACKSON HOUSE INN

A COUNTRY QUILT ACCENTS A GUEST ROOM.

SOME ROOMS FEATURE A FIREPLACE FOR
ROMANTIC EVENINGS.

Vermont

Kedron Valley Stables
South Woodstock, VT

Established in 1971, the Kedron Valley Stables offers lessons, training, boarding, trail rides, and carriage rides. The following write-up about the Green Mountain Inn-to-Inn Ride was written by Deborah Donahue.

Trail riding in Vermont with Paul Kendall is fast-paced and exciting. Paul is an eighth-generation Vermonter and his unsurpassed oral histories of the area are great entertainment, as you trot down a country road or relax at a country inn.

Imagine, riding through the Green Mountains on a beautiful horse that is capable and experienced with the countryside of majestic mountains, rolling hills, roaring brooks, green pastures, beaver ponds and birch groves... a horseman's paradise.

Fortunately, the biggest decision one has to make is the type of holiday they wish at the Kedron Valley Stables. . . . You can choose from a Weekend Riding Vacation, a Weekend Inn-to-Inn Ride, Trail Riding Clinic or a Green Mountain Inn-to-Inn Ride. . . . Each vacation starts with a gourmet evening meal at one of Vermont's finest inns. The following morning you begin your day with a hearty country breakfast, then off to the stables to meet your mount.

The Kedron's staff makes every effort to match you with an appropriate horse and tack. Your two-and-a-half to three hour morning ride will be up and down the Vermont hillsides, past beautiful summer estates, horse farms, new mown hay fields, maple tree groves, stone walls and covered bridges. The picnic lunch-stops on the trail allow for a welcomed rest under a shade tree.

The two to three hour afternoon ride has you meandering down gravel roads, trotting along a soft woods road or climbing a narrow woods trail that leads to a spectacular birch grove with a view of the Green Mountains in the distance. At long last, you arrive at a country inn where your innkeeper is [eagerly] awaiting your arrival. Imagine, afternoon tea on a large porch, a soothing hot bath or a cool swim in an invigorating mountain pond after a day of splendid riding, while the Kedron's staff cares for your horse.

Vermont inns are noted for their friendly charm, authentic furnishings, and scrumptious gourmet dinners. Select one of the wonderful delicacies from the dinner menu while you relive the wonders of the day with your companions. Then seek the comforts of your room, so you may rise the next day and do it all again.

Kedron Valley Stables accepts only strong intermediate and advanced riders. Their Green Mountain Inn-to-Inn Ride (5 nights, 4 days) is a delightful way to walk, trot, and canter one's way through the Green Mountain State, boarding and dining at a select group of historic country inns. Riding distances average 20 to 25 miles a day, and you'll spend up to six hours a day in the saddle.

The Weekend Riding Vacation (2 nights, 1-1/2 days) is a good introduction to the type of trail riding done in Vermont. Families will enjoy this ride, although children must be 12 and able to control a horse at the walk, trot, and canter. Riders are limited to 14 and are split into two or three smaller groups according to abilities. Groups come together for a picnic lunch on the trail.

The Weekend Inn-to-Inn Ride (2 nights, 1-2/3 days) gives you some of the flavor of the Green Mountain Inn-to-Inn Ride in less than half the time.

During the Trail Riding Clinic (5 nights), guests will learn stable management (feeds and feeding, stall care and proper handling of horses, grooming, tacking, bathing, and medicating), ground work, various demonstrations, and some jumping, combined with trail riding over many miles of trails.

Paul Kendall is also a Justice of the Peace performing wedding ceremonies in the surrey "with the fringe on top," sometimes in a birch grove or covered bridge, or in a sleigh on the crest of a hill!

 Address: Route 106, P.O. Box 368, Dept. BP, South Woodstock VT 05071

Telephone: 800-225-6301, 802-457-1480; **Fax:** 802-457-3029

Location: 20 miles from Lebanon, New Hampshire

 **Guest Capacity:** 14 Strong intermediate and advanced riders only.

Season: Year round

Policy on Children, Pets: Children must be 12 and able to control a horse at walk, trot, and canter. No pets.

Rates: Package rates available.

Credit Cards: None

(See Page 303 for Color Photos!)

VERMONT
KEDRON VALLEY STABLES

RIDING THROUGH FALL FOLIAGE.

VERMONT

MOUNTAIN TOP INN
CHITTENDEN, VT

It's the kind of view postcard makers dream about — a spectacular pristine lake surrounded by Vermont's tree-covered Green Mountains. And right in the middle of all that natural beauty lies the Mountain Top Inn and Resort. Known as Vermont's best-kept secret, Mountain Top Inn offers the best of both worlds: an intimate country inn with numerous facilities and superb dining, steeped in the finest New England tradition.

The Mountain Top Inn was originally the barn for an old turnip farm. In 1940, William Barstow, an associate of Thomas Edison, bought the old farm because his wife, Françoise, wanted to open a traditional wayside inn.

Throughout ensuing years, improvements and additions to the original building were made while still maintaining the integrity of its Yankee origin and the beauty and ecology of its natural surroundings. In 1955, the inn was proud to play host to President Dwight D. Eisenhower while he was in the state on a fishing expedition. The inn continues to flourish in the hands of Mike and Maggie Gehan, who now manage and call the inn home.

The inn is situated in the heart of central Vermont's Green Mountain National Forest. In summer, guests can enjoy a wide range of excellent facilities including swimming, tennis, golf, boating, fly fishing, claybird shooting, and horseback riding. In winter, the resort offers cross-country skiing, ice skating, sledding, and horse-drawn sleigh rides that whisk riders through spectacular mountain scenery.

There aren't too many resorts in the country offering riding holidays, and Mountain Top Inn is one of the very few in the Northeast to offer an extensive program of horseback riding vacations. Everyone, from never-ridden-before beginners to advanced horsepeople, can find what they're looking for in an equestrian vacation at Mountain Top Inn. Riders can tailor a personalized riding program to their desires and ability level, making it as varied and intensive as they wish. Each day can be chock-full of riding and lessons, or you can schedule a more leisurely day of part-time riding and full-time relaxing, enjoying the breathtaking vistas and outstanding facilities of the resort. You can explore miles of spectacular trails covering thousands of scenic acres in the Green Mountain National Forest. Choose hour-long and half-day trail rides or combine guided trail rides with specialized individual group instruction in English riding, Western riding, dressage, jumping, or introductory polo.

Mountain Top Inn is the sleeping giant of the cross-country ski world with an astounding variety of trails. Short loops near the lodge get right to the point carrying you just high enough to gaze down into the valley, while several 5- to 10-kilometer loops provide plenty of varied terrain for all ability levels. The ski center, which has snow-making equipment, has been an official training site for the U.S. Nordic Ski Team in past years. And you can bet that you'll receive the finest instruction available from the exceptional staff of the Mountain Top Ski School.

Guests can choose one of 35 well-appointed guest rooms in the inn itself or one of the 22 cottages and chalet units within walking distance of the inn, most of which have a fireplace or wood stove. The dining room offers excellent cuisine, while numerous windows allow you to gaze at one of the most beautiful views in all of New England. The ambiance in the evenings is enchanting, with candles on the tables and a fire softly burning in the fireplace.

Chittenden's Mountain Top Inn is reason enough to visit Vermont any season of the year. Whether visitors are riding, skiing, recreating, or relaxing, all will find what they're looking for at Mountain Top Inn and Resort.

Address: Dept. BP, Mountain Top Road, Chittenden, VT 05737

Telephone: 800-445-2100, 802-483-2311; **Fax:** 802-483-6373

Location: 15 miles northeast of Rutland, 3.5 hours northwest of Boston, 1.5 hours south of Burlington, Vermont

Guest Capacity: 120

Season: Year round

Policy on Children, Pets: Children welcome. No pets.

Rates: $$ Modified American plan. Children's rates.

Credit Cards: Amex, MasterCard, Visa

(SEE PAGE 304 FOR COLOR PHOTOS!)

VERMONT
Mountain Top Inn

A RIDERS' PARADISE.

GETTING SOME POINTERS.

VERMONT

VERMONT ICELANDIC HORSE FARM
WAITSFIELD, VT

Try a most unusual vacation in the Mad River Valley, one of the most scenic areas of Vermont. Tour its mountains, meadows, and forests on one of the oldest and purest breeds in the world. Icelanders describe the experience as "sitting in an easy chair." The tireless and efficient movement of the Icelandic Horse makes it the ideal saddle horse, seeming to effortlessly dance over the earth with lightness and power. The unique gaits of this horse provide horseback riding vacationers with an ideal way to see the landscape of the Mad River Valley in any season.

Since 1988, the Vermont Icelandic Horse Farm, located near the historic village of Waitsfield in the heart of the Sugarbush Resort Area, has offered two- to six-day treks, as well as full and half-day rides on four- and five-gaited Icelandic Horses. On these treks, you will experience the exceptional riding comfort and stamina of this breed.

Brought over by Viking ships, the Icelandic Horse served as the sole source of transportation over Iceland's rough terrain. Isolated by law, nature gifted these horses with no natural predators, but in turn challenged them with sparse forage and a harsh climate.

One thousand years of isolation have preserved in the Icelandic Horse traits once common to all European horses. Among these are two unique gaits. The Tolt, or running walk, is free flowing and effortless, allowing the horse to cover rough terrain swiftly. The Skeid, or flying pace, gives the rider the sensation of floating over the countryside in a way that must be experienced to be understood.

Accomplished equestrians are intrigued by the stamina and speed of the breed and challenged to master the different gaits. Less experienced riders appreciate them for their smooth gaits, surefooted way of traveling, and level-headed temperament. The Icelandic's small size (13-14 hands) does not reflect their excellent strength. They can comfortably carry a rider up to 250 lbs. over great distances.

The Vermont Icelandic Horse Farm offers daily half-day and all-day rides. The farm also has a number of multi-day tours. While riding from one country inn to another, you will pass over beautiful meadows, streams, and lush woods and be able to explore many of the special features of the mountainous Mad River Valley.

The tours feature plenty of opportunity for enjoying the foliage and activities like studying birds and wildlife; there's even fishing if you are so inclined. Fall riders will experience the foliage far away from the tour buses. And in the winter, when most horses are stabled, Icelandics are in their element, confidently crossing icy streams and cantering across snow-drifted fields for a winter adventure beyond compare. All tours are led by experienced horsepersons with a thorough knowledge of the area and its trails.

You'll enjoy comfortable accommodations and gourmet dining at select inns, not to mention a generous helping of Vermont hospitality. All of the inns feature beautiful settings and excellent cuisine. For lunch; a delicious picnic in a pretty spot or stop at a specially selected restaurant.

This is the perfect escape for those who appreciate comfort and fine dining but who still want an active and unusual vacation. You'll return home fit and rested, with the memory of newly found friends and shared adventures. The Icelandic Horse that gave you the overwhelming feeling of floating across the scenic Vermont countryside is likely to keep a special place in your heart forever.

Address: P.O. Box 577, Dept. BP, Waitsfield, VT 05673

Telephone: 802-496-7141; **Fax:** 802-496-5390

Location: 3.5 hours from Boston

Guest Capacity: 12

Season: Inn-to-inn treks summer only; trail rides year round

Policy on Children, Pets: Children must be 10 years old. No pets.

Rates: $$ American plan.

Credit Cards: MasterCard, Visa

(SEE PAGE 305-306 FOR COLOR PHOTOS!)

VERMONT

VERMONT ICELANDIC HORSE FARM

DASHING THROUGH THE SNOW.

VIRGINIA

THE CONYERS HOUSE INN & STABLE
SPERRYVILLE, VA

For the equestrian looking for one of the most unusual horseback riding activities in the United States, you need look no further than The Conyers House in beautiful Virginia. The Conyers House is indeed special, and the fact that it has been written up in no less than twenty-three separate books and has had nine articles written about it in some prestigious magazines and newspapers such as the *New York Times*, *Country Accents*, the *Washington Post*, and the *Baltimore Sun* lets you know you're in for a very special horseback riding vacation indeed.

What makes The Conyers House so special? Well, many things, it is the site of some fast-paced and exciting fox hunting, thought to be found only in places like England.

Fox hunts are held from September through March. Guests are welcome to participate in the sending off of the hounds gala or, even better, to participate in the ride if you're a qualified rider. For those adventurers who are ready to try something new, we suggest you contact Sandra Cartwright-Brown, an experienced fox hunter in her own right, for more information on this exciting sport. There are also miles of beautiful country trails for the less daring.

The Conyers House offers a guided two-hour cross-country ride across vast, privately owned land for an additional fee. Elementary instruction, using English tack, is included. Optional jumping for experienced riders through trails used by local hunts can be arranged for a nominal fee.

The Conyers House was built around 1790 and was known as the Conyers' Old Store in 1815, and then Fink's General Store in 1850. It was bought in 1979 by the Cartwright-Browns and resurrected as a family country house. After a complete redecorating, it was opened as a country inn two years later. Here you can enjoy unpretentious elegance; warm hospitality; eighteenth-century charm; five guest rooms, one suite, and two cottages with private baths and cozy fireplaces; marvelously enthusiastic and compatible guests; Norman's delightful stories and car collection; Sandra's dedication to fox hunting, antiques, and gardening; and their magnificent breakfasts, candlelit dinners, and elegant afternoon teas. There's even a resident "presence!" You'll sleep to the peaceful symphony of the insects, frogs, and birds.

The dogs will take you on a walk through incredible scenery. Hughes River is stocked, and the best fishing hole is just a 15-minute walk away. Old Rag Mountain and White Oak Canyon are nearby and offer great hiking. Crafts and antiques shops dot the countryside. The area offers an incredible array of activities and places to visit. There are moonlight walks; spelunking; spring and fall steeplechases; festivals for apples, mushrooms, and cherry blossoms; the county's annual house tour and dried flower sale; golf; canoeing; tubing; fly fishing; ballooning; hayrides; skeet shooting; archery; wineries; a delightful local movie theater; Skyline Drive; Luray Caverns; and Monticello. You'll be 1.5 hours from Washington, D.C. and within 100 miles of about 80 percent of all of the Civil War monuments.

Come to The Conyers House, a true Virginian country inn and stable for a most unique and refined horseback riding adventure. The fox hunting experience is unmatched, and the Southern hospitality will make you feel more like a close friend than a guest.

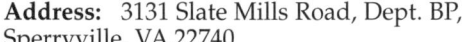

 Address: 3131 Slate Mills Road, Dept. BP, Sperryville, VA 22740

Telephone: 540-987-8025; **Fax:** 540-987-8709; **Web:** http://www.bnb-n-va.com/conyers.htm

Location: 75 miles from Washington, D.C.

Guest Capacity: 20

Season: Year round

Policy on Children, Pets: Children are welcome in cottage accommodations. Dogs permitted with prior permission.

Rates: $ Breakfast included. Riding extra. Pets extra.

Credit Cards: Discover, MasterCard, Visa. Personal checks preferred.

Virginia

The Conyers House Inn & Stable

SPRINGTIME IN VIRGINIA.

VIRGINIA

FORT VALLEY STABLE HORSE CAMPGROUND
FORT VALLEY, VA

This is a campground exclusively for the horse person and his horse with over 80 miles of mountain trails! From a one hour or less ride to a 7 or more hour ride that doesn't travel over the same trail twice, there are trails throughout The George Washington National Forest that surrounds our property as well as trails on over 500 acres of private land bordering the National Forest.

Once you drive your rig into our campground, you will not have to move it again until it is time to go home. All trail heads are easily accessible from the Fort Valley Stable Horse Campground without having to trailer your horse. There are beginner trails which wind over gently rolling hills to exciting, exhilarating trails that climb up the sides of mountains offering the more experienced rider and horse some challenge and excitement. Some of the trails are the very same ones that are used for the Old Dominion Endurance Race, held three times a year. Train on the very same trails used by Endurance Champions, or just have the satisfaction of knowing you have ridden the same trails.

The National Forest Service periodically cuts new trails through the mountains and you may very well be seeing views from the tops of mountain peaks seen by very few people since Civil War troops traveled there to observe the opposing army's troop movements. Or perhaps you are one of the first non-Native Americans to look down on the Shenandoah Valley from that particular area of the mountains. If you enjoy trail riding, we have the trails for you! We will provide you with maps of the different trail's color coordinated.

If along with your trail riding you are looking for a relaxing, restful and tranquil vacation, you will find it at the Fort Valley Stable and Horse Campground. We are located in a valley in the middle of the top of the Massanutten Mountains in Virginia's famed Shenandoah Valley. It is a quiet area of scattered homes and farms, some having been here for over a hundred years. Our Horse Campground offers shaded campsites, some with water and electric hookup, but even if you do not require water and electric hookup, there is readily available water for you and your horse. There is a vault rest room and hot showers. There are also rustic cabins available for those who don't care to camp out.

For your horse, we have available roomy stalls, large paddock area, or you can picket or use a portable corral in the Horse Campground. Hay is available for sale if you need it.

With picnic tables and fire rings at each camp site, and ice and firewood for sale for a nominal charge, evenings can be spent recounting the day's fun and adventures out on the trail over the glow of a crackling fire. As stars sparkle and shine in the darkened sky, you can retire as the quiet of the night is interrupted by the magical sounds of the crickets, frogs, and Whip-R-Wills.

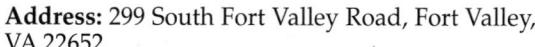

Address: 299 South Fort Valley Road, Fort Valley, VA 22652

Telephone: 540-933-6633; **Toll Free:** 888-754-5771; **Fax:** 540-933-6634; **Web:** www.fortvalleystable.com **E-Mail:** fvs@fortvalleystable.com

Location:

Guest Capacity: 25 horse trailer rigs at one time, 13 with electric hook-up.

Season: Year-round

Policy on Children, Pets: Children welcome. Dogs welcomed but must be on leash.

Rates: $

Credit Cards: None

VIRGINIA

THE INN AT MEANDER PLANTATION
LOCUST DALE, VA

Cradled in the heart of Jefferson's Virginia, The Inn at Meander Plantation offers the traveling horseback rider the rare opportunity to experience the charm and elegance of colonial living and scenic horseback riding through the lovely countryside.

This historic country estate and horse farm, which dates to 1766, returns the horseback riding guest to an earlier, more romantic time when hospitality was a matter of pride and fine living was an art practiced in restful surroundings. The stately three-story white brick Georgian mansion has a twenty-two stall barn and sits majestically atop a hill in the middle of 80 rolling acres of pasture and woodlands—a perfect place in which to pursue equestrian activities. The expansive columned front porch, the columned brick archway believed designed by Thomas Jefferson, and the two-level wrap-around porches on the back side of the house are favorite places for guests to relish the quiet of nature after a morning or afternoon's ride or to view spectacular sunsets over the Blue Ridge Mountains. You can enjoy bird-watching, raft on the gentle Robinson River that borders the Inn, walk nature trails, swing in a strategically placed hammock, or play volleyball or badminton on the lawn. Fox hunts can be arranged during season, and day trip rides up into the mountains are available. Historians might enjoy a walk to the old family cemetery, which dates to the 1700s.

Inside the elegantly furnished mansion, guests can share pleasant conversation in the parlor where Jefferson and Lafayette were frequent visitors, play an impromptu concert on the antique baby grand piano, or search out private nooks for reading, writing, or quiet contemplation. Stroll through formal boxwood gardens or wander woods and fields where wildlife abounds. Watch bluebirds flitter from fence to fence.

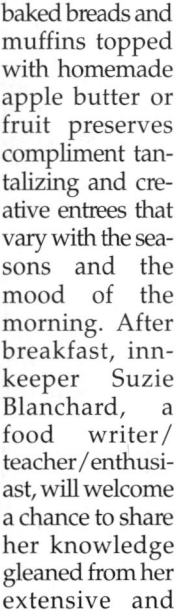

Admire horses grazing on the knoll. Relax in the white rockers that line both levels of the back porches.

The bedrooms, reflective of the elegance and romance of a Virginia country estate, welcome you with warmth and comfort. Four-poster beds piled high with plump pillows beckon you to snuggle beneath down comforters. Each of the eight guest rooms, tastefully appointed and furnished with antiques and period reproductions, offers a private bath.

Join your fellow guests in the formal dining room or under the arched breezeway for a full country gourmet breakfast. Fresh baked breads and muffins topped with homemade apple butter or fruit preserves compliment tantalizing and creative entrees that vary with the seasons and the mood of the morning. After breakfast, innkeeper Suzie Blanchard, a food writer/teacher/enthusiast, will welcome a chance to share her knowledge gleaned from her extensive and eclectic collection of more than 300 cookbooks. Full dinner service is available to guests with advance reservations. For that day of sightseeing, request a take-along picnic basket.

The Inn is the perfect location for your exploration of central Virginia's many attractions. From the Inn, you are only a short drive from Montpelier, home of James and Dolley Madison; the panoramic Skyline Drive; Thomas Jefferson's beloved Monticello and its neighbors, Ash Lawn and Michie Tavern; bustling Charlottesville and the University of Virginia; sobering Civil War scenes at Chancellorsville, Wilderness, and Brandy Station battlefields; and the fasted-paced downhill runs of Massanutten. Local wineries, antique shops, craft stores, country markets, and you-pick orchards abound.

Come to The Inn at Meander Plantation for horseback riding across gentle, rolling hills and mountains of Virginia, the Southern hospitality, and the Southern cooking. Step back into time for your next horseback riding adventure!

Address: HCR 5, Box 460A, Dept. BP, Locust Dale, VA 22948

Telephone: 800-385-4936, 540-672-4912; **Fax:** 540-672-4912

Location: 35 miles north of Charlottesville, 70 miles southwest of Washington, D.C.

Guest Capacity: 28

Season: Year round

Policy on Children, Pets: Children are welcome. Pets only allowed in cottages.

Rates: $ - $$ Full breakfast provided; dinner available with advance notice (extra cost). Stabling of horses extra.

Credit Cards: MasterCard, Visa

WASHINGTON

HIDDEN VALLEY GUEST RANCH
CLE ELUM, WA

Remember what it was like when you were a kid? Every day was a promise, and the most important thing you had to do was have fun. At Hidden Valley Guest Ranch, it's still that way. We offer great opportunities for family and individual activities, from horseback riding and swimming to games on out Sport Court (TM) and hikes along Swauk Creek.

But there's also time here to notice and appreciate the special things: the cry of red-tailed-hawk circling high above the valley; dappled light splashing through the trees along the creek; or the way evening colors grown rich with the essence of each day's weather.

Hidden Valley Guest Ranch is a perfect blend of yesterday and today, of serenity and stimulation. It's a refreshing retreat into the past - and the ideal place to create special memories for the future.

Come to Hidden Valley. Come to where the good guys always win.

Many of our guests think the best thing about Hidden Valley Guest Ranch is what's missing; ringing telephones, schedule conflicts and the crush of the daily routine.

At Hidden Valley there are no alarm clocks, only the cookhouse triangle calling you to a delicious breakfast. Your morning commute may mean a brisk trail ride, hiking around our more than 700 acres, or simply pouring yourself another cup of coffee and ambling down to the swimming pool.

Stop by our recreation center (which resembles a 19th-century hunting lodge with a huge stone fireplace) to test your skills at Ping-Pong or pool.

Go fishing, hiking or mountain biking. During our winter season, enjoy cross country skiing, sledding and old-fashioned winter fun.

Smell the sharp resin of ponderosa pine and hear leather creak as you swing into the saddle. It's easy to imagine what it must have been like to be the first person to ride through this valley. Our wranglers will make your riding experience one you'll remember with pleasure. We raise and train our own horses, and we have mounts suited to riders of almost every size and ability, whether you're an expert or just getting started. We normally offer several rides each day, and we will arrange special group or half-day rides at your request.

Once your day is done, all that's left to do is ease into the hot tub as dusk deepens into twilight and the first evening star brightens above the horizon.

Our cabins and rooms are rustic - much like those used by the original homesteaders - but they're designed for the kind of comfort that's always been welcome after a hard day on the trail. Each one is unique, and all have private entries and baths. Some cabins have fireplaces, none have telephones or televisions, and most have sun decks that overlook the breathtaking beauty of our rugged and secluded valley.

As it was in the Old West, the cookhouse is the center of ranch life. Meals are served family-style, the coffee pot is always on, and there usually are muffins or cookies available for those between-meal-urges.

Appetites are hearty at Hidden Valley, and so is the food. We feature fresh ingredients, homemade breads, and desserts, and ranch cooking as traditional as an open fire. We've improved on tradition, however, with healthful meals that preserve authentic Western character. Outdoor barbecues, chuck-wagon cookouts and other special touches add to the flavor of your ranch experience.

We are happy to accommodate special dietary needs and requests for vegetarian, low salt, macrobiotic or child-pleasing meals. Let us know what you need before you arrive.

 Address: 3942 Hidden Valley Road, Cle Elum, WA 98922

Telephone: 509-857-2322; **Toll Free:** 800-5COWBOY; **Fax:** 50-857-2130; **E-Mail:** brucecoe@Televar.com
Web http://www.ranchweb.com/hiddenvalley

Location: 1.5 hours from Seattle, 12 miles from Cle Elum

Guest Capacity: 25 - 40

Season: Year round

Policy on Children, Pets: Children welcome.

Rates: $ - $$

Credit Cards: MasterCard and Visa

WASHINGTON
HIDDEN VALLEY GUEST RANCH

THE HIDDEN VALLEY GANG.

COME AND PUT YOUR FEET UP.

WASHINGTON

THE K DIAMOND K GUEST RANCH
REPUBLIC, WA

The ranch is located 125 miles north of Spokane near the Canadian border. At an elevation of 2,400 feet, the surrounding mountains are covered with pine, fir and larch trees. With over 1,500 acres of deeded land, we lease an additional 30,000 acres of private and public land for summer grazing of 200 head of cattle.

Guest staying at the ranch are made to feel at home and enjoy the warm hospitality and delightful meals in a relaxed atmosphere. A variety of fun activities and special outings await you:

Horseback Riding — Miles of riding trails with different scenic view on every ride. Lessons given on arrival. With 25 plus horses, there is one for the beginner as well as the experienced rider. All rides are guided.

Fishing — Try catching a colorful rainbow trout in the San Poil River that runs through the ranch. Or try your skill in several mountain lakes along with the well known Curlew Lake just a short distance from the ranch.

Hiking — Walk at a leisurely pace through tall timber trails, grassy meadows and feel the grandeur of the valley below. There is an ever changing view with each foot step. See the wide variety of color in the flowers and landscape. You will be amazed at how blue the sky can be.

Campfires — Feel the warmth and the touch of ranch atmosphere at our campfires. Share tall tales and short stories with other guests.

Hunting — The abundance of wildlife is the great attraction to hunters, small as well as large game.

Photography — A great way to go exploring is with your camera and extra rolls of film. Share your experience with pictures to remember.

Wildlife — You will feel a certain thrill of seeing white tail deer, coyotes or wild turkeys at anytime on the trail or in front of the lodge.

Bird Watching — From the predictable swallows arriving in April, to the amazing "V" formations of the Canadian geese heading north; over 150 species of birds have been identified in the area.

Star Gazing — The clear blue skies put on an evening performance of brilliant planets, stars and constellations. A pleasant surprise is the occasional appearance of the 'aurora borealis', the northern lights.

Golf — A nine hole course known as "The toughest in the West", is only a few miles from the ranch.

A Variety of Things to Do? — Try roping or throwing a lariat, panning in the river for gold, old western video night, cattle roundups-spring or fall, hayrides and sing alongs, learn to build pole corrals, barns or checking a fence line.

All Season Ranch Activities — While the season for vacations is predominately during the warmer months, ranch activities go on throughout the year. As a working cattle ranch we are open all year. Calves and foals are born in the spring, hay harvesting in the summer, feeding hay in the winter and timber harvesting as time permits. Roundups, fencing and maintenance is part of our lives. Each month with its colors and changes is unique in itself. No matter when you come, you are sure to enjoy a distinctive visit. Relax every evening on the front porch watching the sun set, or in front of the fireplace during winter holidays.

Address: 15661 Hwy. 21 South, Republic, WA 99166
Telephone/Fax: 509-775-3536
Web: www.kdiamondk.com
E-Mail: kdiamond.televar.com
Location: 125 miles NW of Spokane, 300 miles E of Seattle
Guest Capacity: 12

Season: Year round (horses on vacation November through March)
Policy on Children, Pets: Children welcome. Kennels are available for dogs, horses boarded
Rates: $$
Credit Cards: None

(SEE PAGE 301-302 FOR COLOR PHOTOS!)

WASHINGTON

THE K DIAMOND K GUEST RANCH

SURROUNDED BY TALL TIMBER.

READY TO RIDE.

WASHINGTON

NORTH CASCADE OUTFITTERS
TWISP, WA

A unique wilderness experience is awaiting you in the North Cascades! For horseback riders and adventurers who are looking for the ultimate horseback riding vacation, try the great Pacific Northwest. North Cascade Outfitters eagerly looks forward to taking you high into one of nature's most beautiful creations.

Owner John Doran packs into the Pasayten Wilderness Area, which encompasses more than a half a million acres of unspoiled mountain country. It has over 500 miles of regularly maintained trails that see relatively little use, which makes for unsurpassed horseback riding. The scenery is spectacular and incredibly diverse in this vast area — from steep valleys to craggy peaks to peaceful high meadows brimming with wildflowers exploding with color. From some points in the Pasayten Wilderness Area, you can see panoramic views of peaks and glaciers from Canada to the North Cascades National Park. The area is abundant with wildlife, and you'll want to bring along a camera with plenty of film.

North Cascade Outfitters features trail-wise gentle horses that will expertly guide you through this amazing country — a true alpine summer paradise. The string of saddle horses is experienced on mountain trails, and the pack mules and horses are trained to carry just about anything. The packer will take care of the horses, pack up the camp gear, and guide you through the mountains. All you have to do is relax on your horse and enjoy!

As you wind along a mountain path, you'll be thrilled by the spectacular beauty of jagged snow-covered peaks and deep, lush green valleys. You'll love the food, cooked the old-fashioned way — over an open campfire. You won't see any freeze-dried food either, only sizzling steaks and fresh vegetables. You'll sleep under the stars or in one of the private sleeping tents, set around a blazing campfire. And the breakfasts? They're so hearty you may need help to get on your horse! (With advance notice, they can accommodate dietary restrictions or preferences.)

One of the favorite destinations is Corral Lake, a beautiful high mountain lake. This trip offers fishing in several lakes, a variety of day rides and hikes, as well as gorgeous scenery and solitude. A minimum of 5 days is recommended for this trip. But you can choose where you want to go and how long you want to stay.

North Cascade Outfitters can also custom design hiking and pack drop trips with customized itineraries. If you are a hiker, the "Walk-n-Pack" trips are perfect. You can walk on selected trails, and they'll carry the gear and meet you in camp. If you like to fish, your trip can be customized to include some excellent rainbow and cutthroat trout fishing in crystal-clear mountain lakes. There is something special about riding your horse and doing some fishing in remote, unspoiled areas.

For a truly magnificent horseback riding adventure, go to the unspoiled Pacific Northwest. Leave the details to North Cascade Outfitters, and you'll have the time to just kick back, relax and enjoy your back-country vacation.

 Address: P.O. Box 395, Dept. BP, Twisp, WA 98856

Telephone: 509-997-1015

Location: 100 miles north of Wenatchee, 200+ miles from Seattle-Tacoma area

Guest Capacity: 12

Season: Summer only

 Policy on Children, Pets: Recommended children be 8 years old for pack trip, 5-6 years old for ridng. Pets not recommended but possible.

Rates: $ - $$ Food included on "first-class" trips.

Credit Cards: None

440

NORTH CASCADE OUTFITTERS

SO PROUD!

WYOMING Dude Rancher's Association

The Wyoming Dude Rancher's Association has 29 member ranches, representing all parts of the state. Each ranch has its own unique Western flavor and a variety of activities and special features. The Wyoming Dude Rancher's Association sets the same high standards as the National Dude Rancher's Association, so you may rest assured you can pick a winner from the following list of members. A horseshoe U beside the ranch's name indicates the ranch has a detailed description and photos elsewhere in this book. Dude ranching in Wyoming is THE family vacation!

Flying A Ranch U

R Lazy S Ranch

Flitner Ranch U

Lazy L & B Ranch U   Two Bar Seven Ranch

Moose Head Ranch

Spear-O-Wigwam Ranch   **Darwin Ranch U**   Crossed Sabres Ranch

Blackwater Creek Ranch U   Brooks Lake Lodge & Guest Ranch

Brush Creek Guest Ranch U

Allen's Diamond Four Ranch

Paradise Guest Ranch U

Bitteroot Ranch   **Triangle C Ranch U**

Red Rock Ranch

Wiggins Fork Lodge   **Triangle X Ranch U**

Klondike Ranch

Gros Ventre River Ranch U   Heart Six Ranch

Kedesh Ranch U

High Island Ranch U   Eaton's Ranch

Absaroka Ranch U

7 D Ranch

T Cross Ranch U

For a free brochure describing these ranches, write to the
Wyoming Dude Rancher's Association, P.O. Box 618, Dubois, WY 82513
Call 307-455-2584 or Fax 307-455-2634.
The WDRA can also be accessed on the Internet at
www.wilderwest.com/wyoming/wdrassoc

WYOMING

A. DRUMMOND'S RANCH B&B
CHEYENNE/LARAMIE, WY

In the heart of Big Sky country, the Drummond's Ranch sits in the middle of some of the prettiest country in southeastern Wyoming. Located on 120 acres of private land, but adjacent to over 55,000 acres of prime riding country, this bed & breakfast caters to either traveling horsepeople who decide to stop for a day (or two or three) or people who want to ride their own horses in the unspoiled Wyoming country.

Featured in *Country Inns, Country Extra,* and *Innviews* magazines, as well as AAA, Mobil, Fodor's and Frommer's guidebooks, Taydie Drummond and her husband Kent provide unique custom-tailored vacation packages to the horseback riding adventurer traveling with a horse. This is a true retreat to both excellent horseback riding and tranquility. Terrycloth robes and outdoor hot tubs are pleasant surprises at the end of a day's ride. Taydie provides complimentary homemade snacks, fresh fruit, and beverages, which are available whenever you may want them. Fine dining is on silver, crystal, and china, while looking out at vistas of the Colorado Rockies.

There are no specific trails in Wyoming's wide open spaces. But there are lots of dirt roads and wildlife trails on 55,000 acres of Medicine Bow National Forest, which is located just 4.5 miles from Drummond's Ranch — all yours for the riding. There are forests with little underbrush; wide-open prairies with deer, elk, and antelope; and rugged hillsides with many creeks and streams. Glorious vistas and breathtaking 100-mile views abound. For those who wish to drive to the Snowy Range, 50 minutes to the west, you will be rewarded with terrific fishing in high mountain lakes. The Drummonds do provide maps of the area, as well as compasses.

Horse facilities at Drummond's Ranch include a 60-ft. x 120-ft. inside arena, one 16-ft. x 12-ft. stall, three 12-ft. x 12-ft. stalls with rubber mats, and four outside runs, one with shelter. Hay and extra bedding are available at a nominal fee, but feed is not provided. Dogs are allowed, but the country has a leash law, so they must be contained at night, either boarded in the barn or in the owners' vehicle.

The bed & breakfast is a no-smoking facility and features Western living at its best. Drummond's can accomodate up to ten people in four bedrooms. The Carriage House Loft has a queen-sized bed, extra-long twin bed, gas fireplace, steam sauna, and a private deck with hot tub. It is a totally self-contained unit with microwave, refridgerator, and gas grill with side burner so that guests have the option of either cooking for themselves or having dinner privately catered in their room; they may even join the other guests in the main house if they choose. The Garden Room comes with a queen-sized bed and private bath, as well as a deck with outdoor hot tub. The Humming Bird Haven has both a queen bed and a youth bed with shared bath. It features an attractive small window seat and 75-mile views to the south to Long's Peak and Rocky Mountain National Park. Kestral's Keep features a double bed and single bed with a sink in the room and a shared bath.

In addition to horseback riding, guests are able to rock climb, hike, mountain bike, llama trek, cross-country ski in the winter, or just relax and enjoy the abundance of natural birds and wildlife in the area. At night, lie in a hammock and marvel at the sky — and know why they call it "Big Sky" country.

Whatever your pleasure, you're sure to enjoy the ruggedness of Wyoming and the thick plush terrycloth robes Taydie provides, not to mention all the homemade goodies she prepares for her guests to eat!

Address: 399 Happy Jack Road, Dept. BP, Cheyenne/Laramie, WY 82007

Telephone/Fax: 307-634-6042

Location: 20 minutes from Cheyenne, 20 minutes from Laramie, 2 hours from Denver

Guest Capacity: 10

Season: Year round. Reservation required.

Policy on Children, Pets: Children are welcome if no other couples are booked. Pets are welcome, but no dogs allowed in house; boarding outside.

Rates: $ - $$ Full breakfast included; lunch and dinner extra. Horses extra.

Credit Cards: MasterCard, Visa

WYOMING

ABSOROKA RANCH
DUBOIS, WY

The Absaroka Ranch is a small, highly personal guest ranch offering solitude and beauty on a grand scale. The ranch is located at the base of the spectacular Absaroka mountain range overlooking the immense Dunoir Valley, ten miles west of the town of Dubois. Wildlife abounds here - elk, deer, pronghorn antelope, badgers, wolves, bears and eagles are often sighted. The valley is surrounded by National Forest and Wilderness. Plenty of peace and solitude awaits guests along with the warm hospitality of the ranch crew.

The Absaroka features a high staff-to-guest ratio which allows guests to tailor their stay to their wishes. Horseback trails offer a variety of terrain, from sage covered hill to grassy meadows, to cool, dark forests with rushing mountain streams. Riding is offered each morning and afternoon along with weekly evening and breakfast cook-out rides and an all-day ride into the high country.

Guided fishing excursions to nearby lakes and streams are offered to all guests. This area features some of the finest mountain trout fishing to be found anywhere. Hiking opportunities are plentiful as well. Guests may wish to hike to the Continental Divide for breathtaking views or any number of high mountain lakes or other spectacular destinations.

Other activities at the Absaroka include slide shows, cowboy music and entertainers, horseshoes, camp-fires, or rousing whiffle ball games after dinner. The cowboy town of Dubois offers a weekly square dance, varied shopping, an interesting museum, and live music on weekends, draws cowboy hats and boots to the dance floor!

A recreation room for the kids, the "Goldpinch Palace," offers a jukebox, soda-pop, foosball table and games. This is a family-oriented ranch where the kids are never segregated from the adults (unless they want to be!)

The Absaroka prides itself on its excellent meals. In addition to popular favorites like beef tenderloin or BBQ ribs, the chefs serve innovative pasta dishes, grilled seafood, and specialties like paella and Greek shrimp. Fresh salads, fruits and vegetables are always available and the homemade breads and desserts bring raves. Wine is offered with dinner and all meals are served family style in the main lodge.

The four guest log cabins are nestled in the aspen trees beside a bubbling stream which runs through the ranch. Cabins are snug and modern, yet rustic and western in decor. Each cabin has two double bedrooms with a bathroom between the bedrooms. All cabins are carpeted, heated, and have pine paneling, and charming rustic touches.

The overall atmosphere at the Absaroka is warm and casual. Owners Budd and Emi Betts and their two school-age kids like to have fun and the staff goes out of their way to make each guest enjoy his/her stay at the ranch as much as possible. With a high rate of return guests the Absaroka books up early each year.

Wilderness pack trips are offered through the Absaroka in addition to the guest ranch. October brings elk hunters to the ranch also.

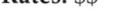

Address: P.O. Box 929, Dubois, WY 82513

Telephone: 307-455-2275

Location: 16 miles NW of Dubois, WY, guests fly into Jackson Hole or Riverton, WY

Guest Capacity: 18

Season: mid-June through the first week in September

Policy on Children, Pets: Children welcome. No pets

Rates: $$

Credit Cards: None

Wyoming

Bill Cody Ranch
Cody, WY

John and Jamie Parson, true Wyoming natives, are proud to share their Wyoming heritage and home with you. Right in the middle of what Teddy Roosevelt called "the most scenic 52 miles in all of America" lies the BILL CODY RANCH, just 30 minutes from both Yellowstone and the famous town of Cody, Wyoming. The historic lodge was built in 1925 and previously owned by the grandson of Buffalo Bill. No matter how much or how little you're looking to do, John & Jamie can build a package that's tailored to fit your family's adventure vacation.

Facilities are nestled in a valley shaded by pine and aspens. Cabins are log sided with western interiors. Some have 2 bedrooms. All are spotless and comfortable and rooms have deluxe private baths, no TV or telephones. All have handmade western furniture and covered porches.

Scheduled or unscheduled riding is the main activity at the ranch. Small groups of no more than 10 go on the ride and there is one wrangler for every 8 riders. Children 6 years and older ride their own horse. The ranch utilizes 16 trailheads in the 2.5 million acres of the Shoshone National Forest. Due to the rugged, mountainous terrain, all the rides are done at a walk. Rides go out morning, noon and evening. Rides either leave from the ranch or are trailered to points near Yellowstone National Park to provide a variety of mountain terrain. On these rides, wranglers cook your lunch over an open fire. Many of the wranglers are accomplished mountain guides that return each year. One is even an author of mountain guide books and western tales!

A hearty, ranch style breakfast is served each day in the dining room. Steaks, trout and nightly specials, served with homemade breads, soup & salad are served five nights a week in the main dining room. Wednesday and Saturday nights there is a creekside BBQ featuring steaks, ribs, chicken and all the fixins. Packed lunches can be provided for your day-trip excursions.

Other activities include guided fishing trips on or off horseback, one or two night packtrips, basketball court and horseshoe pitching. Babysitting is available for children under 6 years old while their parents horseback ride. (Children can join in most ranch activities but are their parent's responsibility.) Weekly musical entertainment could be a western singer or a guest with a guitar and there is a rodeo in nearby Cody every night June through August. Nearby: River rafting, Buffalo Bill Historical Center, Old Trail Town, Yellowstone and Grand Teton National Parks.

The Bill Cody Ranch attracts guests that want to experience a guest ranch and have a home-base while exploring Yellowstone and Cody, Wyoming. Riding is the main activity and is open to the public and you don't have to stay at the ranch to ride but many guests stay a few days, a week or longer!

Address: 2604 Yellowstone Hwy., Cody, WY 82414

Telephone: 307-587-2097; **Fax:** 307-587-6272
Internet: www.wtp.net/BillCody
E-Mail: billcody@cody.wtp.net

Location: 26 miles from Cody, Wyoming airport

Guest Capacity: 60 to 65

Season: May through September

Policy on Children, Pets: Children welcome, must be 6 years old to ride. No pets.

Rates: "Nightly" (alacarte) rates and "All Inclusive". You choose.

Credit Cards: Call

Wyoming

Blackwater Creek Ranch
Cody, WY

Welcome — Blackwater Creek Ranch offers a spectacular mountain dude ranch experience near Yellowstone National Park. Nestled in the scenic river valley between Cody, Wyoming and Yellowstone, our ranch offers a unparalled opportunity for a wide variety of Western experiences your family will cherish forever.

You will stay in an authentic western cabin, rustic from its 1930's vintage. Some of the cabins have fireplaces and for your comfort, all have been updated with modern plumbing and heating for the cool mountain evenings. The summer days at 7,000 feet are quite pleasant. You will enjoy cool evening by the fireplace in our lodge, and warm days of 70's to 80's making your time in the saddle an unforgettable adventure. Riding and ranch activities are well planned for your enjoyment and will present you with a balanced variety of western experiences.

Scenery and Riding — Blackwater Creek Ranch features some of the west's most breathtaking scenery. It's located in America's first National Forest, and just minutes away from Yellowstone National Park. This is the land that our forefathers set aside especially for your enjoyment. Horseback riding is our specialty and rides vary in length and location. All rides take you into the spectacular Absorka mountains and along its creeks which surround the Ranch. You'll be guided by our wranglers and enjoy security of our experienced mountain horses. Every guest is assigned a horse for the week according to the guest's riding ability.

At Blackwater nobody goes hungry, as our hearty ranch style meals are some of the finest in the west. Some of the delicious Western meals will take place around the campfire, among wildflowers on a rolling hillside, while another will be a 10,000 feet, where you are surrounded by a vista of majestic snow capped peaks.

Ranch Activities — Ranch life around our modern lodge is friendly and fun. Games, ping-pong and barbecues, cowboy singalongs, hiking and just plain goofing off make this the most relaxing, rejuvenating vacation of a lifetime and take a refreshing dip in our heated pool.

There are no "musts" at Blackwater. We are here to make this the vacation of your dreams. Soak up the surrounding beauty from your cabin porch and search the mountainside for wildlife, or take a nap under a pine tree. Test your skills at some of the finest fishing with both Blackwater Creek and the North Fork of the Shoshone River both almost within casting distance of your cabin.

In all, absorb yourself in the sights, sounds, and smells of the pristine beauty of nature, the delicate melody of singing birds, the rolling waters of the Blackwater Creek and the fresh scent of the clean mountain air.

A Day in Cody — Cody, Wyoming, the rodeo capital of the world, is truly a western experience in itself. One day each week we offer our guests all the town has to offer, from exciting white water rafting to the outstanding museums of the Buffalo Bill Historical Center.

Visit Historic Trail Town, a collection of historic buildings and gravesites. There's a cabin built for General Custer's scout, Curly, Jeremiah Johnson's grave, historic and prehistoric Indian artifacts and dozens of horse-drawn wagons. Attend the world-famous Cody Nite Rodeo. Great cowboys try their skills on rough stock including broncs and bulls. Skills of roping and racing demonstrate the speed and teamwork of horse and rider. Special events are available for children.

Address: 1516 Northfork Highway, Cody, WY 82414

Telephone: 307-587-5201

Web: http://www.wyo.net/blackwater

E-Mail: bwcranch@wave.park.wy.us

Location: 15 miles from Yellowstone Park

Guest Capacity:

Season:

Policy on Children, Pets:

Rates:

Credit Cards:

(See Page 308-309 for Color Photos!)

Wyoming

Blackwater Creek Ranch

ALL SMILES!

WINTERSCAPE.

Wyoming

Brush Creek Ranch
Saratoga, WY

Brush Creek Ranch is a 6,000acre, family-owned, Orvis-endorsed, working cattle ranch settled by homesteaders over 100 years ago. Guests who explore the many historic log structures and observe our wranglers at work will experience the true flavor of our western heritage.

For those who want to do a little more than take in the breath-taking scenery, we have a variety of activities including horseback riding for every ability level. No 'nose-to-the-tail' format here. With 6,000 acres to explore, our wranglers will take you out for scenic rides, breakfast rides, short rides, or all-day rides plus we generally move the cattle one to three times per week when you can ride along. You may decide that the view is best from the top of your horse. We offer horseback riding for riders of different abilities and comfort levels, from the experienced horseperson to the tenderfoot. We'll fit you to a saddle and take you out across some of the West's most beautiful terrain. The area around here offers every variety of ride experience from soft, gently rolling hay meadows to more challenging mountain trails - let the wranglers take you where you want to go. Because we use a rotational grazing system on the ranch, it's more than likely that you will have a genuine opportunity to help move the cattle from one pasture to another. An additional chance to get involved comes during our annual Roundup Week which occurs once late each summer. Please call ahead for specific dates. This week covers a variety of activities, including gathering and sorting the cattle.

The fishing, too, is unparalleled. We have three private miles of Brush Creek reserved solely for our guests. Stand knee-deep in a mountain stream with a trout tugging at your line, you won't need to be an expert to realize that the flyfishing here is some of the best in the country. That is the reason that Brush Creek became just the second Orvis endorsed lodge in Wyoming. We combine the fabulous fishing resources in this area with an Orvis-endorsed guide who will help you discover the rewarding local waters of the Saratoga and Encampment Valley and the nearby "miracle mile" of the North Platte River. All the classic elements are here in perfect combination: great high mountain lakes feed the cool rushing streams that are tributaries to the legendary freestone rivers, famous among Western flyfishers. As written in the Orvis News last year, the waters around the Brush Creek Ranch "are one of the West's undiscovered venues for the kind of fly-fishing we all like best: big fish on a scenic wild river with nobody else around." Also, to quote the Orvis News, "the experienced, friendly staff provides a unique Western experience for the guests." Our intimate size guarantees a personal touch during your stay and our select Orvis endorsement ensures the quality of every aspect of your time with us.

Please call our toll-free number if you have further questions. We look forward to serving you and welcoming you this year as our guest.Come up to Brush Creek for a genuine Western experience.

 Address: Star Route 10, Saratoga, WY 82331

Telephone: 307-327-5241;
Toll Free: 800-726-2499; Fax: 307-327-5384

Location: Denver International airport is the closest Guest Capacity:

Season: Year round

 Year round Policy on Children, Pets: Children welcome.

Rates: $$

Credit Cards: MasterCard and Visa

(See Page 310-313 for Color Photos!)

Wyoming

Brush Creek Ranch

OUR NATURAL SWIMMING HOLE.

WYOMING

BRUSH CREEK RANCH

FISHING BRUSH CREEK.

Wyoming

Cheyenne River Ranch
Douglas, WY

Cheyenne River Ranch is an authentic working cattle and sheep ranch, owned and operated by the Don Pellatz family. The ranch consists of 8,000 acres of rolling hills covered with sagebrush, grass, and cactus. Trees dot the dry creek beds. The ranch stands amid the Thunder Basin National Grasslands 50 miles northeast of Douglas. This is wide-open prairie country where the wind blows the tumbleweed, and you can see forever. In Wyoming, our nation's history and prehistory are still close enough to touch. Here you can see petrified wood, fossils, and many artifacts that the Plains Indians left behind on their summer hunting grounds.

Like many American West pioneers, Don Pellatz's father homesteaded on the Dry Fork of the Cheyenne River in 1917. For many, this is the Old West — the way it was and still is. If the land seems empty, look closer. You'll see white-rumped antelope grazing on the hills and deer among the cottonwood trees of the Cheyenne River bed. Elk live in the Rochelle Hills and occasionally come down to the lower country. Coyotes, foxes, jackrabbits, prairie dogs, and bobcats live here. Eagles and hawks soar in the big blue sky. As Betty Pellatz says, "The vastness of the prairie where the air is clean and pure, and the smell of sage after a rain is wonderful! This is God's country."

Don, Betty, and their son Chuck have welcomed guests to the ranch since the 1970s, usually only one family or couple at a time to ensure a real hands-on experience. Guests are encouraged to share in daily ranch activities, or just sit and relax. If you come in April or May, you'll get to see new life, for this is calving and lambing time. You might help move sheep to fresh pastures or mend fences. There are always eggs to be gathered, or you might try your hand bottle-feeding an orphan lamb.

There is nothing fancy at all about the Cheyenne River Ranch. The reason guests return is the warm, sincere Western hospitality that the Pellatz family shares. Don was raised in this part of the country. Betty, his bride and partner of over 40 years, came originally from Illinois. Together, they have raised five children and have devoted their lives to each other and the ranching tradition of the West. When you stay with the Pellatz family, their home is your home.

Guests stay in two bunkhouse cabins that have a private bath and queen beds and bunkbeds. Guests can also stay in three bedrooms in the ranch house, one with full bed and private bath, one with full bed, and one with twin beds. The last two share a bath. All of the rooms are immaculate, with handmade quilts on the beds. There's even a sheep wagon for those more adventurous souls. Meals are served family style and nothing compares with country cooking and the aroma of fresh baked bread. The Pellatz family are nonsmokers and non-drinkers, and they especially welcome guests that share these preferences.

The Cheyenne River Ranch also offers five-day cattle drives, May to October for "cowpokes" hankerin' to ride a horse on the vast prairies of the American West. The cattle drives are for beginning as well as experienced riders and are limited to 10 riders. You'll ride 5-6 hours a day and cover a distance of 10 to 12 miles a day moving the cattle from winter range to summer pastures. In October, there's the roundup. Eat hearty western food cooked over an open fire. Sleep under the stars or in a tent.

Most of the horses at the ranch are Quarter Horses or Quarter Horse-Appaloosa crosses; they're horses that are used regularly on the ranch. There are gentle horses for gentle people, spirited horses for spirited people, and for people that don't like to ride, there are horses that don't like to be ridden!

While staying at the Cheyenne River Ranch, you will experience the still hush of the prairie, the multitude of stars in the evening sky, and beautiful sunrises and sunsets. You will discover the true meaning of Western hospitality.

Address: 1031 Steinle Road, Dept. BP, Douglas, WY 82633

Telephone/Fax: 307-358-2380

Location: 50 miles northeast of Douglas

Guest Capacity: 12

Season: May to October

Policy on Children, Pets: Children must be 11 years old for the cattle drives. No pets.

Rates: $$ American plan.

Wyoming

Darwin Ranch
Jackson, WY

The Darwin Ranch has been catering to wilderness enthusiasts ever since 1965 when Loring Woodman began renovating the old log cabins that lay along the original pioneer wagon road into Jackson Hole.

Unstable landslide conditions in the Gros Ventre River canyon just below the Ranch closed down the wagon route in the 1920's. When the first modern highways were engineered into Jackson Hole from the south they were pushed instead down the Hoback Canyon, 25 miles to the west, leaving the Darwin Ranch at the very end of a long, unimproved dirt road, leading nowhere. And nowhere is where we still are - an inholding in Bridger-Teton National Forest, altogether cut off from the rest of Teton County. Though this may have seemed a sad fate at the time, in the long run it has kept us well apart from Jackson Hole's headlong rush to development. We still have the country, but we don't have the people.

In 1983 the area around the Ranch was formally preserved by the U.S. Congress as the Gros Ventre Wilderness. With no other private land in the area, no roads, no dams, no power lines, no nothing, we now have this last, totally isolated section of the Gros Ventre River Valley pretty much to ourselves. riding out in all directions form the Ranch you will rarely see another human being. A few years back a couple, returning for the first time since 1969, had this to say: "We found our first entry in the guest book. Thirty wonderful days and every day a different ride. Now, 23 years later, we find those wonderful rides unchanged and more beautiful than ever - unique in the changing world."

There are guided rides going somewhere in the surrounding mountains almost every day during the summer. Sometimes these trips are as short as several hours, but more often our guests are interested in all-day rides. For beginners we have a number of extremely placid, easy-going animals and can teach you in a day or two to get around more or less comfortably with everyone else.

Riders WE judge to be sufficiently experienced may ride without a guide once they have been riding with us for at least 4 days on guided trips and demonstrated their competence. Riding without a guide is a privilege which must be earned - and retained on a continuing basis. Not everybody can, or should, qualify. You need to work with us patiently, carefully, and respectfully if this is your goal.

Some of our guests come to hike or climb the sedimentary peaks in the Gros Ventre Range rather than to ride. There is also excellent dry fly fishing in the river, though lake fishing is limited. Many of our guests include packtrip as part of their vacation. At present, with only eight rooms or cabins, we can be full with 16, or just 14 guests.

Though the Ranch is not fancy, it has a good deal in the way of creature comforts. The main lodge was built as a private log home by a reclusive Diplomatic Service couple who had endured several winters in Russia during the Second World War. There is a large sitting room with stone fireplace, an informal bar, a piano, lots of books, a dining room and a kitchen. Everybody eats together, including the crew. Food is continental and ethnic, with occasional Thai or Chinese meals by Melody, our Taiwanese cook for over 10 years. Modern plumbing and electricity from a turbine on the creek complete the scene.

Though the Darwin ranch is a true wilderness ranch and not a resort, we like what some guests wrote in the guest book in 1976: "Coming here with expectations of the rough, harsh, and typically Western life, I found class, elegance, taste, and great sensitivity. A vacation to be remembered."

Address: P.O. Box 10430, Jackson, WY

Telephone: 307-733-5588; **Fax:** 307-739-0885
E-Mail: DR@wyoming.com

Location: 30 miles east of Jackson Hole airport, 50 miles north of Pinedale

Guest Capacity: 18

Season: June 15 through September 25

Policy on Children, Pets: No special children's program's, but children are welcome. Dogs are welcome too, but only on best behavior, otherwise they have to be tied

Rates: $$ American Plan

Credit Cards: None

WYOMING

DARWIN RANCH

A SCENIC OVERLOOK AT DARWIN RANCH.

Wyoming

David Ranch
Daniel, WY

Come along on an adult adventure vacation, oriented around smart, well-trained, gentle cow-horses and help on cattle drives, branding, roundup and all the other activities necessary on a working ranch. We show you how our horses are trained and take the time to give you personal instruction on how to really work the particular horse that we pick to be yours during your stay with us but you do not need to have any horseback riding experience to enjoy your vacation here.

You will learn how a real cattle ranch operates, including the problems and rewards of working with horses and cattle, while meeting some of our neighbors. Live an adventure far removed from the hustle of modern day life. The ranch is so remote that we use a diesel-powered generator for our electricity. If you like being up past 10 p.m., we provide a kerosene lamp for your use!

Delicious meals are prepared by the ranch cook and served ranch-style at a common table where the owners, ranch hands and guests all join in lively conversation. The meals are the same type as those served to cowboys years ago. Some of the recipes were used by roundup cooks over 100 years ago and pride is taken in the fact that our meals are cooked from "scratch" rather than prefab concoctions.

Lodging is provided in ample-sized, very comfortable log cabins complete with western decor, gas heat and bathrooms. The bedspreads are homemade quilts.

We have excellent fishing available in nearby streams as well as in our five-acre lake that we keep stocked with rainbow trout.

The beautiful scenery and wildflowers in this area are unsurpassed with an abundance of wildlife such as elk, moose, mule deer, antelope, fox, beaver and coyotes. The antelope can be observed everyday from your cabin. This area is a real haven for the camera buffs!

Our packtrips journey into the nearby Wyoming Range and can be the central focus of your vacation or you may just want to spend a couple of days in the mountains. These packtrips are taken with sure-footed, well-mannered horses. Meals are cooked over campfires in our "Dutch Ovens". Write or call for more information if you have any other questions about our packtrips.

Your hosts, Melvin and Toni David, are lifetime ranchers. Their knowledge of the colorful local history regarding outlaws, ranchers, mountain men, trappers and Indians is extensive and they are happy to share these stories with their guests. It helps you get the feel of the country and the ranch.

We cordially invite you to come and spend a week or more with us. Big-game hunting is also provided in the fall and snowmobiling during the winter. Come along with us for your next vacation adventure.

   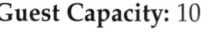

Address: P.O. Box 5, Daniel, WY 83115-0005

Telephone: 307-859-8228; **Fax:** 307-367-2864
Web: www.pinedaleonline.com/davidranch
E-Mail: tdavid@wyoming.com

Location: 90 miles SW of Jackson Hole, Wyoming

Guest Capacity: 10

Season: Year round

Policy on Children, Pets: Children 15 years or older welcome.

Rates: $$

Credit Cards: Call

(See Pages 314 for Color Photos!)

WYOMING

DEER FORKS RANCH
DOUGLAS, WY

Deer Forks Ranch provides horseback riding vacationers with a unique opportunity to ride on a real working cattle and sheep ranch located in the uncrowded, relatively undiscovered part of Wyoming near Medicine Bow National Forest. A limited number of guests can share the relaxed lifestyle and unspoiled and natural wide-open spaces that Wyoming natives have enjoyed for years. This area is abundant with all types of native wildlife. Eighty percent of the world's pronghorn antelope live within 300 miles of Douglas, Wyoming, and the Deer Forks Ranch is home to many of them. You'll delight in seeing them in their natural habitat, so make sure you bring your camera! The road to Deer Forks crosses the historic Oregon Trail, and you'll be able to ride along the ruts left behind by those long-gone settlers. Thunder Basin National Grassland is nearby with the over 22,000 acres of public land to ride, hike, and explore. This is big country!

The Cow Creek unit of Deer Forks Ranch is 85 miles from the home ranch, with rolling plains as far as the eye can see. Here you'll develop an understanding and love of the real West. It's a favorite winter area of the gold and bald eagle as well as various other wildlife species. Many of the cattle and riding activities guests enjoy take place here.

The appeal of this ranch is the relaxed, unscheduled, family-oriented atmosphere. Guests can prepare their own meals and sleep in if they like. Owners Benny and Pauline Middleton have created a haven for people who want to get away from their hectic schedules, come out and relax, and do what pleases them — even if it's doing nothing but enjoying the scenery! Horseback riding is a popular morning event. Afternoons are free for hiking, trout fishing, bird-watching, or touring the many scenic and historic sites around Douglas. Mount Rushmore, Devil's Tower, and Fort Laramie are all located close enough for a day trip. The State Pioneer Museum, Fort Fetterman, and the Natural Bridge are near Douglas. Avid hikers will want to climb Laramie Peak or visit the fire lookout station east of the peak.

Plenty of reading materials and family games are available in the guest houses. You can try horseshoes or volleyball or brush up on your roping. Children enjoy playing in the mountain streams or windmill tanks. They'll also delight in joining in on the daily chores around the ranch, such as feeding the dogs, cats, rainbow trout, lambs, or calves. What makes vacationing at Deer Forks Guest Ranch special is that guests can actually jump in, roll up their sleeves, and take part in seasonal ranch activities such as round-ups, real cattle drives, branding the cattle, shearing the sheep, calving, and lambing.

Another popular option at the ranch is the Winter Wildlife Tour, where guests will be able to view the large elk herd and other wildlife that is more visible at that time of year. Excellent trout fishing can also be arranged in both mountain streams and a privately stocked reservoir. Skiing is available at Hogadon 75 miles away.

Deer Forks Ranch has two modern two-bedroom log guest houses and one guest room with a private bath in the ranch house. One guest house has a loft with bunk beds for four and a master bedroom with a queen-sized bed and additional cot. The second guest house has a queen-sized bed in one bedroom and bunk beds in the second bedroom. Both guest houses are spacious, with kitchens, living rooms, and private baths or showers. The kitchens are fully equipped for housekeeping with everything you'll need except food.

Come and explore the real West, the Big Sky country at Deer Forks Ranch. Come and star-gaze, horseback ride, hike, fish, and create the kind of vacation that other people only dream about!

Address: 1200 Poison Lake Road, Dept. BP, Douglas, WY 82633

Telephone: 307-358-2033

Location: 24 miles southwest of Douglas, 85 miles from Casper

Guest Capacity: 12

Season: May 15 to September 30

Policy on Children, Pets: Children are welcome. Pets allowed with supervision.

Rates: $ Various meal plans available. Children's rates. Horse boarding and riding extra.

Credit Cards: None

WYOMING

DIAMOND L GUEST RANCH
HULETT, WY

The Diamond L Guest Ranch borders the Wyoming Black Hills National Forest and is only 18 miles from Devil's Tower, the nation's first monument. Genuine western hospitality and personalized attention have always been the standard. We limit the number of guests we have each week to a maximum of 16, allowing us to cater to our guests in a very personal way. You may arrive as guests but you will go home feeling like you have made friends here and you can't wait to return.

At the Diamond L our riding program is geared to the individual guest. Our rides are as diverse and interesting as the magnificent landscape we ride in. From high on mountain tops with the whole world laid out at your feet, to valley, canyons, wide open meadows,

country lanes, or grassed-over logging roads made for loping, we have it all. Because we keep our guest to a minimum, we can offer rides that may not be possible at larger guest ranches. We will do everything we can to give you the riding experience that you have always wanted. Our horses, raised in the mountains, are responsive and well trained. We match riders and horses no matter what the skill level, from beginner to old pro. We give riding instructions and provide a video so that when you leave the corrals you are comfortable with your horse and ready to ride.

The Ranch features a cedar log lodge, with a deck suited for relaxing. Upstairs, in the lodge is the large common room which features a floor-to-ceiling rock fireplace, and three of the deluxe guest rooms, (one room has a private bath and the other two share a bath). Downstairs we have the dinning room, gift store, game room and kitchen. To complete the accommodations for the guests, the original bunkhouse has been remodeled into a cabin that features two private guest rooms, each with a wood burning stove and private baths. The ranch has an outdoor hot tub, spring-fed pond and numerous walking & hiking trails.

Activities include at least two daily horseback rides, breakfast and supper rides, an all day ride, fishing, mountain bikes, volleyball, horseshoes, and the best Western home cooking around. The Diamond L is located close to Mt. Rushmore and all the Black Hills have to offer.

In the spring we host "Trail Blazer" week. During the last two weeks of May, our guests are often invited to join our neighbors while they gather their cows and new calves. We also spend some of our riding time getting ready for the summer season, looking for new trails and clearing old ones. We invite adults to join us who enjoy a little work and love to ride. In the fall we offer an "Adults Only" week. If you are looking to spend some serious time in the saddle, this is the week for you. If your idea of the perfect vacation is spending quality time with other adults who share a passion for horses, riding and the great outdoors this is the place.

Address: P.O. Box 70 Dept. BP, Hulett, WY 82720

Telephone: 307-467-5236; Toll Free: 800-851-5909; Fax: 307-467-5486
Web: http//www.diamondlranch.com
E-Mail: info@diamondlranch.com

Location: 120 miles form Rapid City, South Dakota (nearest airport)

Guest Capacity: 16

Season: Year round, horse related activities May through October

Policy on Children, Pets: Children welcome. Pets welcome with prior arrangements.

Rates: $$ American plan, Children's rates

Credit Card: Discover, MasterCard and Visa

(SEE PAGES 315 FOR COLOR PHOTOS!)

WYOMING

HIGH ISLAND GUEST RANCH AND CATTLE COMPANY
HAMILTON DOME, WY

In 1994, Frank Robbins sold the family company business and bought the High Island Ranch, a 130,000 acre cattle ranch in the wide-open spaces of Wyoming. Ranching is not new to Frank. For many years he has owned other cattle ranches in Montana and Alabama. Because of its size and all the country it encompasses High Island offers individuals, couples, families and small corporate groups the freedom to really be cowboys. With their southern hospitality and western traditions, Frank and Karen, together with their two children, Holli and Hank, share their Wyoming paradise and their cowboy way of life with those who want to experience ranch life for a week or more. Here you will see a tremendous diversity of terrain, from the 6,000 foot prairie to 12,000 foot mountains. This working cattle ranch is rugged, remote, beautiful and for those who want ranch life the way it used to be.

Accommodations: The main lodge is located at 6,000 feet on Cottonwood Creek. This lodge has 12 rooms, large den area, large dining area and gift shop, all in western decor. Also, there are 2 private log cabins, four-wall canvas tents, separate men and women shower houses, outhouses and flushing toilets. The lodge on Rock Creek sits at 9,000 feet, has 8 rooms, den area and dining area accented with prints and numerous animal mounts. Also, there are 12 four-wall canvas tents, one private cabin, men and women shower houses and outhouses.

Activities: Branding week; guests get to participate in all the cattle work including branding and doctoring calves. Cattle drives involve 45 miles of moving 200 head of cattle to new pastures. This is the real thing. For those who want to experience long days in the saddle, bedrolls and hearty ranch food cooked over the campfire on the trail, this will be an adventure you will never forget. Unlimited riding available weekly. Trout fishing in crystal clear Rock Creek is excellent. Fish miles of waters that have been protected for years.

Children's Programs: Children 16 years and older go on cattle drives. Children 12 years and older are welcome on branding and roundup weeks, all ages on fly-fishing weeks.

Dining: While on the trail, old-fashioned chuck wagon pulled by two stout mules follow with plenty of hearty food, once at camp everything is cooked on an open fire. Standard western fare at the upper and lower lodges. A non-smoking, non-alcohol environment is promoted. Finale western barbecue.

Entertainment: Listening to the call of the coyotes and cows calling for their young and authentic cowboy music.

High Island is an authentic cattle ranch. Bring your own sleeping gear and come along for one of the greatest weeks of your life.

Address: 3081 Upper Cottonwood Creek Road, Hamilton Dome, WY 82427

Mailing address: 346 Amoretti Suite 10, Thermopolis, WY 82443

Telephone: 307-867-2374; Fax: 307-867-2314
Web: www.GORP.com/highisland/
E-Mail: HIGHLS1994@aol.com

Location: 75 miles S of Cody airport

Guest Capacity: 25

Season: Summer only - June through 1st of September

Policy on Children, Pets: Children welcome but certain weeks are better for children. No pets.

Rates: $$$ American plan. Children's rates available.

Credit Cards: MasterCard and Visa

WYOMING

FLYING A RANCH
PINEDALE, WY

The Flying A Ranch, located some 50 miles southeast of Jackson Hole and nestled at the edge of a valley surrounded by majestic peaks of the Gros Ventre and the Wind River mountains, provides guests with a historic and authentic Western vacation. Built in the 1930s, the Flying A has great historical significance and character. The logs from which the cabins were built were hauled down from the surrounding mountains with horses.

Completely renovated log cabins retain their rustic appearance on the outside while providing luxurious accommodations inside including private bathrooms, handmade pine furniture, kitchenettes, and wood-burning stoves or fireplaces. All of the cabins have private porches for viewing a breathtaking sunset on one of the ranch ponds or the wildlife that abounds throughout the valley.

There's something about the mountain air that makes you hungry, and the Flying A caters to your appetite with a unique combination of hearty Western fare and continental cuisine. Dining at the ranch is served with a casual, yet gourmet, flair. Dinners, which are served in the main lodge, may feature Mexican and Italian entrees along with ranch steaks, charcoal grilled on the deck. Round those out with crisp garden salads, homemade soups, and delicious desserts.

Guests are encouraged to ride, fly fish, hike, and mountain bike — or to sit quietly on the porches of their cabins savoring the spectacular views. The ranch's string of proven trail horses has evolved over the years to include horses that match any rider's skill level. In fact, guests who have stayed at other ranches remark on the quality of the Flying A's horses. The riding program accommodates a variety of levels, and instruction is offered for guests who wish to improve their skills. Half-day and all-day rides are planned, on over twenty different trails. Riding groups are kept small so guests can fully appreciate the majestic views available on the trails. Through grassy meadows, ponderosa pine forests, aspen groves, or steep mountain trails, there's a new spectacle to be savored and stored at every turn. Wildlife abounds and moose, bear, antelope, coyote, elk, sand hill cranes, and deer can often be spotted around the next run in the trail or on the far side of the meadow.

Rainbow and brook trout roil the surfaces of the ranch's three lakes and fly fishing lessons are available for guests who want to try their hand. For the more advanced angler, trips with an expert guide can be arranged to nearby mountain lakes, streams, and rivers, including the Green, New Fork, and the Snake.

Hiking and mountain biking are always available and a great way to explore the high country of the surrounding national forest. Evenings are tranquil — a time for peaceful relaxation and good conversation with other guests or your hosts. Of course, there might be a volleyball game or a hayride lurking around the corner!

Each month at the Flying A offers guests a different vacation experience. In June, herds of elk can be seen moving to their summer pastures. July is resplendent with wildflowers — each week seemingly promoting a different color. August is warm with clear nights and meteor showers. And in September the aspens paint the hills with color and the elk begin to bugle.

This area of Wyoming offers plenty of attractions. Yellowstone and Grand Teton national parks are a short distance away. Jackson Hole is just an hour's drive away with lots of shops and museums. Pinedale itself offers beautiful mountain lakes, shopping, and the Museum of the Mountain Man.

The Flying A, under the management of Debbie Hansen and her husband Keith Dagel, offers the discerning adult the most distinctive of western vacations.

Address: Route 1, Box 7, Dept. BP, Pinedale, WY 82941

Telephone: Summer: 307-367-2385; Winter: 800-678-6543;

Fax: Summer: 307-367-2385; Winter: 605-336-8731

Location: 50 miles southeast of Jackson Hole, 27 miles northwest of Pinedale

Guest Capacity: 12-14

Season: mid-June to October 1

Policy on Children, Pets: Children welcome if guest books entire ranch. No pets.

Rates: $$ - $$$ American plan.

Credit Cards: None

WYOMING
FLYING A RANCH

THE WIDE OPEN SPACES OF WYOMING.

THE FLYING A HAS SOME OF THE FINEST HORSES IN THE WEST.

WYOMING

GROS VENTRE RIVER RANCH
KELLY, WY

Millions of people travel to Jackson Hole and Yellowstone Park every year but most stop only briefly. The Gros Ventre River Ranch offers its guests a chance to slow down and really experience the area. It is a special place to us and we would enjoy sharing it with you.

We are located eighteen miles northeast of the town of Jackson in a valley that was frequented by trappers and explorers 150 years ago. The ranch is bordered by National Park and Forest land and at 7,000 feet altitude we enjoy the panoramic view of the Tetons. The main lodge, guest cabins and corrals are set amidst meadows, pines and aspens. The Gros Ventre River flows through our land. The Ranch is only fifteen miles from the Jackson Hole airport.

We can accommodate 34 guests in our log cabins and lodges. All rooms are carpeted, electrically heated with private baths and decks. The lodges with living rooms have wood burning stoves and kitchenettes. Ranch life is casual, but comfortable. You can curl up by the fire with a book, nap under a down comforter or in a hammock, or enjoy a glass of wine on the deck as you watch the sun set behind the Tetons.

Riders at the Gros Ventre River Ranch have access to 45,000 acres of trails in the Teton National Forest and Wilderness area. We assign each guest a horse for the week, compatible with their riding ability. Our wranglers guide you through fields and meadows full of wildflowers, across creeks and up to ridge tops overlooking the entire Jackson Hole valley. Tuesday's ride takes you to Lucy Hill for a cook-out and, for the adventurous, all-day rides into the Gros Ventre Wilderness to look for moose, big horn sheep, elk and deer. We keep our rides to 6-8 people and break them out into various riding levels and interests. We offer morning and afternoon rides all week or signup mid-week for all-day rides into Bridger-Teton National Forest or Gros Ventre Wilderness.

In the evening we gather in the dining room overlooking the river for carefully prepared meals that may include prime rib, fresh trout, grilled chicken, fresh garden vegetables and homemade breads and desserts. We often serve meals on the decks or have cook-outs around a roaring fire in the yard. After dinner, it's time for toasting marshmallows and cowboy songs and poetry.

In mid-summer, evenings may find some guests fly fishing in our two stocked trout ponds or stalking native cutthroat trout along the mile-and-a-half of river running through the ranch. Guided trips on the legendary Snake River are also popular with anglers.

Some guests take a day away from the ranch to explore the area. Day trips to Teton and Yellowstone Parks offer opportunities for hiking and wildlife viewing. If you head into the town of Jackson, you will find galleries, museums and shops of every description. Evening rodeos bring cowboys to test their skills and golfers can choose from two championship courses with views of the Tetons!

The Gros Ventre River Ranch is a wonderful place to enjoy the valley's natural beauty and peace. We hope you will come and spend some time with us in this unique and beautiful part of the world.

Address: 18 Gros Ventre Road, Kelly, WY 83011;
Mailing Address: P.O. Box 151, Moose, WY 83012
Telephone: 307-733-4138; **Fax:** 307-733-4272
Web: http://www.jocksonwy.com/lodging/gurr/.htn
E-Mail: grosventreriverranch@compuserve.com
Location: 12 miles from Jackson Hole airport

Season: Mid-June through mid-September
Policy on Children, Pets: Children welcome but those under 7 years of age do not go on trail rides. No pets.
Rates: $$
Credit Cards: None

(SEE PAGES 316 FOR COLOR PHOTOS!)

WYOMING

HIDEOUT AT FLITNER RANCH
GREYBULL, WY

The Hideout at Flitner Ranch is located in Greybull, Wyoming on the west slope of the Bighorn Mountains, and is one of the oldest and most impressive cattle ranches in Wyoming. The Flitner Ranch is home to The Hideout, one of the most beautiful and remote ranch retreats in the world. The five-generation cattle ranch was founded in 1906. The ranch cattle graze on vast range lands between the Bighorns and Rocky Mountains. This spectacular landscape ranges from 4,000 to 10,000 feet in elevation. Guests have many options: you can relax in the rustic setting of the main lodge with its many amenities, fish in our nearby streams, or ride with the cowboys while they tend to a large herd of our Black Angus cattle.

As a guest of the Hideout, you will have the opportunity to participate in the team-building activities of a working ranch. Check, move or doctor 1,000 head of cows. Repair fence lines or scatter salt. Enjoy a Flitner Ranch exhibition roping and test your skills in competitive horseback events in our custom-built rodeo arena. Fish in nearby Shell Creek, enjoy a chuckwagon barbecue or take a scenic trail ride into the wilderness. Whether you want to "rough it" or just relax is our rustic elegance, a vacation to the Hideout is one you and your family will never forget.

The setting of The Hideout is spectacular. The Hideout facility is only a few short miles from Shell Canyon. It includes a luxurious 5,000 square-foot lodge, complete with decks and rocking chairs, that surround the lodge and offer guests the perfect place to put their feet up and enjoy the beauty that surrounds them. From your porch you will see deer roam on nearby pastures while watching beautiful sunsets. The cozy cabins and main lodge are decorated in a luxurious western style and offer the latest modern conveniences. There is a large dining room, a lounge area with wet bar where the owners invite you for cocktails, a library with satellite TV and conference rooms. The Hideout is a perfect setting for incorporating a business or family retreat with a unique western flavor in spectacular surroundings and …the food is outstanding!

The ranch arena is perfect for calf roping and barrel racing. You can try your luck at throwing horseshoes or trap shooting, or relax in the hot tub. And there is plenty of fishing in nearby streams. You may want to visit nearby Cody for a summertime rodeo, or go up to Yellowstone National Park.

Our remote but modern mountain cabins are available for hunting and fishing, as well as family reunions or small retreats. They are decorated in a rustic western motif and each room has a separate entrances and a private bathroom. The Hideout is the ideal setting for your next conference or corporate event.

The Hideout is a family business where attention to detail and personal service is of the utmost importance. Whether you come for a cowboy adventure, a honeymoon, or small corporate meeting, every effort will be made to make your stay a most memorable one.

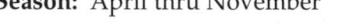

Address: 3208 Beaver Creek Road, Greybull, WY 82426

Telephone: 307-765-2080; **Toll Free:** 800-354-8637; **Fax:** 307-765-2681; **Web:** thehideout.com
E-Mail: hideout@tetwest.net

Location: 70 miles E of Cody

Guest Capacity: 10 to 50

Season: April thru November

Policy on Children, Pets: Children welcome (with restrictions). No pets.

Rates: $$$

Credit Cards: Visa, MC, Amex

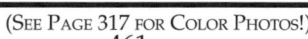

WYOMING

KEDESH RANCH
SHELL, WY

Kedesh Ranch is nestled at the entrance to Shell Canyon on the "Western Slope" of the Big Horn Mountains. The Shell Canyon is known worldwide for its extraordinary beauty, beautiful geological formations and recent dinosaur discoveries, the Shell Canyon Waterfall, Indian petroglyphs, plus fossils and otherwise interesting rocks. The ranch sits alongside a clear-running stream at an elevation of 4,300 feet. The surrounding mountains rise to 13,000 feet, providing a stunning backdrop for the ranch.

Horseback riding is the main attraction at Kedesh. Experienced wranglers match each guest to a horse according to riding ability. Groups are limited in number, and the riding is varied enough for both the experienced and the inexperienced rider. Rides are scheduled twice a day. The ranch is best suited for children who are old enough to ride since children are able to ride with their families. Competent wranglers guide the riders on these daily excursions either on gentle terrain or on mountain trails. The view of the breathtaking Big Horn Mountains, Chimney Rock, and White Canyon are all part of the territory!

Guests can do as little or as much as they want. There is trout fishing for both the beginner and experienced angler in Shell Creek, which runs through the ranch. Guests can try their hands at hooking the abundant rainbow and brown trout and Rocky Mountain whitefish. Serious anglers should bring along their own gear. Wading, swimming, and tubing in the creek are popular pastimes. The ranch even provides a rafting trip on the Shoshone River near Cody.

Hikers will find many trails leading from the ranch to the Big Horn Mountains. In addition, there are outdoor games of volleyball, basketball, and horseshoes and indoor games of table tennis, pool, and cards.

Evenings after a delicious meal, guests may just want to relax. A hot tub visit may soothe those weary bones after an active day. Or, guests might want to participate in the weekly talent night. There are also Western dancing, videos, and good conversation. The ranch's old wagon with straw bales makes great fun for the old-fashioned hayride.

Depending upon guests' interests, weekly sightseeing trips are provided to some of the local attractions. Spend a day in the beautiful Big Horn Mountains, enjoy the view from the top of Big Horn Canyon and a pontoon boat ride up the canyon when available. This is the third-largest canyon in the United States. The Medicine Wheel and the mysteries it still holds of our early Indian heritage can be found on the top of the Big Horn Mountain. A trip to Cody for the "Cody Night Rodeo" will show you some real cowboy and cowgirl fun. A visit to the Buffalo Bill Historical Center will give you a real feel for the Western life.

The cabin accommodations are a combination of rustic log exteriors and modern interiors. They're nestled among cottonwood trees, and some back up to Shell Creek. The screened porches are especially nice places to enjoy the evening quiet. The three-times-a-day, home-cooked meals are served family style and are all you can or want to eat. They truly are an event! A Western steak cookout with all the fixings is held once a week.

Yellowstone National Park is just two hours from the ranch. You may want to add a visit there either before or at the end of your week at the ranch.

The Kedesh Ranch is casual, friendly, and laid back. The three generations of the Lander family are eager to share the beautiful Big Horn country, their home, and their way of life.

Address: 1940 Highway 14, Dept. BP, Shell, WY 82441

Toll-free Telephone: 800-845-3320; **Telephone/Fax:** 307-765-2791

Location: 70 miles east of Cody, 140 miles from Billings

Guest Capacity: 20

Season: Last week of May to first Saturday in October

Policy on Children, Pets: Children are welcome; special activities available but no daily supervision. No pets.

Rates: $$ American plan. Children's rates.

Credit Cards: Mastercard, Visa

Wyoming

Kedesh Ranch

RED ROCK WONDER.

WYOMING

LAZY L&B RANCH
DUBOIS, WY

Lee and Bob Naylon with their young daughter Piper have owned and operated the Lazy L & B Ranch since 1993. Their beginnings in the horse recreation business began in 1972 with many summers outfitting guests in the Eastern High Sierra Mountains. Winters were spent working in the ski industry and enjoying mountaineering. After 5 long years looking for the "right" ranch they purchased the Lazy L & B from long time dude ranchers Leota and Bernard Didier. Keeping the reputation of a "rider's ranch", Lee and Bob invite you to share a week of western fun while enjoying the spectacular beauty and drama of Wyoming's vast country.

The lodge has had its original structure added to over the years with the last addition being a new dining room, front porch and extended kitchen in 1994. The lodge also has a cozy fireplace, library and game tables where guests enjoy evening entertainment of cowboy poetry and songs.

The cabins and the lodge are shaded by the towering cottonwood trees along the river. Ranging in size from the old 1-room schoolhouse for one or two to large family cabins that house up to eight, they all have private bathrooms, electric heat and comfortable western furnishings.

The Lazy L & B riding program is based on the exploration of the varied scenic country we have surrounding the ranch. From open meadows and ranch land, canyons and red rock country, to the green river bottoms and alpine forests we cover all kinds of spectacular terrain. There are trails leading up the mountain ridges but the rest of the riding is open enough that we invite you to spread out and explore.

The major philosophy of the riding program at the Lazy L & B Ranch is to provide a safe enjoyable and exciting horseback vacation to all guests. We want to include everyone in our program that wished to enjoy the pleasures of ridding in the west regardless of age or ability level. The program includes walking rides for beginners and young children to faster rides with trotting and loping for the more experienced riders. For riders of all levels our wranglers are always guiding each ride. Instruction and safety are part of all rides.

As our guests arrive, we encourage them to get to the corrals as soon as possible. At that time, each guest is introduced to his or her preassigned horse by a wrangler. The wrangler then assesses the rider's ability and comfort around the horses and leads the guest to the main arena where he is mounted on the horse and stirrups are adjusted. There is always a wrangler in the arena to instruct and assist each rider as they become more comfortable with their horse. Each group or family is then taken on a 1 to 1-1/2 hour ride around the valley. Rides at the Lazy L & B consist of no more that 7 guests in a group. This first-day ride is a good chance for the staff to evaluate each rider.

Sunday night after dinner, all guests are required to attend a safety demonstration at the corrals. We explain: how we tie up our horses, cinching up in the morning basic riding skills (uphill climbing, crossing rivers and ditches, downhill riding) as well as many safety tips. We like our guests to become involved in saddling, birdling and general care of the animal. Although we don't require it, we feel learning the basics make for a safer experience.

The Lazy L&B is famous for its ranch meals with home made breads and desserts. Start the day with a hearty buffet breakfast with homemade muffins and cinnamon rolls, eggs and bacon or pancakes. IF you prefer, enjoy fresh fruit and cereal. Our lunches are served either at a buffet picnic on the lodge porch or at a scenic location on an all day ride. Four nights each week guests enjoy supper in our beautiful dining room, the other evenings outdoor BBQ's are the rule. Special diets—just a little advance notice.

Address: 1072 East Fork Road, Dubois, WY 82513
Telephone: 307-455-2839; **Toll Free:** 800-453-9488; **Fax:** 307-455-2634; **Web:** www.jimbalzotti.com
Location: Jackson, WY airport 2 hours away
Guest Capacity: 35

Season: May 31 through mid-September
Policy on Children, Pets: Children welcome
Rates: $$
Credit Cards: None.

(SEE PAGE 319 FOR COLOR PHOTOS!)

WYOMING

Lazy L&B Ranch

NESTLED IN THE VALLEY.

SNOW-CAPPED PEAKS.

Wyoming

Lozier's Box "R" Ranch
Cora, WY

This 1,600-acre working cattle/horse guest ranch is located just 60 miles southeast of Jackson Hole and Teton and Yellowstone national parks. It's nestled between two glacial lakes at an elevation of 7,500 feet on the western slope of the Wind River Mountain Range. Here you'll find resident herds of moose, antelope, deer, and coyote. The Box "R" borders the Bridger Teton National Forest — 840,000 acres of unfenced and undeveloped country — and is the only deeded land in Wyoming to border the state's wilderness lands! Riding will take you through plush meadow bottoms, quaking aspen hillsides, pine timber with open parks, and the above-timberline granite peaks, while being surrounded by glistening alpine lakes and streams.

Horseback riding is the main activity, and riding options vary as the week progresses. The Box "R" has over 100 horses and mules, all finely trained and managed to ensure you of a top-notch riding vacation. Experienced wranglers match guests with their own personal horses and tack and instruct everyone on the basics as well as how to rein, stop, walk, trot, and gallop a horse safely and efficiently. After that, weekly options may include moving 25 to 1,000 head of cattle, all-day rides that climb from 7,500-ft. to 11,000-ft. elevations and dozens of other scenic rides on the 840,000 acres of the Bridger Teton National Forest. A sunrise breakfast ride and a moonlight ride provide fun diversions.

The ranch offers 3-day (or more!) pack trips that can be combined with a stay on the ranch. Guides will escort you deep into the heart of the Rocky Mountains, where you can sightsee, hike, and fish. Hundreds of clear sparking lakes, six varieties of trout, and awesome alpine vistas 13,800 feet high await you.

Or, if you're more adventurous, you can be a true-to-life cowboy! Join the folks at the Box "R" as they move over 9,000 head of cattle 65 miles during three consecutive cattle drive weeks from private ranch land to the summer range in the Bridger Teton National Forest. You'll be assigned your own personal horse and tack for the week and will learn how to work with your four-legged partner. You'll spend several days on the trail moving the cattle, chowing down on delicious campfire grub, sleeping in tents or "roughing it" a little. No experience is necessary, although a desire to spend long days in the saddle will increase your overall experience.

Back at the ranch, accommodations include one-, two-, and three-bedroom log cabins, each with its own private bath and daily maid service. The three-bedroom cabins feature a recreation/social room, bar, fireplace, pool table, darts, card table, TV and VCR.

The Box "R" prides itself in providing robust family-style meals that vary from Western steak and potatoes to a Mexican chicken quesadilla casserole to chile relleños. Sunrise breakfasts, beach cookouts, and barbecues provide even more great times. There is excellent trout fishing both on and off the ranch. The ranch has six ponds, and Willow Creek runs the length of the property. You can fly fish or spin cast at numerous nearby lake and rivers. Hiking is always an option, and there are horseshoe pits, a pool table, darts, and roping dummies to practice on.

Yellowstone and Teton national parks and the town of Jackson Hole are the main attractions in the area and are all within a 2-hour drive of the ranch. You might want to add a visit to one of these at the beginning or end of your vacation at the Box "R." There's a rodeo in Jackson, and if you're interested, transportation can be arranged. Nearby Pinedale has a number of shops, dancing at local saloons, and the Mountain Man Museum.

Join Irv and Levi Lozier on this second-oldest family-owned operation in Wyoming, and form friendships that will last for a lifetime!

Address: P.O. Box 100-AAA, Cora, WY 82925

Telephone: 800-822-8466; **Fax:** 307-367-6260

Location: 65 miles southeast of Jackson Hole, 245 miles northeast of Salt Lake City

Guest Capacity: 20

Season: late May through mid-September

Policy on Children, Pets: Children must be 6 years old (8 recommended). No pets.

Rates: $$ American plan.

Credit Cards: Discover, Mastercard, Visa for deposits only. Cash, cashiers check, traveler's checks accepted.

(See Page 320 for Color Photos!)

Wyoming
Lozier's Box "R" Ranch

GUESTS ENJOYING THE VIEW.

LASTING MEMORIES.

WYOMING

PARADISE GUEST RANCH
BUFFALO, WY

Paradise Guest Ranch, nestled in a valley in the Big Horn National Forest in northern Wyoming, is one of the oldest guest ranches in the West, and, if the number of returning guests is any indication, it's also one of the best. As reported in the *Ladies Home Journal*, "Paradise Guest Ranch at 7,600 ft. is surrounded by over a million acres of the Big Horn National Forest. The ranch encompasses the traditional values and activities of dude ranching. But along with this, as Jim says, 'We keep an ear to the ground as to what modern-day guests need and expect.'"

Riding is the main activity, and Paradise is known for its strong horse program. You'll find plenty of guidance and experience here; there's one wrangler for every seven guests. They will match you to a horse that suits your riding ability, whether you're a raw beginner or an expert rider. And that horse is "yours" for the length of your stay. Adults and children may ride together or separately. The riding program is flexible enough to meet guests' varied needs. Nine to twelve separate rides go out each day, and guests can choose walking, trotting, or loping rides. Both youngsters and adults are welcome to join the ranch rodeo at the end of the week.

There's fishing too! Paradise has fishing opportunities that can be enjoyed by guests of all ages. A stocked fishing pond for kids as well as French Creek, which flows right through the ranch and is chock full of rainbows and brookies. The ranch's resident fishing guide is available for instruction and organizes the many different fishing trips offered both on horseback and by vehicle. The ranch promotes catch-and-release fishing, but you can bring your fish back to the cook, who will prepare it for you. A 4-day fishing-oriented pack trip, which could be an add-on to your stay.

A supervised children's program is available. There are different activities for different age groups and include such things as games, arts and crafts, hikes, different overnight campouts for the various ages, pony rides for children under age 6, and much more. Children age 6 and up can participate in the horse program, riding on the trails with the wranglers. Babysitting is provided from 8:30 to noon everyday for children under age 3. Teenagers have their own program, including a teen social each week. Teens also have a trip into town one morning, followed by a pack trip that evening.

Paradise Guest Ranch has eighteen individual new and renovated luxury log cabins. Each cabin has a living room area — complete with a fireplace, full bath, and even a porch where you can relax and read. Most have washers and dryers.

Our excellent food is a perfect complement to the Wyoming hospitality you'll find at Paradise. You'll look forward to three meals a day, served family style (all you can eat) whether in the dining room or at our various western cookouts along mountain trails. We even bake our own breads and desserts. A highlight is the chuckwagon dinner on Friday night, where riders out on the trail meet up with the mule-drawn chuckwagon and the haywagon or buckboard carrying those who chose not to ride. The ranch has a liquor license and the French Creek Saloon in the lodge is well stocked. Evening entertainment is varied and includes a singalong bonfire, talent show, adult night, and our ever-popular square dance.

The first two weeks of September are a special time at Paradise when all the kids have gone back to school. Because there are fewer guests on the ranch during this time, the riding program is more flexible and includes the very enjoyable breakfast ride. Blue skies, cool temperatures, beautiful fall colors, and no children make these weeks very popular with adults.

The latch string is always out. Join hosts Jim and Leah Anderson and do as much — or as little — as you wish.

Address: P.O. Box 790, Dept. BP, Buffalo, WY 82834

Telephone: 307-684-7876; **Fax:** 307-684-9054

Location: 46 miles south of Sheridan, 123 miles north of Casper, 173 miles south of Billings, Montana

Guest Capacity: 75

Season: late May to mid-September

Policy on Children, Pets: Children of all ages welcome. No pets.

 Rates: $$ American plan. Children's rates.

Credit Cards: None

Wyoming

Ranger Creek Guest Ranch
Shell, WY

"Where the Spirit of the Old West Remains"

Since 1918, Ranger Creek Guest Ranch has been welcoming guest to enjoy the spectacular scenery of the Big Horn Mountains. Located within the Bighorn National Forest, the Ranch started as a "small cabin and tent spike camp." Today, the Ranch consist of an old historic mountain lodge and guest cabins. Restored and furnished with antiques and family heirlooms, the Ranch reminds one of days gone by.

Situated at an elevation of 8,300 feet and with access to more than 1.1 million acres of pine forests, sparkling lakes, blue-ribbon trout streams, historic sites and spectacular views within the Bighorn National Forest, the Ranch offers both summer, fall and winter activities in a rustic setting.

The summer horseback riding season starts in early June and is centered around an Old West theme with plenty of western style horseback riding. The horseback riding program is designed to make even the most novice rider feel comfortable in the saddle. Horses chosen for you by the Ranger Creek wranglers are ridden for the duration of your stay. Relax and enjoy a short ride to Shell Creek or for the adventurous at heart, an all day ride into the Cloud Peak Wilderness Area.

Children of all ages can enjoy the "Little Pardoner's Program" designed to keep even the most restless little cowboy or cowgirl busy with crafts, nature hikes, picnics, games, animal studies, and riding in the corral with the help of the Kid's Counselor and wranglers. The Kid's Counselor and Kid's Corral provide a safe place for the younger rider to experience the thrill of being on horseback for the first time. Helping around the corral, feeding and taking care of the animals in the Kid's Corral will be the highlight of their vacation.

Your Old West Adventure will include wagon rides, black powder shooting, fishing, hiking and our nightly entertainment, along with many more activities. Take a trip to Historic Cody and visit The Buffalo Bill Historical Center, a trip to the site of Custer's Last Stand at the Little Big Horn, visit the Medicine Wheel (a prehistoric structure) within the Bighorn National Forest, or spend an extra week and visit the wonders of Yellowstone National Park.

In early fall, hunters take to the mountains to pursue a wide variety of big game, including black bear, mountain lion, elk (Wapiti), mule and whitetail deer, moose and Big Horn Sheep.

Winter brings on the adventure of experiencing the mountains on snowmobiles or cross-country skis. The Ranch is located on the Big Horn Mountain/ Northcentral WY trail system, a system of trails (387 miles of trails), of which 281 miles are groomed snowmobile trails. Downhill skiing is also available at nearby Antelope Butte.

Address: P.O. Box 47, Shell, WY 82441

Telephone: 888-817-SPUR
Web www.netset.com/~billc

Location: 66 miles west of Sheridan, Wyoming

Guest Capacity: 25 to 30

Season: Year round

Policy on Children, Pets: Children welcome.

Rates: $ - $$ American Plan

Credit Cards: None

Wyoming

Rimrock Ranch
Cody, WY

The Fales family has been operating Rimrock Ranch for more than 40 years and through two generations. They take great pride in providing a seven-day Western "happening."

Your hosts, Gary and Dede Fales, are continuing the family tradition of receiving guests that began in 1956. Gary grew up on the ranch and met Dede there. For 20 years, they have operated an outfitting business taking guests on wilderness pack trips and big game hunts. Now they are following in the footsteps of Gary's parents, Glenn and Alice, who are still involved in the daily operation of the ranch.

The Rimrock Ranch is named after the rock formations surrounding the property. Just 26 miles from the Yellowstone National Park's east entrance and at the edge of the Shoshone National Forest. This family-owned and -operated ranch is located on Canyon Creek, one mile from the North Fork of the Shoshone River.

At Rimrock, there's something different going on every day. Horseback riding, of course, is the main attraction, and Rimrock has a fine string of 130+ horses. After getting settled in and meeting the other "folks," there is an excellent horse orientation, and one horse is assigned to each guest. That horse is exclusively the guest's for the week, with instruction for all levels of riding ability expertly given by the experienced wranglers. There are plenty of rides offered for different skill and fitness levels. The rides vary from one hour to half day to all day. There are often as many as three separate rides going in different directions and at different paces. You may ride along the lowlands, the river, or up into the high ground, which surrounds the ranch. In all, the rides are wonderful opportunities for camera enthusiasts.

The week's entertainment is varied. One night after dinner, the whole ranch goes to the Cody Night Rodeo, where you'll get to see some real cowboys and cowgirls and cowkids "strut their stuff." There are softball games and "River Float," a family whitewater rapids trip on the Shoshone River. Fishing opportunities are available in the river as well as in the ranch pond. There's even a fishing derby one day where everyone gets involved! There's a full-day escorted trip to Yellowstone National Park. You'll get to see Yellow- stone Lake, Old Faithful, the Geyser Basins, the Grand Canyon of the Yellowstone, and the varied wildlife of the Park. If you've already seen Yellow-stone, you can spend the day in Cody at the Buffalo Bill Historical Center with its Western art gallery, firearms museum, Plains Indians museum, and of course the Buffalo Bill memorabilia. Riding is an option for those who prefer to remain at the ranch. Toward the end of your week, there's the ranch gymkhana, where guests can compete against one another for ribbons in the horseback games. The log ranch house is a gathering spot at the end of the day and exudes rustic hospitality. Many of the guests like to stretch out here, reading or letter-writing, while enjoying the ranch's spectacular views.

Guests stay in one of eight comfortable cabins, which are situated along the banks of Canyon Creek. The meals are very satisfying, hearty, and varied with a hearty cowboy breakfast or lunch over a campfire along the trail, a steak dinner cookout, picnics on the lawn, or a sit-down dinner in the Lodge.

While the ranch is open for horseback riding May to September, it is available in the winter for snowmobile trips into Yellowstone National Park. Wilderness pack trips, which are separate from the ranch activities, can be arranged.

To paraphrase a guest's comments, "The Rimrock Ranch is a place of the warmest of welcomes, ready smiles, a feeling of home and friends kind and true clasping your hand in true Western style."

Address: 2728 Northfork Route, Dept. BP, Cody, WY 82414

Telephone: 307-587-3970; **Fax:** 307-527-5014

Location: 26 miles west of Cody, 26 miles from east entrance of Yellowstone National Park

Guest Capacity: 32

Season: May to September 1; winter snowmobiling

Policy on Children, Pets: Children must be 6 years old to ride. No pets.

Rates: $$ Sunday to Sunday. American plan. Children's rates. Pack trips extra.

Credit Cards: MasterCard, Visa

(See Page 321 for Color Photos!)

Wyoming

Wyoming

ROCKY MOUNTAIN HORSEBACK VACATIONS
LANDER, WY

With Rocky Mountain Horseback Vacations, you won't just visit the West, you'll live it! Owner Ed Dabney offers guests a number of exciting horseback riding vacations. These include an authentic 1800s Oregon Trail Adventure Ride cross-country in the high plains, complete with Indian attacks and outlaw hold-ups; a stay at a working cattle ranch with roundups and branding; and high mountain pack trips.

You awake to the clip-clop of horses' hooves as the wrangler rounds them up and brings them into camp to get ready for a day of riding the high country. The air is crisp and clean, the camp cook has a crackling fire going to take the chill off the morning mountain air, and the aroma of cowboy coffee and a hearty breakfast beckons you out of your warm cocoon. It's morning in Wyoming and you're experiencing the horseback riding vacation of a lifetime!

Rocky Mountain Horseback Vacations hosts families, couples, and individuals who are seeking a unique blend of relaxation and excitement in the spectacular mountain West. Rocky Mountain's specialty is offering customized summer horseback riding vacations, trout fishing trips, and big game hunting trips, geared to small groups. It operates over a huge tract of the beautiful Shoshone National Forest in the Wyoming Rocky Mountains near the Yellowstone area and on the high plains of the Oregon Trail country.

This is a small but top-quality, professional operation. It provides gentle, excellent horses; a friendly, congenial staff; and first-rate food and equipment. There are even hot showers with real toilets! Each outfitter and guide has over 20 years of experience serving customers, guiding pack and hunting trips, and handling horses. Rocky Mountain owns and trains all its own horses and does an excellent job of matching up a horse and rider, whether you are a non-rider or an expert rider. There are no "follow the leader, stay on the trail" rides. This is big, open country, so you'll just take off and ride!

You may choose the High Mountain Pack Trip, where you'll ride on horseback into the majestic mountain back country. You'll set a leisurely pace along meandering snow-fed streams through deep rock canyons to pristine high altitude forests and flower-filled meadows. Along the way, you're likely to see an abundance of wildlife. You may fish for trout in the nearby alpine lakes. On the Oregon Trail Adventure Ride, you'll explore the area where the historic trails came together to cross the Continental Divide. On the Outlaw Trail Ride, join a five-day horseback trip along the trails used by Butch Cassidy and the Sundance Kid and visit their Hole-in-the-Wall hideout! This is a historically accurate re-enactment, so keep your eyes peeled for the sheriff's posse or for other outlaws, and be prepared for a gunfight! During the five-day Cattle Drive, you'll spend a lot of time in the saddle as you help round up and move herds of cattle along the historic stock trails that cross the ranch. Camp along the trail each night with the herd in the spirit of the Old West. Finally, you might want to try the Working Ranch Stay. Spend a week on the ranch, working with the cowboys and getting a taste of the real cowboy life.

Since Rocky Mountain caters to small groups, many items on the basic itinerary can be adjusted to guests' desires. You may choose to ride, take photographs, nap, hike, eat, and enjoy the scenery to your heart's content.

At Rocky Mountain Horseback Vacations, this is more than a vacation, it is a spirit, a tradition of Western hospitality, warmth, honesty, family, and natural beauty. Experience the natural beauty and tranquility of the outdoors with Rocky Mountain Horseback Vacations.

Address: 7696 Highway 287, Dept. BP, Lander, WY 82520

Telephone: 800-408-9149; **Fax:** 307-332-2686

Location: Statewide locations depending upon the trip chosen.

Guest Capacity: 15

Season: May 1 to September 30

Policy on Children, Pets: Children must be 6 years old. No pets.

Rates: $$ - $$$ American plan. Children's rates.

Credit Cards: None

WYOMING

ROCKY MOUNTAIN HORSEBACK VACATIONS

BEAUTIFUL COUNTRY.

HEADING INTO THE PASS.

WYOMING

SAVERY CREEK THOROUGHBRED RANCH
SAVERY, WY

Lying in a beautiful part of Southern Wyoming that was once a sanctuary and hunting grounds of the Ute Indians is an unpretentious ranch dedicated to the horse and riding. The area is almost unchanged since Indian and pioneer days, except for the old ranches along the dirt road, producing cattle, sheep, horses and hay. Savery Creek is still a free flowing stream, and tall Cottonwood trees still tower beside it and along the tranquil meadows.

The Ranch caters to enthusiastic riders, both English and Western. We can usually go in three directions without seeing other people. There is much trail riding over sage covered hills, through valleys and meadows, and in the Medicine Bow National Forest with its aspens and pines. We have cross country practice jumps over different obstacles scattered through towering 200 year old cottonwoods, as well as a jumping area with homemade jumps. Guests may also wrangle and drive cattle.

Non-riders can enjoy fly fishing, swimming in the creek, playing tennis, hiking and, of course, just plain loafing. Wildlife is abundant: elk, deer, antelope, beaver, badger, bald eagles, and wild horses in the nearby Red Desert. There are rodeos and local rodeo practice, and cutting horse competitions. It is a photographer's/painter's/writer's paradise.

Neighboring towns provide a wealth of wild-west history and chances to find interesting memorabilia. This is Wild Bunch country and Butch Cassidy's cabin can still be visited in Baggs, just a few miles from the ranch.

Our lifestyle is relaxed and low-key. Accommodations are in the ranch house upstairs and include antiques and Queen size beds. The River Room is large, with private bath, cedar post bed, and overlooks Savery Creek. Two smaller bedrooms share a bath, are attractive, but little more than comfortable places in which to lay down weary bones at the end of the day. Meals are plentiful with quite a lot of picnics, and cookouts on the bank of Savery Creek. Dinners at times are by candlelight with wine. We serve many fresh vegetables, crisp salads and fruit; some of which come from our own garden. While not entirely eliminated, we try to avoid cholesterol producing and fattening foods as much as possible.

Because we are a working horse ranch that, like Topsy, "just growed," (and is still growing in the sense that there is still much improving to do - landscaping, adding new jumps etc.) we only accommodate about six guests. Each guest has much personal attention, and we have a very high percentage of returnees. In view of the limited guest quarters, we are not able to provide separate rooms for smokers, so we ask that guests do not smoke indoors. While we often serve wine with dinner, guests usually bring their own spirits. Everyone usually gathers for refreshments and conversation before dinner.

In general we are particularly suited to advanced and intermediate riders, partly because we have a number of horses that are 'too much' for the novice. But the inexperienced should not be daunted as we have several confidence-inspiring, gentle horses. There is no charge for lessons for beginner, if given before and during rides. Fast riders and novices usually do not ride together unless. for example, parents want to ride with their children. Experienced, thoughtful riders may ride alone if they prefer. There are exceptions, but usually we do not encourage riding for those weighing more that 200 lbs.

Address: Box 24, Savery, WY 82332
Telephone/Fax: 307-383-7840
Location: West of Cheyenne
Guest Capacity: 6
Season: Mostly Spring, Summer, and Fall

Policy on Children, Pets: Children old enough to ride are welcome. No pets.
Rates: $$ - $$$
Credit Cards:

Wyoming

SAVERY CREEK THOROUGHBRED RANCH

SOARING TOGETHER.

BEAUTIFUL COUNTRY — FRIENDLY HORSES
AWAIT YOU AT SAVERY CREEK.

T CROSS RANCH
DUBOIS, WY

The T Cross Ranch is located in a secluded valley at the base of the spectacular Absaroka Mountain Range. The outside world is all but forgotten once you enter the T Cross gate. You'll step back in time on this 160-acre piece of history. T Cross was homesteaded in the late 1800s by a fugitive from the Johnson County cattle wars and has been operated as a dude ranch since 1920. The main lodge and cabins were gradually handcrafted by tiehacks from the ranch's own lodgepole pine. Unique in its authenticity, T Cross remains a true vision of the frontier West.

T Cross maintains an informal atmosphere that offers a colorful blend of the Old West and the creature comforts of home. Guests reside in eight cozy log cabins nestled in the pines, decorated with lodgepole furnishings, Indian rugs, and artifacts. Each cabin is outfitted with wood stove, fireplace, hot shower, and its own porch enticing one to sit back while enjoying the peaceful solitude and dramatic views of the surrounding wilderness.

The main lodge, filled with authentic Western charm, is a fun gathering spot for meals and socializing. Its grand stone fireplace wards off the chill of the mountain mornings and evenings, while the spacious front porch is perfect for relaxing with a good book, chatting with new friends, or simply embracing the splendid seclusion T Cross has to offer.

The horses at T Cross are hand-picked by manager/owner Ken Neal, who has worked ranches and horses since he was knee high to a grasshopper and fully understands the importance of sound, congenial animals. Each guest (6 years or older) is assigned a horse for the week, in accordance to ability. (You can bring your own horse if you wish.) Riding instruction is available for those who want to "brush up." Half-day, full-day, and picnic rides offer a variety of scenic options. The deep timber, stands of aspen, tranquil meadows, and breathtaking vistas of the Washaki Wilderness are just out the back gate. And don't forget the elk, deer, moose, porcupine, or coyote you'll probably see along the way.

Fishing has become a popular pastime on picturesque Horse Creek, which winds through T Cross. There is a healthy population of brook trout and mountain white fish, and an annual program has been initiated with the addition of hundreds of pounds of 10- to 14-inch rainbow and cutthroat trout. T Cross subscribes to catch-and-release, barbless hook fishing. However, fish are occasionally kept for breakfast or dinner enjoyment. A Wyoming fishing license is required and may be purchased at any sporting goods store in the state. Fly-casting lessons and tactics for catching trout are available during the season.

Independent vacationers will appreciate the freedom to arrange their own days. Activities are informally organized, and the friendly, knowledgeable staff are always ready and willing to cater to special requests. T Cross takes pride in not only offering a relaxed, unstructured vacation, but also numerous opportunities to experience the wonders of nature and wildlife in many capacities, be it riding, hiking, fishing, campfire singalongs, tubing down Horse Creek, or just pitching horseshoes. Guests are constantly surrounded by the tranquil, pristine beauty that nature has assembled in this remote Wyoming wonderland.

Your hosts, Ken and Garey Neal, have been mastering the dude ranching operation for over 30 years. The relaxed atmosphere that permeates T Cross stems not only from the Neals' great rapport with guests but also from their innate understanding of the need for flexibility. Over the years, they have developed and nurtured relationships with guests who continue to return to T Cross year after year. Some weeks, you'll find three generations of one family making more unforgettable memories.

Why not start your own memories at the T Cross Ranch.

Address: P.O. Box 638, Dept. BP, Dubois, WY 822513

Telephone: 307-455-2206; **Fax:** 307-455-2720

Location: 2 hours from Jackson Hole

Guest Capacity: 24

Season: June through October

Policy on Children, Pets: Children should be 6 years old. No pets.

Rates: $$ American plan. Minimum 7-day stay July and August, 4 nights June and September.

Credit Cards: None. Personal checks, cash, traveler's checks accepted.

Wyoming

Triangle C Ranch
Dubois, WY

Just as important as workin' for the good life, is findin' a place to enjoy it. The answer is The Triangle C Ranch, where activities are limited only by your imagination.

Folks up here spend all day workin' up mountain sized appetites and sooner or later it leads to the Lodge's dining room. You can feel free to sit back, let your belt out a notch and watch the sky turn Crimson over the Dinnacles with cookin' that is second only to the view.

Let us wake up the Lil' Cowboy inside you with our horseback riding program. We run about 150 head of horses and mules and you won't have to hurry - there's time to do everything or do nothin'. Maybe you'll watch the hands bring in the horses from the pasture and pick out your mount and ride back where you won't see another soul.

If you can't remember when was the last time you wandered off into the woods before breakfast, bound for nowhere, then its time to slow down and unwind in our cozy log cabins way back in the woods, and relax, sort out the day and snuggle down.

The Garnick family is one of the most colorful dude-ranch families in the Jackson Hole area. Over the past 30 years the family has been involved in Dude Ranching and entertainment. Son, Carmen, who has a wonderful family himself, with eight children, has 17 movies to his credit and operates the Jackson Hole Playhouse. In 1994, the family bought the Triangle C Ranch, a ranch with a colorful past and as the National Register of Historic Places. What drew the Granicks to the Triangle C Ranch was its old-time charm, magnificent views and proximity to Yellowstone and Grand Teton Parks and the most famous Wyoming western towns, Jackson Hole and Dubois.

The main lodge is on a bluff, overlookin the Wind River and out across the Absaroka Mountains. Some of the cabins date back to 1920 and radiate that Old West Charm. The new Deluxe log complexes have King and Queen Suites, some with fireplaces and some with Jacuzzis.

The Garnicks have one of the most famous children's programs in the country for kids 1 to 18. Your youngsters are always in good company with Cameron's eight ranch-raised youngsters. With our naturalist, outdoorsman and wrangler programs for youngsters along with a teepee village overnight.

Fishing is great in the Wind River, that runs through the ranch, as well as swimming to join the Polar Bear Club. They take you river rafting, Yellowstone touring and square dancing.

The landscape at Triangle C unfolds like a storybook tellin' tales of Outlaws, Indians, Mountain men and wagon trails and there are places where you'll have to hold your breath so you won't startle the big game in the clearings.

Their friendly ranch hands have put the "Wild" back in Wild West! They open their ranch gate and promise to send no one home without a fresh spring to their step and a treasured memory. Their ranch hospitality makes you feel like family comin' home.

The Triangle C is what the West used to be!

Address: 3737 US Hwy. 26, Dubois, WY 82513
Telephone: 307-455-2225; **Fax:** 307-455-2031
Location: 18 miles West of Dubois
Guest Capacity: 35 - 50

Season: June — March (closed April and May)
Policy on Children, Pets: Children welcome.
Rates: $$
Credit Cards: Amex, Visa and Mastercard

(See Page 322-323 for Color Photos!)

Wyoming

Triangle C Ranch

HORSE WORK.

COMING DOWN THE STREAM.

Wyoming
Triangle C Ranch

BLAZIN' NEW TRAILS.

WYOMING

TRIANGLE X RANCH
MOOSE, WY

The Triangle X Ranch has been owned and operated by the Turner Family for 70 years. Four generations of this well know family been welcoming guests for fun filled vacations in Jackson Hole.

The Triangle X offers one of the most compete recreational package of any facility in the Rocky Mountains and elsewhere in the Western states. Guests enjoy horseback riding, river float trips, fishing, cookouts, square dancing, hiking, scenic tours with many photographic opportunities. A special children's, "Young Wrangler" program for children from ages 5 to 12 is a special feature of the ranch.

All riding levels, from beginner to experienced are accommodated. Guests can choose from a slower, scenic ride to one with more energetic riding. As the ranch is very family oriented, the Young Wranglers and the Teenagers have their own riding groups with their own wranglers. Each guest has his or her own horse for the week, chosen from a string of well trained saddle horses.

The maximum guest capacity is 80 and there will be an average of 65 to 70 people at the ranch during the summer months. Our guests meet other guests from throughout the country and from other countries as well. Our family style dining set-up allows our guests to get to know one another in a relaxed and congenial atmosphere while they enjoy the beautiful view of the Teton Range from our sun porch dining room. Two nights each week there are cookout suppers (one on the ranch and the other at a site by the Snake River). At these cookouts you will enjoy being serenaded by talented campfire singers.

Rates, which depend on the number of people in your family or group, range from $785 per week, per person to $1,120 per week. The rates include, lodging, all meals and riding. The ranch accepts guests from late May through the first week in November.

Also, instead of, or before or after a stay at the ranch our guests can also enjoy the unforgettable experience of a Wilderness pack trip. Get away from it all in the high country and "rough it" in style in one of America's largest and most spectacular wilderness areas. Guests will enjoy riding into the peaceful, magnificent scenery. Fields of wildflowers, excellent fishing and just relaxing all add to the experience. Guests can sleep in sturdy tents or under the stars in our cozy sleeping bags on comfortable foam pads. Our expert guides will spoil you with three delicious meals each day.

This horseback camping trip (minimum four days and we do suggest at least five days) is perfect for families or a group of friends. The daily itinerary is tailored to your interests, whether it be just riding, photography, fishing, exploring for wildlife or just relaxing.

Pack trips are offered from mid June until early September. Reservations are necessary and should be made as soon as possible prior to the season to ensure getting the preferred dates.

Address: Triangle X Ranch, Moose, WY 83012
Telephone: 307-733-2183; Fax: 307-733-8685
Web: Trianglex.com
E-Mail: Johnturnes@Blissnet.com
Location: 25 miles north of Jackson Hole
Guest Capacity: 80

Season: May 25 through October 31
Policy on Children, Pets: Children welcome. No pets.
Rates: $$
Credit Cards: None

(SEE PAGES 324-325 FOR COLOR PHOTOS!)

WYOMING

TWO BARS SEVEN RANCH
TIE SIDING, WY

Two Bars Seven is a fourth generation Schaffer family owned, real cattle spread established in the days of territorial government. Our ranch family raises stock horses and Beefalo cattle, a buffalo hybrid, hardy for our rugged winters. Their meat is natural, leaner and more nutricous that domestic cattle.

The ranch sits on 7,000 acres of secluded mountains, meadows and forest beauty. It lies on both sides of the Colorado and Wyoming boundary, along the eastern slope of the Rockies. We carry on our ranch operation free from the pressures of modern commercial enterprises.

We invite you to come share the pleasant relaxation of a truly Western experience!

Two Bars Seven saddle horses carry the blood lines of famous working cattle horses. They are carefully trained for both ranch work and pleasure riding.

You may join us with our cattle and ranch work, the degree of physical challenge is easy for you to control, not all ranch work is rough riding, a lot of it is quiet-time work.

Are you a rodeo fan? Schedule your vacation around our nearby and famous world-class rodeos: Greeley Stampede, Laramie Jubilee Days, Cheyenne Frontier Days and several others.

Many guests just want to go riding with each other in this beautiful unsettled area. We plan our rides to safely suit their ability. The varied terrain allows riders the opportunity to trot smoothly across the meadows, ride carefully into valleys and draws and canter across the open spaces.

We choose a saddle horse suited in nature to your special taste and riding experience. Your Corral Boss (western from the beaten up Stetson to the toes of well worn boots), will see that you have able and kindly riding instructions.

For evening entertainment there are the western night lights of Laramie and Cheyenne. Here at the homestead, we have our own dancehall with a pool table, an antique juke-box and player piano. Relax in the hot tub under the stars where you are often serenaded by distant coyotes. We have a cozy lounge with a beautiful fireplace or there might be an ice cream making party on the verandah.

At Two Bars Seven Ranch cooking is bountiful and delicious, prepared with fresh ingredients, much of which are home grown. Healthy balanced nutrition is basic to our meal planning.

The awesome beauty of the land's varied terrain rejuvenates the heart and brings out the adventurous explorer in most everyone, or some may want to enjoy it in stillness. They can spend a relaxing vacation quietly communing with nature.

Having counted 56 bird species here, L.J. Strickman, Ornithologist, stated; "This is one of the best areas for bird watching that I have ever found."

Two Bars Seven filled with wildlife, small creatures and predatory animals, as well as deer, elk and antelope. (no poisonous snakes) Hunting seasons and license information available on request.

 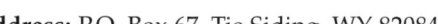 

Address: P.O. Box 67, Tie Siding, WY 82084

Telephone: 307-742-6072; Fax: 303-265-9046
Web: www.twobarsevenranch.com
E-Mail: lschaffer@vcn.com

Location: 1 hour from Laramie airport, 2-1/2 hours from Denver airport

Guest Capacity: 20

Season: late May through early November

Policy on Children, Pets: Children welcome, must be 7 years or older to ride.

Rates: $$

Credit Cards: Visa and Mastercard

Wyoming

Two Creek Ranch
Douglas, WY

Since 1987 guests have been coming to the Two Creek Ranch for an authentic "cowboy experience." The biggest events of the year are the two 75-mile cattle drives — one in May and one in October — which show in part what a real working cattle ranch is all about. But the work is year round and, from February to October, other facets of a cowboy's life are open to the public: calving, branding, gathering, pregnancy testing the heifers, roundup, moving the cattle, dehorning, trailing, and marketing the steer calves. After expert instruction, the guests are invited to work beside the ranch hands doing *real* cowboy chores.

Two Creek Ranch is located on 25,000 acres of land on the North Platte River. Part of the Oregon Trail is nearby, and on the drives the cattle move along a short piece of this historic route.

Each "wannabe" cowboy or cowgirl is issued his or her own horse, saddle, blankets, and bridles on the first day and it is the cowpoke's responsibility to keep track of and care for that tack for the duration of the adventure. Both beginning and advanced riders are welcome — and every level in between.

In April, there are a number of chores that have to be done before the spring cattle drive. First, the cattle pairs (mother cow and calves) must be gathered from the winter range and moved about 17 miles to the spring range along the Oregon Trail. Since the calves are so young, this usually takes several days and a lot of patience! The week before the cattle drive is spent gathering and branding the calves, a process that has been done in the West for over one hundred years.

The spring cattle drive usually takes place at the end of May. You and fellow cowboys and cowgirls will trail about 1,000 head of cattle from the spring range at an altitude of 5,000 feet to the summer range at an altitude of 7,000 feet on the Laramie Plains. The trail follows the old Fort Fetterman freight road. The first few days are spent collecting the cattle and then driving them along the route. The next four nights are spent under the stars, sleeping in bedrolls by the campfire or in provided trailers and tents. Once you reach the Laramie Plains, you can relax a bit at the summer cow camp, which the Daly's call the "Little House on the Prairie." Here there are modern amenities but no telephones or TVs.

In the summer, there are a number of "chores" that have to be attended to on the Laramie Plains — changing pastures with the cattle; moving the bulls to different pastures; pregnancy testing the heifers; weighing and vaccinating the calves; separating the steers from the heifers; dehorning. Guests can take part in any of these activities.

At the end of September, it's time to get prepared for the fall cattle drive. The steer calves and their mothers must first be trailed to the shipping pen. Then the mother cows must be collected and trailed back to the pasture on the Laramie Plains.

The 75-mile fall cattle drive is at the end of October. This is the spring cattle drive in reverse. The weather on this cattle drive is a little unpredictable. It can be anything from bright, sunny Indian summer to rain to snow to perhaps a blizzard. But the feeling of satisfaction after completing the task is one that is not to be forgotten. It may even bring you back year after year.

On the drive, accommodations are on the trail and at cow camp, other activities are at the ranch with nights spent in the ranch house or a nearby cabin. All meals are hearty Western fare — plentiful and all you can eat. Beef is almost always on the menu, which is only natural since Two Creek Ranch is a beef producer! Sorry, no special diets.

So if the movie *City Slickers* is your idea of a perfect vacation, then Two Creek Ranch is an experience you won't want to miss.

Address: 800 Esterbrook Road, Route 6, Dept. BP, Douglas, WY 82633

Telephone: 307-358-3467

Location: 60 miles southeast of Casper, 10 miles south of Douglas

Guest Capacity: 12

Season: Year round

Policy on Children, Pets: Children must be old enough to take care of their own horses. No pets.

Rates: $ American plan.

Credit Cards: None

CANADA

WARNER GUIDING AND OUTFITTING LTD.
BANFF, ALBERTA

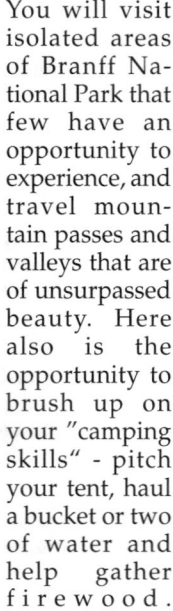

For the past 37 years, Warner Guiding and Outfitting Ltd. has offered the best in providing a holiday on horseback through exploring the Banff National Park located in the Canadian Rocky Mountains.

Leave behind deadlines and schedules and trade them for fresh mountain air, a good horse and a trail that leads you to one spectacular view after another. Wake to the bugling call of a bull elk or the tantalizing aroma of pan fried trout. We'll take you to the perfect spots for once-in-a-lifetime photos, mountain stream fishing and viewing wildlife in its natural habitat.

The combination of mountain air and a day out on the trail usually stirs up a mean appetite and our cooks are ready to meet the challenge. Breakfast and dinner are sit-down affairs and lunch is served picnic-style along the trail. Belly up to the table and enjoy the fresh, mouth-watering, ranch style menu. But don't forget to leave room for our ample selection of home-baked desserts!

Our rides are perfect for people of all ages and riding abilities. With over 300 head of horses, we'll provide you with a sturdy, sure-footed mount to match your experience. Our professional cowboy guides will give you an inside look at western culture and provide you with enough good times to create a week full of memories.

Warner Guide and Outfitting offers trips to accomodate almost every type of rider, from folks who are looking to take a little bit of home into the backcountry to more challenging and rugged trips. For people who want to experience the beauty of the Canadian Rockies but want some of the creature comforts, we offer trips where you will spend your evenings in a cozy lodge that has ten sleeping rooms, a large country kitchen and a cozy living room area where you can curl up by the woodstove and lose yourself in a good book. We call that roughing it - the civilized way!

The most challenging and rugged trips we offer are the Adventure Expedition rides designed for the experienced rider and camper. New camps are established daily as the expedition penetrates further into the mountains. You will visit isolated areas of Branff National Park that few have an opportunity to experience, and travel mountain passes and valleys that are of unsurpassed beauty. Here also is the opportunity to brush up on your "camping skills" - pitch your tent, haul a bucket or two of water and help gather firewood. Guest participation is a very important part of these trips. A maximum of eight guests are allowed on the moving trips and duffel is kept at a weight not exceeding 30 pounds.

We also offer four, five and six day wilderness tenting trips that explore some of Branff National Park's most spectacular backcountry. Simple luxuries abound: washstands and hot water; a fire pit for evening socializing; a large kitchen tent where meals are cooked and served; and plenty of "A" frame canvas tents equipped with wood floors.

Come and experience the hospitality, fun and incredible views of a holiday on horseback. Let us show you what we do best!

Address: Box 2280, Banff, Alberta, Canada TOL 0C0

Telephone: 403-762-4551; **Fax:** 403-762-8130
E-mail: warner@horseback.com
Web: www.horseback.com

Location: Calgary International Airport is located 75 miles east of Banff.

Guest Capacity: 100

Season: Year round

Policy on Children, Pets: Children welcome. No pets.

Rates: $$

Credit Cards: Amex, Visa, MC

(SEE PAGE 326 FOR COLOR PHOTOS!)

CANADA

WARNER GUIDING AND OUTFITTING LTD.

CANADA

SADDLE PEAK TRAIL RIDES
COCHRANE, ALBERTA

A horseback riding vacation with Saddle Peak Trail Rides is a riding adventure in the Canadian Rockies that you'll never forget. For over seventeen years, owner Dave Richards has been taking guests from all over the world, and from all walks of life, on pack trips high into the majestic Canadian Rockies. Guests are guided through some of the most spectacular scenery along the Eastern Slopes of the Rockies, including the renowned Devil's Head Mountain. They don't just offer you a holiday, but a retreat, an outdoor experience by horseback into Alberta's Rocky Mountains.

Saddle Peak offers two-, three- and four-day packages. One of the more popular trips, aptly called "The Ultimate," is a three-day wilderness experience into the mountain country of Devil's Head and Johnson Creek. This region of the Rockies is known for its wild horse herds, and it's where bighorn sheep can be seen. Fishing is also available as well as a chance at taking some excellent pictures. See Alberta the way it was meant to be seen, on horseback.

On departure day, guests meet at the foothills ranch at 9:30 a.m. and are transferred by vehicle to the ranch's Lesseuer Creek Trail Head Camp. There you'll meet your guide and horses. As your gear is being packed on the pack horses, you'll start to feel your excitement build. After packing up, you'll mount and have a leisurely 12-mile ride to the base camp, stopping along the trail to each lunch. Over the next two days, you'll be able to fish, hike, take photos, and just generally enjoy being in some of the most remarkable country you've ever seen. Every morning, you'll wake up to a fresh cooked hearty breakfast served Canadian style — plenty and delicious! On these trips your guides will share with you the history of the area, and you may get involved as much or as little as you wish with the care of your horse.

Saddle Peak also offers a four-day combination ride. During the first three days you'll ride the high country of the Eastern Slopes of the Rocky Mountains. On the third night, you'll stay at the ranch in a warm, rustic cabin and enjoy a ranch-style dinner served in the main ranch house. On the fourth and last day, you'll do an all-day ridge ride that crosses the Ghost River and follows the Side Hills Trail high above Lesseuer Creek. The foothills ride offers a panoramic view of the mountains from Devil's Head to Kananaskis Country, with the Ghost River below.

Saddle Peak's most popular package is the five-day Explorer trip, which combines the three-day trail ride, along the Eastern Slopes of the Canadian Rockies within the valley of the Ghost River, with two full days of whitewater rafting on the Red Deer River. The Red Deer was twice chosen for the Canadian National White Water Championships. For this all-inclusive memorable experience, all you need to provide is a sleeping bag, clothing, and personal effects. Saddle Peak provides everything else — from delicious home-cooked meals, equipment, and experienced guides to round-trip transportation from the rendezvous point. Saddle Peak also offers similar but shorter trips.

Saddle Peak Trail Rides also offers hourly rides, all-day rides, horse boarding, barbecues, and a variety of different trips and activities out of the foothills ranch.

Just about an hour's drive southeast of the ranch is the city of Calgary, home of the greatest outdoor show on earth — the Calgary Stampede. If you want to combine your vacation with the chance to take in the Stampede, you'd better book early! Saddle Peak is also just a 45-minute drive away from the mountain resort town of Banff.

Whatever package you choose, you'll end up knowing why we say you'll have a "holiday in the Canadian Rockies you'll never forget!"

Address: P.O. Box 1463, Dept. BP, Cochrane, Alberta, Canada T0L 0W0

Telephone: 403-932-3299

Location: 43 miles west of Calgary

Guest Capacity: 16

Season: mid-May to mid-September

Policy on Children, Pets: Children must be 9 years old for pack trips, 6 years old for trail rides. No pets.

Rates: $ American plan.

Credit Cards: None

CANADA

SADDLE PEAK TRAIL RIDES

PACK TRIP.

GETTING AWAY FROM IT ALL.

CANADA

SPRINGHOUSE TRAILS RANCH
WILLIAMS LAKE, BRITISH COLUMBIA

The Springhouse Trails Ranch is a total fun and adventure vacation for the whole family. In British Columbia's big sky country of the Cariboo, you'll experience views that go on forever, and you'll smell air so fresh that you will feel an exhilaration all day long and sleep like a log each night. In the heart of the rolling landscape is Springhouse Trails Ranch, tucked away but easily accessible — the best of both worlds when you want to get away from it all on your horseback riding vacation.

Springhouse Trails Ranch is a summer resort that offers active families a wide variety of fun activities. Naturally there are horses, lots of them! And the horses range from mounts suited for the beginner rider to more spirited, tireless animals that are a better match for the experienced rider. The wranglers at Springhouse Trails pride themselves on  choosing the perfect horse for you, and they're so good at it that you'll want to take your horse home with you!

The trails around the ranch are many and varied. The scenery in every direction is beautiful, and you don't even have to be on horseback to enjoy it. Whether you're on a one-hour or an all-day ride, you'll marvel at the towering trees and big country. For something a little different, try a night ride or a carriage ride.

Bring your hiking boots and explore the many hiking trails and paths. There's a bird sanctuary nearby that is a favorite watching spot. And don't forget to bring your fishing pole and try your luck at some of the local fishing spots.

For the super-adventurous, the river rafting excursions are always popular. After getting wet, you'll be glad to come back to the ranch and dry off, put your feet up, and enjoy a hot beverage! You can explore nearby gold rush towns, go to a local rodeo, visit the Williams Lake Stampede, or even take a 20-lake airplane tour. So whether you are the type of person who likes to be on the go all day, hunting up all kinds of new adventures, or the type who likes to put his or her feet up and relax, you will find that the Springhouse Trails Ranch will be the perfect ranching holiday.

Accommodations are comfortable and spotlessly clean. They range from rustic cabins, to apartment-type lodging, to rooms with fireplaces. For those who prefer to camp out, Springhouse Trails has a campground and RV site with full hookups, complete with showers, bathrooms, and laundry facilities. There's even a playground for the kids and a trampoline.

The dining room serves up hearty home cooking, sure to please even the fussiest palate! Twice a week there are barbecues outside, over a cookfire. After dinner, if you can still move after all that good food, try a game of horseshoes or kick up your heels on the dance floor in the steakhouse.

So, if you want to escape the hectic pace of day-to-day life, visit the big sky country of British Columbia and have a horseback riding vacation like no other.

Address: Box 2, RR 1, Dept. BP, Williams Lake, British Columbia, Canada V2G 2P1

Telephone: 250-392-4780; **Fax:** 250-392-4701

Location: 20 kilometers south of Williams Lake on Dog Creek Road

Guest Capacity: 45

Season: May - September

Policy on Children, Pets: Children welcome. Pets allowed with prior permission.

Rates: $ American plan.

Credit Cards: MasterCard, Visa

(SEE PAGE 327 FOR COLOR PHOTOS!)

CANADA

THE HILLS HEALTH AND GUEST RANCH
100 MILE HOUSE, BRITISH COLUMBIA

Can you imagine horseback riding on 20,000 acres with 6 different lakes, fantastic ranch food, knowledgeable cowboys and guides, and your very own friendly horse? Well that's only part of what you get at The Hills Health & Guest Ranch.

The Hills Health and Guest Ranch, which opened in 1985, was designed by Pat and Juanita Corbett as Canada's first year-round health and fitness vacation resort. Here you can refresh your mind, slenderize your body, and renew your spirit. The fabulous outdoors compliments the indoor facilities and allows guests to take advantage of a large Western-style riding stable, which offers all-day cookouts and rides, plus one- and two-hour rides over thousands of acres of rolling Cariboo ranchlands. In the midst of an inspiring landscape with some of the largest cattle ranches in Canada, you'll ride through a land of a thousand lakes, rolling meadows and heavily forested hills and mountains.

You'll enjoy the fun of going on an old-fashioned horse-drawn hayride in the dawn of evening, heading out to Willy's Wigwam. Once inside this large Indian teepee, you'll join in an old-fashioned singalong around the big campfire, and be delighted by one of The Hills' many entertainers. It's a memory you'll cherish forever.

Returning under the starlight skies — where often the Northern Lights dance before your eyes — the rhythmic jingle of the bells on the horses' backs reminds you that you're in a different world. You'll have a smile to your face, and a smile in your heart, knowing that you've not had this much fun in a long time! Chatting with the teamster, listening to the hoof beats of the horses, it's all different, it's all fun! On another evening, you'll enjoy the Texas line dancing parties — learning together, laughing together, making new friends. The smiles of the instructors, their encouraging approach, it's night time at The Hills Health & Guest Ranch.

In addition to some of the best horseback riding offered in Canada, The Hills Health and Guest Ranch offers one of the largest groups of wellness professionals of any resort in Canada. These health professionals include a kinesiologist, a nutritionist, a physio-therapist, a physician, a nurse, a number of certified fitness instructors, and gourmet chefs who are members of the prestigious Canadian Federation of Chefs de Cuisine. The Hills combines education, physical testing, and self-responsibility into a winning package. You may come here to experience the wonderful horseback riding, but you'll also be able to take advantage of the wellness centre, which delivers the most supportive, informative, personal wellness program you'll find. The ranch's wellness centre encourages health and wellness through medically supported and research-based lifestyle changes using an integrated program of exercise and nutrition guidance, stress management, relaxation, lifestyle education, and body treatments.

The Hills Spa cuisine is world renowned. The chefs are well trained and take pride in creating appetizing, low-fat, and calorie-controlled meals. Your nutrition education continues with nutrition seminars and shopping and cooking workshops. The Hills water is extraordinary and comes clear and cold from a deep underground source. Drink, swim and bathe in it; you'll cleanse your body inside and out.

The air in the Cariboo region of British Columbia is clean and pollution free. The dry climate and the ranch's 3,400-foot elevation are very complimentary to your health and are especially beneficial for many respiratory disorders.

Food, water, and air — three crucial fuels that can revitalize your body at The Hills. Come to The Hills Health and Guest Ranch for a horseback riding vacation. When you leave, you'll find yourself physically, emotionally, and spiritually renewed.

Address: C-26, Dept. BP, 108 Mile Ranch, BC, Canada V0K 2Z0

Telephone: 250-791-5225; **Fax:** 250-791-6384; **E-mail:** thehills@netshop.net

Location: 55 miles south of Williams Lake

Guest Capacity: 200

Season: Year round

Policy on Children, Pets: Children and pets are welcome.

Rates: $$ All meals included.

Credit Cards: American Express, MasterCard, Visa

CANADA

THREE BARS GUEST AND CATTLE RANCH
BRITISH COLUMBIA

Who says you can't be a real ranch, have world class facilities, great accommodations, fantastic food, real ranch atmosphere, cows, cowboys, tennis court, hot tub and a heated pool all at the same time? We think you can have the best and we do!

Mountains reaching 10,000 feet into bright blue skies, rushing clean water beaming with trout, wildlife, cattle, horses and friendly people - Welcome to the Canada and Three Bars Cattle and Guest Ranch.

The Beckley family has been in the ranching and tourism business for 20 years, opening Three Bars Cattle and Guest Ranch to the public in 1992. This combination working cattle ranch, horse training facility and guest ranch manages to merge all three aspects of the ranch into a unique experience for their guests.

Jeff and April Beckley and their two sons are ever present and involved with the ranch. Jeff is a professional reining horse trainer and offers the weekly riding lessons to the ranch guests as well as two horsemanship weeks per summer. Tyler Beckley is the activities director and Jesse works with the horses. April oversees the lodge operations and is who you are apt to get if you phone the ranch office.

While the ranch atmosphere is about western Canadian hospitality and great service, the facilities are up to the standards you would expect from a much larger venue. Guests are housed in nicely appointed log cabins, each with a private porch, full bath and custom made log furniture. Boardwalks weave through landscaped gardens connecting the tennis court, cabins, lodge and pool facility. The ranch's indoor riding arena, outdoor arena and horse barn are in close proximity to the cabins, keeping with the real ranch atmosphere. The ranch's lodge serves as a gathering place with a library, watering hole, game room and dining room.

Great location and wonderful facilities are nice, but you want to have a great time on your holidays. That's what Three Bars is really about - having a wonderful holiday. The Beckley family and staff have made a pledge to make a Three Bars experience one that each of their guests will long remember as a truly great holiday experience. To this end, the activities are set up at the ranch to keep you as busy as you want to be. The ranch experience includes all the riding you want on your own personal horse for the week, guided trail riding, river rafting, campfire programs, riding lessons, fly fishing lessons, archery, volleyball and more.

If location is important to your holiday decision, Three Bars is hard to beat. The ranch is located in the beautiful Canadian Rocky Mountains south of Branff National Park and just north of Montana's Glacier Park. Guests often fly into Calgary and drive to the ranch taking in Banff, Lake Louise and Jasper. Another choice is to fly into Kalispell, Montana and see Glacier Park on the way to the ranch. Cranbrook is also serviced with connector flights out of Calgary, Alberta and Vancouver, BC daily. From the Cranbrook airport, it is a fifteen minute drive to the ranch. The ranch is happy to provide van service to the ranch. However you get here, you'll be glad you came!

Address: S.S. 3, Site 19-62, Cranbrook, B.C., V1C 6H3
Telephone: 250-426-5230; **Fax:** 250-426-8240
Location: Call
Guest Capacity: 45

Season: Call
Policy on Children, Pets: Call
Rates: Call
Credit Cards: Call

IRELAND

BLACK VALLEY EQUESTRIAN
KILLARNEY, CO. KERRY

Travellers all over the world agree on the beauty of the Irish landscape and it is generally accepted that the county of Kerry is the jewel in the crown. The best way to see this beautiful country is on horseback at Gene Tagney's Black Valley Equestrian.

From the edge of the European continent and at the remote end of the Killarney National Park in the county of Kerry, Black Valley Equestrian operates quality, personalized, guided riding vacations.

Gene Tagney lives where he grew up - in the heart of Ireland's most magnificent landscape. An avid horseman and mountaineer, Gene has travelled widely and he knows that his homeland has a beauty rarely equalled around the globe. Black Valley Equestrian is his dream creation, motivated by his desire to expose discerning visitors who share his passion for horses to the Ireland he knows so intimately.

The county consists mostly of rocky promontories that extend further west into the Atlantic than any other part of Europe. The sea is never far away and the Gulf stream ensures that even here at latitudes comparable to Moscow and Labrador, lush tropical vegetation thrives. It is a wild beauty recently described by one writer as a place of "mountain and ocean with a little land in between".

The southwest of Ireland is dominated by the country's highest mountain range, the McGillycuddy Reeks. Along the southern flank of this lofty ridge lies the most remote and beautiful glen in the country. Mighty glaciers crawl through here, carving from the sandstone mass that masterpiece of nature known as the Killarney National Park. Waterfalls tumble from the crags and remnants of ancient oakwoods, spared destruction by their remoteness. Corrie Lakes, giant boulders and tiny farmsteads fit tightly together. On a wet and cloudy day in winter the place is self explanatory - it is Black Valley.

On magnificent Irish mounts, you will gallop over miles of open sands alongside Atlanic waves. You will take ancient byways over rugged passes and through tranquil glens as you learn and are entertained on the subject of Ireland and it's people.

This is a land where you will trek through Kerry's finger-like peninsulas with names like Beara, The Ring of Kerry and Dingle. You will discover the seafood to be the best and when you visit a pub or two and sign a few tunes, the day will seem perfect after your time in the saddle.

Gene Tagney was born into of of the dozen or so families in the Black Valley. His is a small, specialized and highly personalized business, operated by Gene, his wife, Joanne, and their son, Feidhlim, from the spectacular stone and thatched house that they built as their home. As well as being a highly accomplished horseman, Gene Tagney is also a keen mountaineer who has climbed and travelled in some of the worlds most remote regions. The magic and beauty of the Black Valley, together with Gene's unique way of life and his rare understanding of his place and people, makes time spent riding alongside him a very special experience.

Address: Beaufort, Co. Kerry
Telephone: 353-64-37133; **Fax:** 064-37133
E-mail: bvalleyequestrian@tinet.ie
Web: www.kerrygems.ie/bvalleyequestrian
Location: West of Killarney

Guest Capacity: Call
Season: Year round
Policy on Children, Pets: Children welcome
Rates: $$
Credit Cards: Call

(SEE PAGE 328-330 FOR COLOR PHOTOS!)

MEXICO

SADDLING SOUTH
LORETO, BAJA

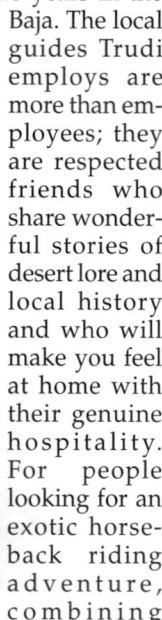

For a terrific horseback riding adventure, you have to go south of the United States — Mexico to be exact. Trudi Angell runs Saddling South, a horseback riding vacation that explores thousands of miles of unfenced country in Baja California, Mexico. You'll follow original, historic missionary trails that wind along rocky desert ridges, through cactus garden valleys, and into lush palm springs. With more and more people looking for a "different" horseback riding vacation, Saddling South offers something most travelers never see — a fascinating look into a way of life that has barely changed since the early years of Spanish settlement in Baja. On surefooted mules and horses, you'll boldly head into red, rugged canyons of the Sierras, on trails barely visible beneath swaying carpets of floral colors.

You will sleep by peaceful oases. Around the campfire at night, after the traditional ghost stories, Trudi and her guides — who are real Mexican cowboys born and bred in the mountains like their fathers and grandfathers before them — will enchant and delight you with stories rich in their cultural heritage.

Because of the rugged terrain, most of the locals use mules and burros for cross-country travel. For Saddling South tours, they have selected the best, desert-trained animals from the area in which you will be riding. Trudi's guides have had a hand in raising and training the stock and will match your skill level with an appropriate animal.

Another unique offering of Saddling South is the opportunity to occasionally stay at isolated ranches. You'll feel that you have gone back in time to the seventeenth century. Trudi offers special excursions into rugged, hidden canyons where you can view ancient Indian murals and cave paintings. There's also a separate springtime ride through the desert for women only, which includes a visit to the mission of San Javier. You might sleep under open palm thatch porches and eat homemade ranch food including rice, beans, fresh homemade tortillas, stew, enchiladas, and fresh-roasted ranch coffee.

Trudi is an experienced trip organizer and speaks fluent Spanish, a result of living the last 18 years in the Baja. The local guides Trudi employs are more than employees; they are respected friends who share wonderful stories of desert lore and local history and who will make you feel at home with their genuine hospitality. For people looking for an exotic horseback riding adventure, combining horseback riding in remote and beautiful areas, this will be an adventure you will never forget.

For those of you who are interested in kayaking, you'll be happy to hear that Trudi also runs Paddling South, 6- to 9-day adventures that may be combined with the horseback riding vacation. No previous kayaking experience is necessary, as Trudi will introduce you to the basic skills used in Baja-style touring. On these trips, you will kayak 5 to 10 miles a day to one of many beautiful beaches. There you'll spend the day snorkeling and exploring the new environment. Starry skies and the sound of the sea will lull you to sleep at night. The next day, you'll be back in the kayak, following the coast and exploring wilderness beaches and canyons before ending your trip at a small fishing village and remote roadhead. Combined with Saddling South, this might just be one of the most memorable horseback riding vacations you'll ever take!

Mailing Address: 4510 Silverado Trail, Dept. BP, Calistoga, CA 94515

Telephone/Fax: 707-942-4550

Location: Trips leave from Loreto, Baja, Mexico.

Guest Capacity: 8. Private trips arranged.

Season: October to April

Policy on Children, Pets: Children over 12 welcome on most rides. No pets.

Rates: $$ American plan.

Credit Cards: All major credit cards accepted.

Mexico
Saddling South

ON THE MEXICAN TRAIL.

GETTING AWAY FROM IT ALL.

Saddling South

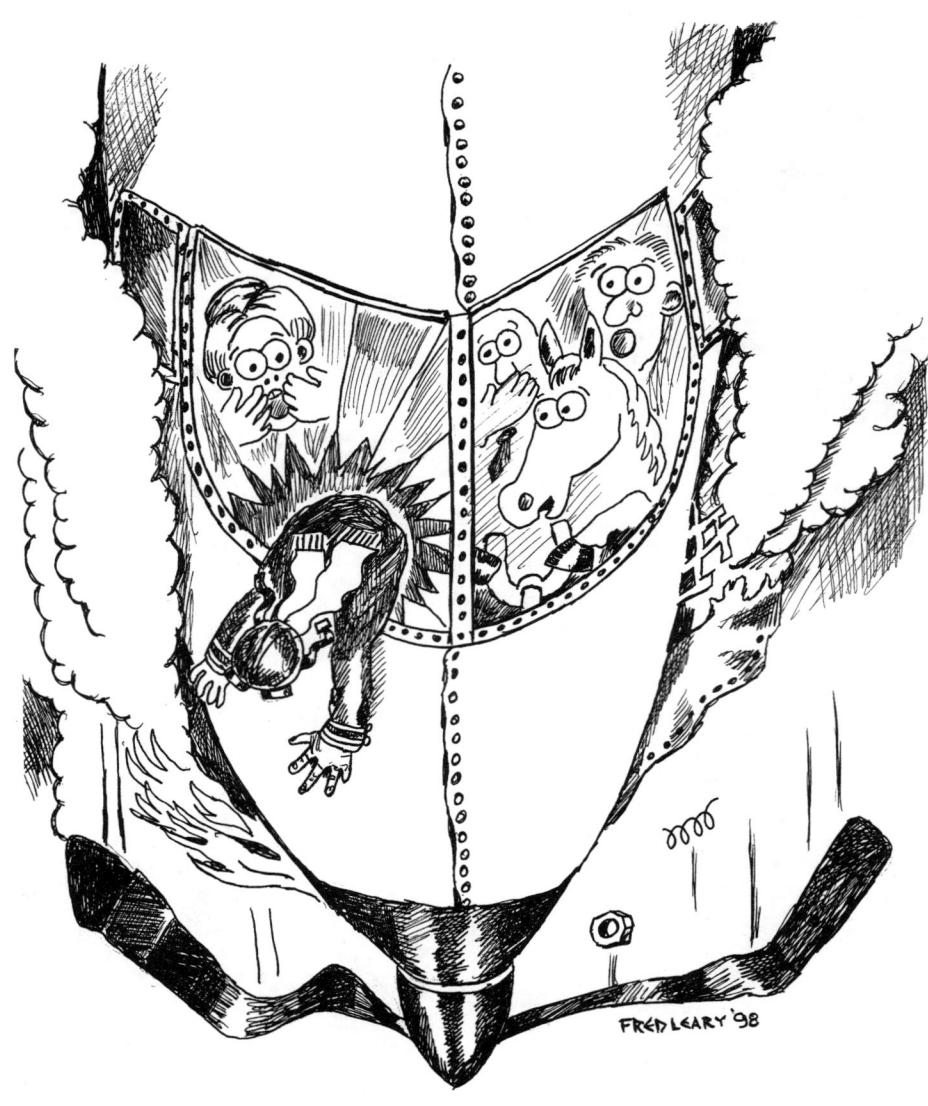

But Then Trigger Grabbed The Controls And Saved The Plane.

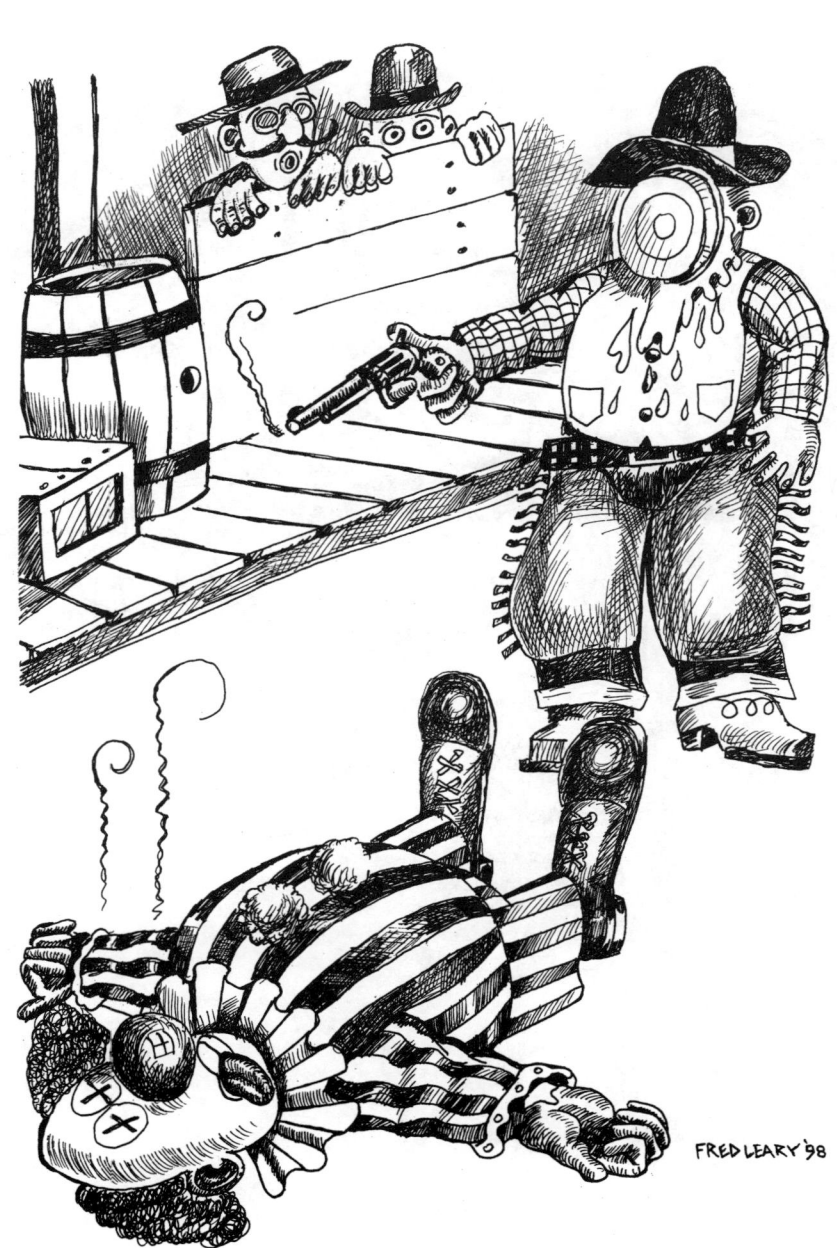

The Fight Was Over, But No One Could Decide on Who Was The Winner!

"Hold On There, Honey! I Promise I'll Never Forget Our Anniversary Again!!"

WHAT HORSES WEAR TO COSTUME PARTIES

"Gosh, You Always Manage To Catch A Few Bullets., Ed. Why Is That?"

Why Horses Are So Nervous In The Morning.

Most People Never Heard Of The Famous Battle Of Cowboys And Outer Space Invaders.

Medical School In The Old West.

"Horse Stand-up Comedy Clubs!"

FRED LEARY

About the Cartoon Artist!

Fred Leary started drawing at the age of three and has never put down the pencil.

His illustrations and cartoons have been published in many Boston area publications over the last twenty years.

He has started a small comics business called Belfry Comics- thusly named because the main office is in a church belfry built in 1866 and saved from demolition by Fred and rehabbed into an office at his house.

He does comics both comical and political and illustrations with pencil and pen and ink for any occasion.

He attended the University at Bridgeport in Connecticut.

He has a background in newsprint and journalism.

He lives in Pembroke, Massachusetts with his wife, four daughters and two cats.

INTERVIEW WITH A FARRIER

By Ed McCollum

You've heard of "Interview with a Vampire". Well, after one full year of searching through stables, farms and even circuses, I was finally able to locate one of New England's most elusive and sought after farriers, Joe DeLowery. Joe is legendary in New England for his knowledge of horses and his ability to keep them on their feet. Joe is known for his down-home sense of humor and low-key style. He is also known as a miracle worker because of his ability to get "lame" horses back on their feet and out riding again. He is the farrier for the internationally famous Big Apple Circus. We had to hobble Joe long enough to conduct this interview, but it was worth it. This is the first time Joe has agreed to a personal interview.

Ed: Joe, how did you come to be a farrier?

Joe: My grandfather was a farrier way back when not that many people made a profession out of it. In fact, I still have and use some of his tools. My folks came to the United States from Ireland and always had horses. My father shoed horses after WWII and I remember he used to charge a buck a foot. I came out of the service and used the G.I. Bill to go to the Martinsville School of Farrier back in 1970. After school, I hooked up with a Scotsman and stayed with him for five years before branching off on my own. He was a great guy and a great farrier who taught me a lot.

Ed: What's the secret of your legendary success working with horses? I'm sure not all of them are eager to be shod.

Joe: No, not at all. First of all, it's important to understand the animal so I try to be one of them. I start my day very early in the morning. I muck out my ten horses, turn them out and make sure they're doing fine. Then I clean my barn and do my shop work. By the time I get to my first appointment, you might say I smell like one of the herd. Because if they feel you're one of them, they'll stand around and trust you and feel they're in good hands. And that is what's most important. I may be getting paid by the client, but the horse is the one that I'm working for and they get all my attention. Their sense of smell is so strong. I want to smell like an old friend before I shoe that first horse. It may sound strange to the novice, but it's an old trick. I knew an old smithie that had one shirt that he shod in and he shoed seven or eight horses a day in that one shirt. At the end of the day, he took that shirt off and put it in the truck. The next day he'd come and put that same shirt on. You couldn't stand to be around this guy, but the horses loved him and he never had a problem shoeing a horse. Sometimes, it only takes a bad after-shave lotion or a strong deodorant or even some medicine that makes them think you're a veterinarian that causes a problem.

Joe DeLowery, Farrier
68 Bowen's Lane, Rochester, MA 02770
508-763-1741

Ed: How has the industry changed in the past 25 years?

Joe: The amount of work has gone up considerably. Even though there are shoeing schools pumping out farriers, the number of horses out there is growing. Consider that just in the U.S. there are six million thoroughbreds registered as a fresh crop on January first. There are more quarter horses in this country than thoroughbreds not to mention all the Morgans, Arabs and other breeds. Just in the state of Tennessee there are over 40,000 registered Tennessee Walking horses. You have 70,000 registered quarter horses in the state of Texas and that's not counting all the working horses the cowboys use. I see people in their 30's and 40's who never had a horse but dreamed of it are picking them up, starting to enjoy riding and forming riding clubs.

Ed: This new group of farriers, are they sticking it out or finding it's tough work and giving up?

Joe: They're finding out it's not an easy dollar and it's a long dollar. If they're in it to make a quick buck, they'll find out what hard work it is. You have to be shoeing alot of horses to make a living at it. For me, it's peace of mind because I just enjoy doing this. It's the secret of a good life to enjoy what you do for work.

Ed: What advice to you give people who are trailering their horses. Should the horses be shod or should they pull the shoes?

Joe: If someone is traveling for a day or to a riding spot, usually they'll have their horses shod in a nice flat steel.

They'll want shoes on their horses because when they come off the trailer, most likely they be riding them right away. But you want to be careful about shoeing your horse with any kind of shoe that's apt to get hung up in the trailer. Now, if you are using a national transportation company, ask their advice, but you may want to ship them barefooted. Also, check the condition of the trailer to make sure there are no cracks or crevices that the horses can get hung up on. I like rubber all around myself.

Ed: Are steel shoes and rubber mats a bad combination? Will the horses slip on the surface?

Joe: Oh no, steel and rubber are excellent. Bare hoof on rubber is also great. There is a suction factor there. The heat of the horse combines with the rubber while they're standing and it makes for a nice suction cup on the rubber mat. But if they're going to ride the horses when they get out of the trailer in the mountains or in the desert, they'll need a set of shoes on them. I recommend going flat and never mind the traction devices when you're traveling because it can be a booby trap and you're asking for an accident to happen.

Ed: Do you recommend wrapping the legs to keep them from skinning themselves up inside the trailer?

Joe: Absolutely. A good wrapping with just regular cotten is better than nothing at all, but they're making some great shipping boots now that will do the trick. They'll cover the vital parts of the leg, the cannon bone, the ligaments and the lower vascular system. A good pair of shipping boots is a lot cheaper than replacing your horse.

Ed: Do you recommend that people use a hay bag when traveling?

Joe: Always! I always like to keep a bag tied high, not where they can get their legs caught up in it. It gives them something to do and it's a pacifier. It calms them down.

Ed: If a rider is on a trail ride and a shoe comes loose, what should they do?

Joe: Well, there are some tools that horsemen can buy and they should ask their farrier how to use them. There are simple little clinching devices that can help reclinch a shoe and tighten some nails that have been sprung. I don't recommend doing any kind of sharp paring or removing and sole, but these new clinchers can tighten up the shoe and get you home.

Ed: What if the shoe can't be clinched?

Joe: That same tool is also a pair of pulloffs that you can very gently pry the shoe off and if you are carrying a neoprene pad you can make a temporary shoe that cowboys have been making for years that will get you home. Just fasten it with some vet wrap or good adhesive tape and off you go. A good idea is to ask your farrier to cut you a set of pads that match your horse and keep them with you when you ride. It's a quick fix for a shoe that's been sprung or been pitched or left in the mud.

Ed: What about these temporary boots I see on the market?

Joe: There are alot of brands out there like Easy Boots. They are a great invention, but make sure you get them to fit properly. Also, if the apparatus is made of hard rubber, then it should be made to fit one size larger than the horse's normal size. For instance if the horse wears a size one, fit them to a size two. It's a quick fix, and you're not going to use it for very long, but it will get you through the trail ride until you get to the next blacksmith's shop.

Ed; How do you tell your horse's shoe size?

Joe: You can ask your farrier, but it is stamped on the hoof bearing side of each shoe. You'll see a number set in the steel.

Ed: You hear alot about corrective shoeing. Is this relatively new? Did they do corrective shoeing back in the 1880's?

Joe: Absolutely. Some of the old shoes I have in my collection have a piece of steel welded on one side of it to correct a fault in a horse where he was dragging his hind leg or where he was a plow horse and you might see a cock welded on the inside heel to correct a horse that dragged his toe. In these creatures, there's no two alike and there's no horse with perfect limbs and perfect hooves so the smithies would try to help the guy who owned the horse and he could read from the old shoes where the horse was dragging. It would be worn in a certain area so he could take his new shoe and maybe double it over and make it twice as thick on one side. But corrective shoeing has been around since the first burros carried Mary and Joseph. Horse shoeing has be3en documented back to 900 B.C. The first shoes that have been found in Europe were of Celtic nature. They figured out that the only reason Hanibal was able to conquer as muchas he did was because he had his blacksmiths and armor makers put steel on the hooves of his horses to protect their feet to cross the Alps. He conquered tribes that couldn't compete with the distances Hanibal covered in conquering Europe. There are old shoes to be seen in the Smithsonian and they are basically the same as we are using today. A shoe is just iron ore that's been hammered out, nails punched and nails made. Nailed the same way, through the hoof capsule and bent over in the form of a clinch.

Ed: How about the Native American Indians? Did they use iron shoes?

Joe: Not to our best knowledge. They covered the horses hooves and used a lot of deer hide, bear hide and buck skins. They had the right idea but they didn't have the implements. They didn't do a lot with steel. The were always careful to try to protect the hoof capsule. They knew that with the foot protected, they would get 30 years out of their horses if they were also fed right and used right.

Ed: Are horses really used for work anymore?

Joe: Sure, but not like they used to be. You'll still see the plow horse here and there, and of course the working cow horse but you have to remember one hundred years ago horses were driving this country. The Teamsters had leather in their hands, not steering wheels. Horses were pulling wagons in every little town in America.

Ed: Has the horses physical condition changed since then?

Joe: Back then the horses work load was immense. You could always see ribs. If you looked at pictures of those horses, you'd say they weren't fed, and by today's standards that's probably right. We over feed them. The vets say founder is way up. We're feeding our horses like beef cows. In the old days, the horses were out working. The horse would be taking the father to work in the morning and bringing him home. The kids would probably use the horse to go visit their friends and at night they might harness up to go out to church or someplace to socialize. The hay and oats he was being feed was being worked off. Nowadays, it seems we're working for the horse. We've got automoibiles in the driveway so we don't need them for transportation. So we over feed them out of kindness and end up killing them with kindness.

Ed: For the new horse owner, how would they go about getting comfortable with their new horse, especially if it's on the frisky side?

Joe: It's just time. If you got one that isn't finished yet and isn't quite sure of everything in its environment, you just can't quit. Give that horse two or three hours a day of your time if you can. It's a daily routine that I highly recommend to these folks. You'll gain that horses confidenece and see a big improvement when he goes out on the trail or is in the trailer.

Ed: How do you shoe a horse that is big, mean and ornery?

Joe: Just had one the other day. If you lose your temper and start physically manhandling the horse, you'r going to lose. In the old days, you'd apply a Yorkshire Twitch which was put on her nose and gave her something to think about while you'd go about getting her shod. Most new horse owners now don't like that method, so if a feed bag won't do it you have to call your friend the veterinarian. There is a new drug on the market called Demozodan that I highly recommend. It's safe to use and will make the heaviest and nastiest horse comply without any problem or long term effects. As the horse gets used to being shod, you might try something milder like granules of Promosine that is sold at feed stores. Very important, take the time getting your horse used to someone handling their feet by making it a part of the grooming process.

Ed: You shoe for the famous Big Apple Circus when it comes to Boston. What's that like?

Joe: I really look forward to it every year. As you know, the Big Apple Circus is probably the best circus in the United States now. They are based in Florida and every spring come to Boston. Only the cream of the crop in talent gets hired as performers. The Big Apple was Max Shuman's circus. He's a Danish horse trainer and circus owner from Europe, and now his daughter Katja and her husband Paul Binder run it. They have the best jugglers, the best dog acts, the most amazing trapeze, and of course, their world famous horse act.

Ed: What kind of horses do they use?

Joe: Max Shuman has a set of Liberty horses that are the most skilled horses I've ever seen perform in the United States. Max Shuman is the best horse trainer that's out there. There's never a line attached and they do everything at his command. Usually they'll have eight to ten horses and to see them running in a circle the way they do shows how Max gets these horses to show their true intelligence to mankind. I was lucky to hook up with them. They're all a great bunch of people.

Ed: Joe, being a farrier for all these years, have you ever been hurt?

Joe: Oh yeah! In fact, the old expression of "having the crap kicked out of you" is about blacksmiths! You take your kicks and bites, go home chew on some aspirin and jump into a streaming hot bath. I've never had a broken bone though. Got a few cracks. Got a few kicks in the femurs and tibia that warranted taking an Advil on a cool day. It can be a dangerous profession. We had a fellow farrier that got kicked and was found dead under a horse. The horse still was on cross ties when they found this poor smithie. So you have to be careful. It's a good idea to shoe with someone around.

Ed: Last question. What's your favorite breed of horse?

Joe: Well, Ed, it's like women. I love 'em all.

INTERVIEW WITH A VETERINARIAN

By Jim Balzotti

In this edition, I was thrilled to ask one of New England's most prominent vets some of the questions most of our readers ask time and time again, namely what do I do if... Dr. Bruce, as he's known here in the northeast, is both a teaching and practicing veterinarian who is at the top of his field. Here are some of his answers to questions every rider wants to know.

Jim: If while on a trail ride, my horse suffers a cut or puncture wound, what should I do?

Dr. Chase: The first step in any wound is to control bleeding, which is generally accomplished by the application of direct pressure. When hemorrhage is controlled, rinse the wound with water to remove contaminants followed by the application of topical antiseptics. The wound should be evaluated as to whether sutures are appropriate. It is generally best to make the decision to suture or not during the first twelve to twenty four hours after the injury. If the wound is on the lower leg, bandaging should be considered.

Jim: While riding and possibly miles away from my trailer, my horse begins to limp a bit. Is it safe to ride him back to the trailer or should I get off and walk back?

Bruce E. Chase, DVM, PC
66 East Grove Street, Middleboro, MA 02346
508-947-9400 Fax 508-947-0030

Dr. Chase: At the first suspicion of a possible lameness problem, it is always wise to evaluate the foot for injury or the presence of a foreign object. A quick review of the rest of the leg for obvious injuries or swellings is also important. Whether to continue the ride back or not would depend on the cause and severity of lameness. It is always wise to err on the side of caution and not ride a lame horse due to the potential for further injury or the risk of stumbling resulting in the fall of the rider.

Jim: What are the signs that show whether or not you are overworking your horse?

Dr. Chase: The most accurate means of evaluating exhaustion has generally been to evaluate the time required for the pulse and respiration rates return to the normal resting level. This first requires an appreciation for the normal pulse and respiratory rate values. If there is a concern of exhaustion, the horse should be stopped and monitored every five to ten minutes. For most horses a proper recovery to resting values occurs within ten minutes. If recovery extends to more than twenty or thirty minutes, excess stress has occurred.

Advanced exhaustion is reflected by extreme weakness, staggering and incoordination. This represents an extreme emergency and requires immediate medical attention.

Jim: What's the best way to get your horse into shape, especially if he/she hasn't worked all winter.

Dr. Chase: A routine conditioning program by monitoring pulse and respiratory rate recoveries is the most reliable means of assessing your horse's condition. This system allows the evaluation of each horse's individual needs in a conditioning program.

Jim: Do horses get cold?

Dr. Chase: Yes. Generally hypothermia or decreased body temperature would be reflected by shivering similar to the human population. The horse is blessed with an excellent form of insulation and therefore hypothermia is a relatively rare condition. Monitoring the horse's rectal temperature is the most important step in the diagnosis and management of the problem.

The greater concern for horses is the occurrence of hyperthermia or heat exhaustion. During extended periods of work, the process of metabolism releases energy in the form of heat which must be dissipated through sweating and other cooling processes. During extensive work on a warm day a rider should be constantly aware of the horse's body temperature and condition.

Jim: What are the signs of a horse in distress and what should I do about it? What if I am miles away from home?

Dr. Chase: The most common medical emergency experienced by the horse is the syndrome of colic. A variety of activities can be associated with colic such as loss of appetite, sweating and restless behavior including pawing the ground, lying down, rolling and other signs of abdominal pain.

The most important step is to attempt to stabilize the horse and send for assistance. If at all possible, getting the horse to a location for treatment is the most important consideration.

Jim: Do you recommend bringing along medical supplies on an extended trail ride? What items? What about a day's ride?

Dr. Chase: A first aid kit would take into consideration that the two most common equine emergencies involve lacerations or the occurrence of colic. Bandaging materials for leg injuries and first aid therapy for colic as suggested by your veterinarian are therefore the most important items in a first aid kit.

Jim: What is the best way to relieve your horse from pain?

Dr. Chase: There are several medications available for pain relief, however, phenylbutazone or "Bute" is still an economical and reliable anti-inflammatory. Colic pain can be relieved by several medications used on a first aid basis. One of the most popular medications at this time is flunixin meglumine or "Banamine" which comes in an oral paste as well as an injectable form and can be administered safely for a wide variety of medical conditions.

Jim: Do horses really need electrolytes?

Dr. Chase: Yes. The electrolytes required in greatest amount for the proper metabolic performance are sodium, potassium and chloride. Sodium chloride is also known as common salt. These electrolytes are often found in adequate amounts in commercially mixed rations and are rarely found to be deficient. They can be lost in significant amounts through sweating and therefore, horses that sweat extensively in their performance should be supplemented. Supplementation can occur by the addition of trace mineralized salt to the diet or the addition of a commercially prepared electrolyte mixture.

We want to thank Dr. Chase for taking the time out of his busy schedule to talk to us. The one thing that you do not want to learn the hard way, is if your horse appears hurt or ill, do not delay in calling your vet. Unfortunately, some permanent injuries and deaths can be prevented by calling your vet promptly. When in doubt, call.

Jim Balzotti

A Cowboy's Guide to Life

If you find yourself in a hole, the first thing to do is stop diggin'.

Don't squat with your spurs on.

Don't interfere with something that ain't botherin' you none.

If it don't seem like it's worth the effort, it probably ain't.

Never ask a barber if you need a haircut.

If you get to thinkin' you're a person of influence, try orderin' somebody else's dog around.

The easiest way to eat crow is while it's still warm.
The colder it gets, the harder it is to swaller.

It don't take a genius to spot a goat in a herd of sheep.

The biggest troublemaker you'll probably ever have to deal with watches you shave his face in the mirror every morning.

Timing has a lot to do with the outcome of a rain dance.

Don't worry about bitin' off more'n you can chew;
your mouth is probably a whole lot bigger'n you think.

Always drink upstream from the herd.

Generally, you ain't learnin' nothing when
your mouth's a-jawin'.

Tellin' a man to get lost and makin' him do it are two entirely different propositions.

If you're ridin' ahead of the herd, take a look back every now and then to make sure it's still there with ya.

Good judgement comes from experience,
and a lotta that comes from bad judgment.

When you give a personal lesson in meanness to a critter or to a person, don't be surprised when they learn their lesson.

Lettin' the cat outta the bag is a whole lot easier than puttin' it back in.

The quickest way to double your money is to fold it over and put it back into your pocket.

Never miss a good chance to shut up.

ON THE ROAD AGAIN

TRAILERING TIPS FOR SAFE HAULING
By Jim Balzotti

What can be so hard about trailering a horse? That's what I asked myself many years ago. I remember the first time I had to transport a horse...

Years back, I had bought a large Tennessee Walker who was a little to rambunctious for his owner. I had borrowed my friend's horse trailer and headed up to the stable to pick up the horse. I dropped the ramp, then started to lead him into the trailer. Easy enough, right? Wrong. He reared, squealing as if he had been shot, and rammed his head on the trailer roof-hard enough to leave a sizable dent that my friend will never let me forget. The horse then bolted into traffic on a nearby major highway. There were a few anxious moments but, luckily, we were able to get him back to the trailer.

That was my first experience in horse transportation but far from my last. Over the years, I have transported horses many times throughout almost every state in the country. From my own experiences and those I have heard from other horse owners, there isn't too much that would surprise me about trailering horses.

Once, when finishing a 2-day trail event, one of my horses, a calm, older, bomb-proof horse I usually reserved for friends, panicked for no apparent reason after being led into the trailer. He wedged himself under the butt-bar. Only the quick availability of a hammer to dislodge the jammed holding pin saved the horse. Still, he peeled enough skin off his back that he couldn't be ridden for many months.

In trailering horses, nothing beats experience. Years after those first incidents, I bought a ranch in Arizona and began to transport horses there from my home in M<assachusetts. I made horse transportation a science. I talked to many veterinarians, horse feed manufacturers, and consulted any and all material I could get my hands on concerning horse transportation. Simple questions such as what to feed, when to feed, how often to rest the horses, and even where to spend the night became extremely important.

The key to trailering and traveling with your horse is preparation—for both you and your horse. When I am traveling across the country, I try to plan my trip to the last detail, from beginning to end. When I know which day I will be leaving, I schedule my veterinarian to pay a farm visit about a week before that day, to do a Coggins test and provide a health certificate for the horse. It is important to carry and have within easy reach a negative Coggins test, a current health certificate, and I would recommend, ownership papers complete with a bill of sale. Some states require other specific health tests for the horse, but if you have the papers mentioned, they will satisfy most any inquiry.

Next is an equipment check. I make sure my truck is serviced. Belts should be checked, fluids topped off, and the oil changed. There is absolutely no worse time to have a frayed belt snap than when traveling down an interstate while towing horses. Tires should be checked for wear and inflated properly. This is also a good time to check on the condition of your spare tire and jack. A good investment to make is a portable floor jack that will lift a truck safely and quickly. Insurance papers, registration, and driver's license should be properly stored in the truck. I am a big believer in joining AAA, which provides emergency roadside service, up-to date information on the best routes to take, and accurate state maps.

I also suggest investing in a CB radio for all your long trips. Tune into Channel 19, the nationwide truckers' band, for information on highway conditions and breakdowns. If you travel often, you may want to invest in a cellular phone for additional security.

Often overlooked during preparation is the horse trailer itself; it is essential that it is also properly prepared for a long trip. Pull off the wheels and pack the bearings with grease. Check the trailer tires for wear and proper inflation. Check the lug nuts to make sure they are not "frozen" by rust. Once, while traveling across New Mexico in a biting snowstorm, my horse trailer blew a tire. It took me 30 minutes just to free a frozen lug nut. I was on the shoulder of an interstate highway with huge tractor-trailers whizzing by—not a safe or comfortable situation. So before leaving on a trip, loosen and then retighten each lug nut. Every horse trailer should be equipped with safety chains in case the hitch fails.

Inside the trailer, check for any sharp objects or loose metal that may injure your horse. I put down plenty of shavings for bedding and use rubber floor mats for some shock absorption and more comfortable traveling for the horse. You should also have emergency gear such as a fire extinguisher, flashlight, spare leads, rope, road flares, hammer, and a knife in case you need to free a tangled horse. Keep these things in an easily accessible place. In the event of an emergency, you don't want to be hunting around for this equipment.

If you are not familiar with towing a horse trailer, do not wait until your departure date to learn. Hook up your empty trailer and take some practice runs. Get the feel of how much distance you need to turn, pass other vehicles, or brake. Remember that with a loaded trailer and extra weight you have to allow even more distance than with an empty trailer. I like to drive to a nearby shopping center early in the morning, when the parking lot is empty, to practice turning, parking, and backing up with the trailer. Sooner or later you will find yourself in a tight spot; for example, at the gas pumps when someone pulls in front of you and someone behind you. You will be thankful for the extra practice.

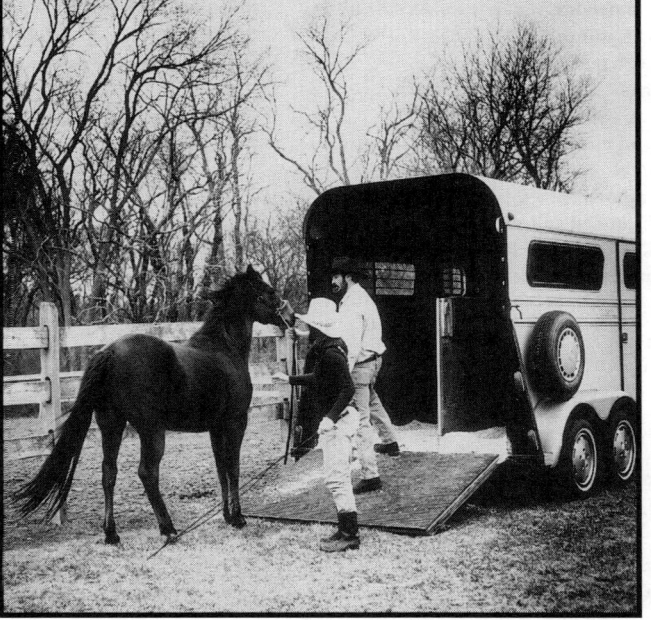

If you are traveling with only one horse and pulling a two-horse, straight load trailer, always load him on the left side. The reason being most highways, especially two-lane highways, have a very slight crown in the middle to help facilitate runoff of rain and melting snow. You want the weight of the horse toward the crown of the road to help stabilize the trailer. The same is also true if you must pull onto the shoulder of the road, which drops off to the right. For the same reasons, load the heavier horse on the left side if you are hauling two horses.

When I am traveling any distance, I try to stop every two hours or so. This gives me a chance to stretch my legs and gives the horses a chance to relax. I give the truck and trailer a quick visual inspection to make sure everything is still properly connected—the hitch and the safety chains. I usually try to offer the horses water at this point, but do not take them out of the trailer. Most horses can remain on a trailer for a full day as long as you are providing them with breaks, and if the horses are urinating. Some horses are reluctant to urinate in a trailer if they cannot stretch out to do so.

I carry along two 6-gallon containers filled with the water the horses are used to drinking. I also pack as much of the horse's regular feed and hay as I can store. If you are going to be traveling often, you may want to get your horse accustomed to a feed that is easily found across the country. Purina Mills, for example. Some horsepeople do not feed grain when they are hauling, and horses can certainly go a few days without grain. If you are worried about your horse binding or colicing, you may want to feed them a nice watery brand mash with a little molasses in it. That'll be sure to keep them loose.

As for hay, be aware that different hay is available in different parts of the country. Take into consideration what type of hay your horse is used to and what part of the country you are heading to. If you make a sudden change in your horse's hay, he could colic. For example, in the East, timothy hay is most common. If you are heading west, alfalfa is common. Take along enough of what your horse is used to so you can gradually mix it with the type of hay you can find along the way.

If you do unload your horses during the day while trailering, do so in a safe location, such as a rest area if regulations allow. Do not let your horses graze. Many roadsides are sprayed with weed-killers that can be very toxic to horses. People often ask me how far they should travel in a day. This really depends on a number of things. I know from experience that I can cover 500 to 600 miles a day on interstate highways. I always like to get an early start and stop every two hours or so for a short break. It is always important from me to stop for the day well before dark. I always try to avoid arriving at a boarding stable after dark because I want to inspect the facility for cleanliness and safety, and also to let them run around and get a little exercise before they are put in for the night. If the stable has a round pen or an arena, you might ask if you can use it so your horses can get a little exercise. After all, they've been cooped up in your trailer all day.

If you do not get an early start, your mileage could go down considerably. Road conditions and bad weather can also slow you up. It's best to leave yourself plenty of time so that you don't push yourself to a dangerous point of fatigue. Stop when you are tired.

Our U.S. Stabling Guide that lists overnight and/or short term boarding can be indispensable. Do not hit the road assuming you will find a stable along the way. Invariably, you won't ; then you run the risk of spending the night on the side of the road with your horses.

One of the most horrific stories I've heard was about a woman traveling cross-county with four horses. One night, unable to find accommodations for her horses, she put them in a portable corral off the side of the highway. Sometime during the night, one of the horses put his nose under the bottom of the corral and it collapsed. The horses ran onto the highway and all were struck and killed by a passing tractor-trailer. Put yourself and your horse up somewhere safe for the night. If worse comes to worst, leave the horses in the trailer while you catch a few hours sleep. Safety first, for both you and your horse.

CAMPFIRE RECIPES

Ranches are famous for their wholesome, stick-to-your-ribs cooking.
Guests have been delighted for years after coming back from a horseback ride
to the mouth-watering smells coming out of the kitchen.
We were able to coax some of these famous, secret recipes for your enjoyment.

MAIN DISHES

SPINACH LASAGNA WITH SPAGHETTI SAUCE

Lasagna:
2 (10-oz.) pkgs. frozen chopped spinach cooked and drained
1 T. chopped parsley
2 cloves garlic, minced
2 lbs. sm. curd cottage cheese
2 eggs
1/4 cup butter, softened
1/3 tsp. ground nutmeg
1 tsp. salt
1/4 tsp. pepper
8 oz. lasagna noodled, cooked and drained
1 lb. Monterey Jack cheese, shredded
1 cup grated Parmesan cheese
1 qt. spaghetti sauce (see below)

Squeeze all excess moisture from spinach with your hands. Combine spinach with parsley, garlic, cottage cheese, eggs, butter, nutmeg, salt and pepper. Preheat oven to 350 degrees. In a buttered 13 x 9 x 2-inch baking pan, layer 1/3 noodles, 1/2 spinach mixture and 1/3 each cheese. Repeat layering. On top, place last 1/3 noodles and cheeses. Bake, uncovered, for 30 to 40 minutes or until hot and bubbly. Let stand for 15 minutes before cutting. Serve with heated spaghetti sauce on the side. Yields 12 servings. Excellent lasagna!

Spaghetti Sauce:
1 lb. ground meat (or hot sausage)
2 to 3 oz. mushrooms
1/4 cup chopped fresh parsley
1 tsp. garlic salt or 1 clove fresh garlic, minced
1 tsp. salt
1 tsp. oregano
1/4 tsp. pepper
1 (6-oz.) can tomato paste
1 (8-oz.) can tomato sauce
1 (16-oz.) can tomatoes

Brown meat and add mushrooms to saute. Add the remaining ingredients and simmer for 1 hour. Serve over hot pasta or with spinach lasagna.

— from Flying A Ranch in Wyoming

RATTLESNAKE RECIPE

Skin rattlesnake and cut in 3" and 4" pieces.

Roll in mixture of flour, corn meal, milk and egg. Salt and pepper. Deep fry in hot oil. Serve hot.

— from Shannon Ranch in California

PORK TENDERLOIN ROAST WITH MUSTARD SAUCE

2 (3-4 lb.) pork tenderloins
1/2 c. soy sauce
1/4 c. water
2 cloves garlic, minced
1 T. grated onion
1 T. white vinegar
1/4 tsp. ground red pepper
2 slices bacon

Place tenderloins in a shallow dish. Combine soy sauce, water, garlic, onion and white vinegar; stir well. Pour marinade mixture over tenderloins. Cover and marinate in refrigerator for 8 hours, turning occasionally. Remove tenderloins from marinade, reserving marinade. Place on a rack in a shallow roasting pan. Place 1 slice of bacon on top of each tenderloin. Bake, uncovered, at 325 degrees for 1 hour and 10 minutes or until thermometer inserted in thickest part of the tenderloin reads 160 degrees, basting frequently with marinade. Yields 4 servings.

Mustard Sauce:
1/3 cup mayonnaise
1/3 c. sour cream
1 T. dry mustard
1 T. grated onion
1 1/2 tsp. white vinegar
1/4 tsp. salt

Combine all ingredients in a bowl, stirring well. Cover and chill. Yield: 3/4 cup. Serve with pork tenderloin.

— from Flying A Ranch in Wyoming

ELK, VENISON OR ANTELOPE CURRY

2 lbs. elk boned, cut in cubes
2 tbsp. olive oil
2 large onions, chopped
3 garlic cloves, minced
2 celery stalks, chopped
1 tbsp. worcestershire sauce
1 tbsp fresh curry
2 cups water
2 beef bouillon cubes

Brown meat in oil, add onion, garlic and celery. Cook until crisp and tender. Add worcestershire, curry, water and bouillon cubes. Stir, cover and simmer for 30 minutes or until meat is tender. Serve over rice.

— from Circle Bar Ranch in Montana

POLLA ALLA DIAVOLA
(Devil's Chicken and Wine Rice)

Yield: 6 servings

Devil's Chicken:
Six 6 oz. skin-on chicken breasts
1/2 cup lemon juice
1 tablespoon coarsely ground black pepper
3 tablespoons olive oil
2 teaspoons salt

Wine Rice:
3/4 cup finely minced onion
12 tablespoons butter
4 1/2 cups uncooked
3 cups chicken stock
3 cups dry white wine
3 cups water
3 parsley sprigs
1 bay leaf
1/2 teaspoon thyme
ground black pepper & salt to taste

Place the chicken in a deep dish. Pour the lemon juice over the chicken. Add the pepper and olive oil. Let marinate for 2 to 6 hours, stirring the chicken once every hour. Remove the chicken from the marinade. Sprinkle with salt. Place on a grill, skin side down and cook, using the marinade to baste the chicken.

To prepare the wine rice, slowly saute the onion in butter, until it is soft, not browned. Blend in the uncooked rice. Stir constantly over moderate heat, until the mixture takes on a milky appearance (it will first turn translucent). Combine the chicken stock, white wine and water in a pot and bring to a boil. Pour the boiling liquid mixture over the rice mixture. Add in the remaining ingredients. Place the entire mixture into a casserole. Simmer until all the liquid is absorbed.

Serve the chicken and rice with au natural steamed sugar snap peas.

— from Rawah Ranch in Colorado

DROVERS STEW

2 T. olive oil
4 boneless, skinless chicken breast halves
(1 lb.), cut into 1 inch pieces
1 c. chopped onion
1/2 medium green bell pepper, chopped
1/2 medium yellow bell pepper, chopped
1 tsp. chopped garlic
2 (14-1/2 oz.) cans stewed tomatoes
1 (15 oz.) can pinto beans, drained and rinsed
3/4 c. picante sauce
1 T. chili powder
1 T. ground cumin
1/2 c. shredded cheddar cheese
6 T. sour cream

In a large stockpot, heat olive oil over medium heat. Add chicken, onion, bell peppers and garlic and cook until chicken is no longer pink. Add tomatoes, beans, picante sauce, chili powder and cumin. Reduce heat to low and simmer for 25 minutes or up to 2 hours. Place in individual serving bowls and top with cheese and sour cream. Serve with cheddar muffins.

— from Flying A Ranch in Wyoming

BARBECUED VENISON MEATBALLS

1 egg, beaten
1/3 cup milk 1 cup soft bread crumbs
1/2 cup chopped onion
1 tsp. salt & 1/8 tsp. pepper
1 tbsp. worcestershire
1-1/2 lbs. ground venison
1 tbsp. olive oil

Combine above ingredients (except olive oil). Mix together and shape into small (tablespoon size) meatballs. Brown in olive oil.

Sauce:
1/2 cup catsup
1/2 cup water
1/4 cup brown sugar
2 tbsp. vinegar
1/4 tsp. cayenne pepper
1/2 tsp. salt
1 tbsp. worcestershire sauce
1/2 tsp. pepper
1/2 tsp. celery seed (optional)

Heat sauce ingredients to a boil, pour over meatballs and bake 350 degrees for 25 minutes. You can also use either elk or antelope.

— from Circle Bar Ranch in Montana

WILD DUCK IN ORANGE SAUCE

1 Wild duck, cut into 1 inch pieces
1/4 c. olive oil
1 cup beef broth
1 cup orange juice
3 tbsp. grated orange rind
1/2 cup dry vermouth
1/3 cup brown sugar
Salt & pepper to taste

Season duck with salt & pepper, brown well in olive oil. Transfer to crockpot. Add remaining ingredients. Cook on low for 8 hours. Thicken juices with flour or corn starch if desired. Serve over wild rice.

— from Circle Bar Ranch in Montana

GRILLED CHICKEN BREAST
With Southwestern Prickly Pear Fruit Salsa

Yield: 10 servings

10 - 6 oz. chicken breasts, double lobe
1 med. cantaloupe (sm. dice)
1 med. honeydew (sm. dice)
1/2 cup red onion (fine dice)
1/2 cup lime juice
1/4 cup cilantro (chopped)
1 mango (diced)
1/2 cup red pepper(fine dice)
1/2 cup green onion (fine dice)
1/2 tsp. ground cumin
1/2 cup prickly pear syrup

Mix gently. Grill chicken breasts until done. Spoon a heaping amount of salsa over chicken and serve.

— from Lazy K Bar Ranch in Arizona

PAULA'S MEATLOAF

2 lbs. lean ground beef
1 medium onion, chopped fine
2 eggs
1 to 1-1/2 cups crushed saltines or bread crumbs
2 tsp. salt pepper to taste
3 Tbsp. worcestershire sauce
a little garlic powder
about 1 cup heavy whipping cream

Mix and shape loaf. .Put in foil lined baking pan. Bake at 375 degrees for approximately 1 and 1/2 hours.

— from Rawah Ranch in Colorado

HERB BUTTER FOR STEAKS

1 lb. butter
2 tsp. basil
2 tsp. oregano
1 tsp. garlic powder
1/4 tsp.thyme

Mix well, shape into log, wrap in plastic wrap, freeze, until ready to pack in the cooler. Slice butter and place on steaks. It's fabulous!

— from Triangle C Ranch in Wyoming

BREADS AND BISCUITS

CHEDDAR MUFFINS

2 c. flour
1 T. baking powder
1 tsp. salt
1/2 tsp. paprika
2 T. sugar
1 egg, beaten
1/2 c. milk
1/2 c. sour cream
1/3 c. butter, melted
1 c. cheddar cheese, shredded
1/3 c. chopped onion, sauteed in 2 T. butter
1/4 c. fresh parsley

Stir together flour, baking powder, salt, paprika and sugar. Add egg, milk, sour cream, butter and cheese. Stir just until blended. Fold in onions and parsley. Bake at 400 degrees in greased muffin tins for about 20 minutes or until golden. Makes 1 dozen. Serve with a hearty bowl of Drovers Stew.

— from Flying A Ranch in Wyoming

COVERED WAGON BISCUITS

2 cakes yeast
1/4 cup lukewarm water
1 tbsp. sugar
4 cups flour
2 tsp. salt
4 tbsp. shortening
1 tsp. baking powder
1-1/4 cups buttermilk

Sprinkle yeast into bowl, add sugar and lukewarm water. Stir and let stand. Sift flour and salt into separate bowl, cut shortening in small pieces. Stir baking powder into buttermilk. Add at once into yeast mixture with dry ingredients. Knead lightly until smooth. Turn onto floured board. Roll out 1/4 inch thick. Cut into biscuits. (You can use a floured juice glass for cutting). Place biscuits on greased pan about 1/4 inch apart. Brush melted butter on tops. Prick with fork; let rise 40 minutes. Bake at 425 degrees for 12 to 15 minutes.

— from Rankin Ranch in California

HERB FOCACCIA

1-3/4 cup warm water
3 T. extra virgin olive oil
2 tsp. active dry yeast
3 cups of bread floor
3/4 cup all-purpose flour
2 tsp. salt 1 tsp. ground black pepper
1 T dry oregano
1 tsp. garlic powder
2 tsp. dry basil
1 tsp. crushed rosemary needles

Combine water, olive oil, & yeast. Stir until yeast is dissolved. Combine bread flour and seasonings, add to water/yeast mixture and stir well. Add remaining 3/4 C of all-purpose flour, add a little more flour if needed to make a moderately stiff dough. Knead until smooth and elastic. Transfer to a slightly oiled bowl and cover with plastic wrap. Rise till doubled. Punch down and rise 20 minutes more. Punch down again. Place dough in a greased, and lightly sprinkled with corn meal, 10 inch iron skillet. Brush top of loaf with olive oil and sprinkle lightly with Kosher salt. Cover with plastic wrap rise till doubled. Bake at 400 degrees for 20-25 minutes until browned. Remove from skillet and cool on wire rack. Yield is one loaf. This is a high altitude recipe. Wrap with plastic when cool.

— from Triangle C Ranch in Wyoming

TO HELL WITH TEXAS CORNBREAD

2-1/2 cup cornmeal
2-1/2 cup all purpose flour
2 cups sugar
4 eggs
2 T. baking powder
2 T. salt 1 pint milk

Mix 15 seconds in mixer and add 1/2 cup liquid shortening. Put in 13" x 9" sheet pan. Cook for 45 minutes at 325 degrees.

— from Wilderness Trails Ranch in Colorado

Desserts

SOUR CREAM COFFEE CAKE

1 cup butter softened
2-3/4 cups sugar, divided
2 tsp. vanilla
4 eggs
3 cups all-purpose flour
2 tsp. baking powder
1 tsp. baking soda
1 tsp. salt
2 cups (16 ounces) sour cream (can use plain yogurt)
2 tbsp. ground cinnamon
dash nutmeg
1/2 c. chopped pecans or walnuts

Cream butter and 2 cups sugar until fluffy. Add vanilla. Add eggs one at a time beating well after each addition. Combine flour, baking powder, soda and salt; add alternately with sour cream, beating just enough after each addition to keep batter smooth. Spoon 1/3 batter into greased 10-inch tube pan. Combine cinnamon, nuts and remaining sugar; sprinkle 1/3 over batter in pan. Repeat layers two more times. Bake at 350 degrees for 70 minutes or until cake tests done. Cool for 10 minutes. Remove from pan to a wire rack to cool completely.

— from Circle Bar Ranch in Montana

"APPLE" ZUCCHINI PIE

Use big zucchinis that are still tender enough that you can pierce the skin easily with your thumbnail. Peel, cut into quarters lengthwise, remove the seed and slice crosswise.

Toss together:
4 cups sliced zucchini, cooked until tender crisp
2 tbsp. lemon juice
dash salt

Mix in a bowl:
1-1/4 cup sugar
1-1/2 tsp. cinnamon
1-1/2 tsp cream of tartar
dash nutmeg
dash ground cloves
3 tbsp. flour

Add zucchini and mix well. (Mixture will be runny; that's ok). Dump into a 9 inch piecrust, dot with butter and add top crust. Bake 40 minutes in 400 degree oven until golden brown.

— from Circle Bar Ranch in Montana

PEANUT BUTTER BARS

2 saran pkgs. graham crackers
3/4 lb. peanut butter
1/2 lb. powdered sugar
1/4 lb. softened butter
1 lb. chocolate

Mix all ingredients (except chocolate) well in mixer. Spread in 9" pan. Melt chocolate. Spread on top and chill. Bring to room temperature for easier cutting.

— from Wilderness Trails Ranch in Colorado

T CROSS ICE CREAM PIE WITH CHOCOLATE DECADENCE SAUCE

Awesome and easy to make! You can buy a chocolate crust to make it even easier. The sauce makes this more than just an ice cream pie. The chocolate sauce makes great sundaes, too. We prefer the orange whisper of Grand Marnier over the plain vanilla version. 8 - 10 servings

Crust:

25 chocolate wafers, crushed (about 1-1/2 cups)
6 tablespoons butter, melted

Filling:

1 pint coffee ice cream
1 pint chocolate ice cream
1/2 cup chopped pecans
1/2 cup chopped chocolate-toffee candy bars

Pull ice cream from freezer to soften. Preheat oven to 350 degrees. To make the crust, crush the cookies to fine crumbs and mix with butter. Press into a 10" pie plate and bake 10 minutes. Remove and completely cool. While the crust is cooling, make the chocolate sauce (below).

When the ice cream is softened, mix ice cream, pecans and chopped candy in a large mixing bowl. Pour into cooled crust and freeze until hard. To serve, cut pie into 8 to 10 slices, place on a chilled dessert plate and drizzle pie and plate with chocolate sauce.

Chocolate Decadence Sauce

4 ounces semi-sweet chocolate, chopped
1 cup heavy cream
1 cup sugar
3 tablespoons butter
1 tablespoon vanilla or Grand Marnier

Over simmering water in a double boiler, melt chocolate and stir in cream. Mix to combine, stirring constantly. Add sugar and butter and stir until sugar is dissolved. Remove from heat and stir in vanilla or Grand Marnier. Allow to cool at room temperature (refrigerating will cause sugar to crystallize and you'll have a grainy sauce).

— from T Cross Ranch in Wyoming

PECAN PIE

3 eggs
2/3 cup of sugar
dash salt
1 cup dark corn syrup
1/2 cup melted butter
1 cup pecans
1 (9inch) unbaked pastry shell

Beat eggs with sugar, salt, cornsyrup and melted butter. Add pecans. Pour into unbaked shell. Bake at 350 degrees for 50 minutes or until knife inserted halfway between outside and center of filling comes out clean. Easy, but soooo good!

Traveling with Your Horse? Don't Forget Your Stabling Guide!

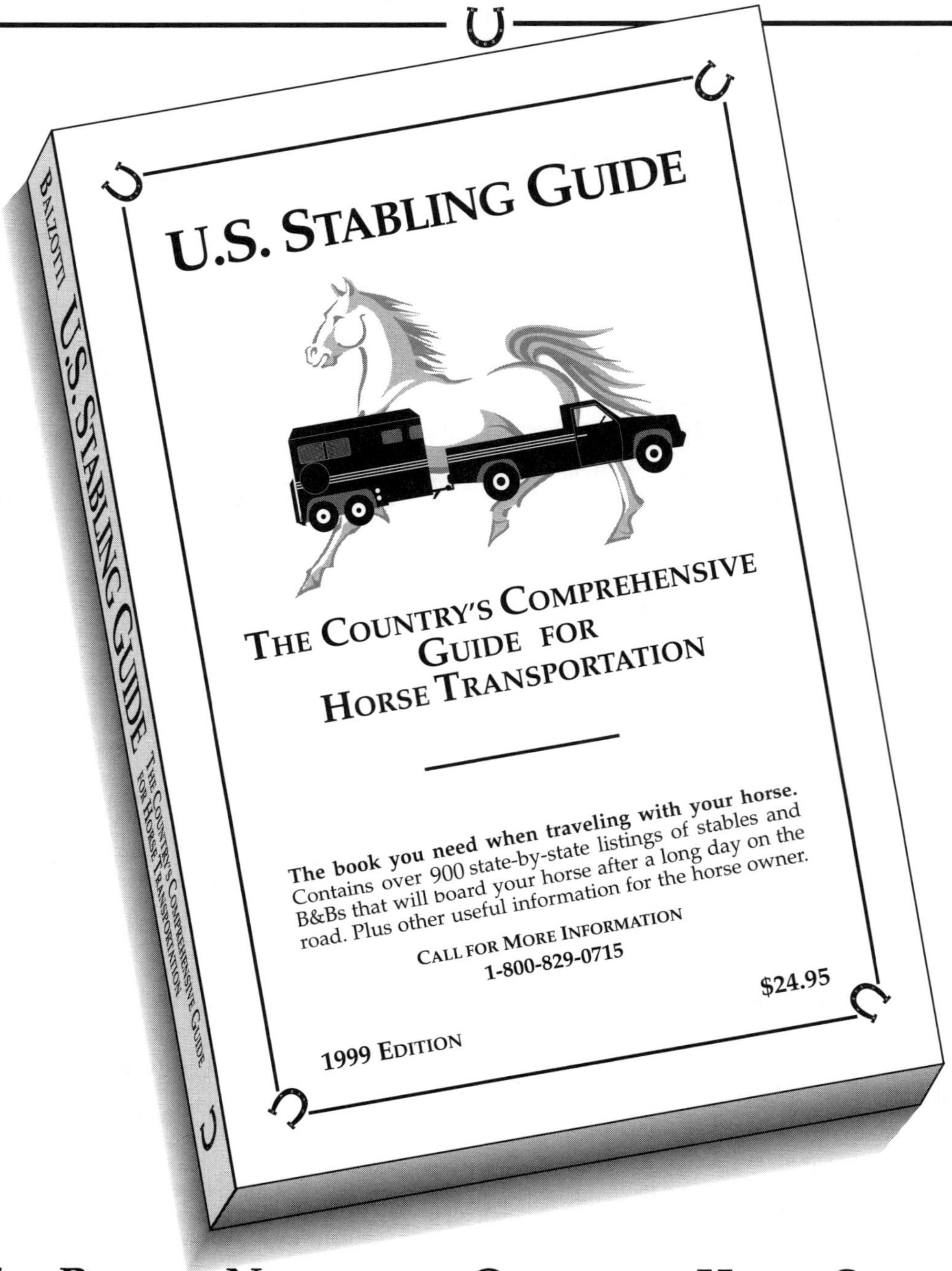

U.S. Stabling Guide

The Country's Comprehensive Guide for Horse Transportation

The book you need when traveling with your horse. Contains over 900 state-by-state listings of stables and B&Bs that will board your horse after a long day on the road. Plus other useful information for the horse owner.

Call for More Information
1-800-829-0715

$24.95

1999 Edition

The Bible of Nationwide Overnight Horse Stabling Carried by AAA Offices Throughout the U.S. and Canada!

- A Unique Gift for Yourself and Friends with Horses!
- $24.95 plus Shipping and Handling.
- We Accept MC, Visa, and American Express.

For More Information
Call: **1-800-829-0715**   Fax: **1-781-829-9087**

Notes and Reminders

Notes and Reminders

Notes and Reminders

Notes and Reminders

Notes and Reminders

Notes and Reminders

Notes and Reminders

NOTES AND REMINDERS